# Lecture Notes in Artificial Intelligence    9896

Subseries of Lecture Notes in Computer Science

More information about this series at http://www.springer.com/series/1244

Friedhelm Schwenker · Hazem M. Abbas
Neamat El Gayar · Edmondo Trentin (Eds.)

# Artificial Neural Networks in Pattern Recognition

7th IAPR TC3 Workshop, ANNPR 2016
Ulm, Germany, September 28–30, 2016
Proceedings

 Springer

*Editors*
Friedhelm Schwenker
Ulm University
Ulm
Germany

Neamat El Gayar
Cairo University
Giza
Egypt

Hazem M. Abbas
Ain Shams University
Cairo
Egypt

Edmondo Trentin
Università di Siena
Siena
Italy

ISSN 0302-9743                      ISSN 1611-3349   (electronic)
Lecture Notes in Artificial Intelligence
ISBN 978-3-319-46181-6           ISBN 978-3-319-46182-3   (eBook)
DOI 10.1007/978-3-319-46182-3

Library of Congress Control Number: 2016950420

LNCS Sublibrary: SL7 – Artificial Intelligence

Printed on acid-free paper

This Springer imprint is published by Springer Nature
The registered company is Springer International Publishing AG
The registered company address is: Gewerbestrasse 11, 6330 Cham, Switzerland

# Preface

This volume contains the papers presented at the 7th IAPR TC3 Workshop on Artificial Neural Networks in Pattern Recognition (ANNPR 2016), held at Ulm University, Ulm, Germany, during September 28–30, 2016. ANNPR 2016 followed the success of the ANNPR workshops of 2003 (Florence), 2006 (Ulm), 2008 (Paris), 2010 (Cairo), 2012 (Trento), and 2014 (Montreal). The series of ANNPR workshops has served as a major forum for international researchers and practitioners from the communities of pattern recognition and machine learning based on artificial neural networks.

The Program Committee of the ANNPR 2016 workshop selected 25 papers out of 32 for the scientific program, organized in regular oral presentations. The workshop is enriched by two IAPR invited sessions: A Spiking Neural Network for Personalized Modelling of Electrogastography (EGG) given by Prof. Nikola Kasabov, Auckland University of Technology, New Zealand and Learning Sequential Data with the Help of Linear Systems presented by Prof. Alessandro Sperduti, University of Padua, Italy.

This workshop would not have been possible without the help of many people and organizations. First of all, we are grateful to all the authors who submitted their contributions to the workshop. We thank the members of the Program Committee and the additional reviewers for performing the difficult task of selecting the best papers from a large number of high-quality submissions. We hope that readers of this volume will enjoy it and get inspired from its contributions. ANNPR 2016 was supported by the International Association for Pattern Recognition (IAPR), by the IAPR Technical Committee on Neural Networks and Computational Intelligence (TC3), by the University of Ulm, Germany, and the Transregional Collaborative Research Center SFB/TRR 62 Companion-Technology for Cognitive Technical Systems. Special thanks to the people responsible for local organization, in particular to Markus Kächele, Viktor Kessler, Sascha Meudt, and Patrick Thiam. Finally, we wish to express our gratitude to Springer for publishing our workshop proceedings within their LNCS/LNAI series.

July 2016

Friedhelm Schwenker
Hazem M. Abbas
Neamat El Gayar
Edmondo Trentin

# Organization

## Organization Committee

| | |
|---|---|
| Friedhelm Schwenker | Ulm University, Germany |
| Hazem M. Abbas | Ain Shams University, Egypt |
| Neamat El Gayar | Cairo University, Egypt |
| Edmondo Trentin | Università di Siena, Italy |

## Program Committee

| | |
|---|---|
| Shigeo Abe | Kobe University, Japan |
| Amir Atiya | Cairo University, Egypt |
| Erwin M. Bakker | Leiden Institute of Advanced Computer Science, The Netherlands |
| Mohamed Bayoumi | Queen's University, Canada |
| Daniel Braun | Ulm University, Germany |
| Ludovic Denoyer | Université Pierre et Marie Curie, France |
| Mohamed M. Gaber | Birmingham City University, UK |
| Eric Granger | École de Technologie Supérieure, Canada |
| Mohamed Abdel Hady | Microsoft, USA |
| Markus Hagenbuchner | University of Wollongong, Australia |
| Barbara Hammer | Bielefeld University, Germany |
| Hans A. Kestler | Ulm University, Germany |
| Jonghwa Kim | University of Augsburg, Germany |
| Simone Marinai | University of Florence, Italy |
| Marco Maggini | University of Siena, Italy |
| Nadia Mana | Fondazione Bruno Kessler, Italy |
| Heiko Neumann | Ulm University, Germany |
| Günther Palm | Ulm University, Germany |
| Luca Pancioni | University of Siena, Italy |
| Stefan Scherer | University of Southern California, USA |
| Eugene Semenkin | Siberian State Aerospace University, Russia |
| Ah-Chung Tsoi | Macau University of Science, SAR China |
| Zhi-Hua Zhou | Nanjing University, China |

## Local Arrangements

Markus Kächele
Viktor Kessler
Sascha Meudt
Patrick Thiam

## Sponsoring Institutions

International Association for Pattern Recognition (IAPR)
Technical Committee 3 (TC3) of the IAPR
Ulm University, Ulm, Germany
Transregional Collaborative Research Center SFB/TRR 62 Companion-Technology for
Cognitive Technical Systems, Ulm University and Otto-von-Guericke University
Magdeburg, Germany

# Contents

**Invited Papers**

Learning Sequential Data with the Help of Linear Systems. . . . . . . . . . . . . 3
  *Luca Pasa and Alessandro Sperduti*

A Spiking Neural Network for Personalised Modelling
of Electrogastrography (EGG). . . . . . . . . . . . . . . . . . . . . . . . . . . . 18
  *Vivienne Breen, Nikola Kasabov, Peng Du, and Stefan Calder*

**Learning Algorithms and Architectures**

Improving Generalization Abilities of Maximal Average Margin Classifiers . . . 29
  *Shigeo Abe*

Finding Small Sets of Random Fourier Features for Shift-Invariant Kernel
Approximation . . . . . . . . . . . . . . . . . . . . . . . . . . . . . . . . . . . 42
  *Frank-M. Schleif, Ata Kaban, and Peter Tino*

Incremental Construction of Low-Dimensional Data Representations . . . . . . . 55
  *Alexander Kuleshov and Alexander Bernstein*

Soft-Constrained Nonparametric Density Estimation with Artificial Neural
Networks . . . . . . . . . . . . . . . . . . . . . . . . . . . . . . . . . . . . . . 68
  *Edmondo Trentin*

Density Based Clustering via Dominant Sets . . . . . . . . . . . . . . . . . . . 80
  *Jian Hou, Weixue Liu, and Xu E*

Co-training with Credal Models . . . . . . . . . . . . . . . . . . . . . . . . . . 92
  *Yann Soullard, Sébastien Destercke, and Indira Thouvenin*

Interpretable Classifiers in Precision Medicine: Feature Selection
and Multi-class Categorization . . . . . . . . . . . . . . . . . . . . . . . . . . 105
  *Lyn-Rouven Schirra, Florian Schmid, Hans A. Kestler,
  and Ludwig Lausser*

On the Evaluation of Tensor-Based Representations for Optimum-Path
Forest Classification . . . . . . . . . . . . . . . . . . . . . . . . . . . . . . . 117
  *Ricardo Lopes, Kelton Costa, and João Papa*

On the Harmony Search Using Quaternions . . . . . . . . . . . . . . . . . . . . 126
  *João Papa, Danillo Pereira, Alexandro Baldassin, and Xin-She Yang*

Learning Parameters in Deep Belief Networks Through Firefly Algorithm . . .   138
*Gustavo Rosa, João Papa, Kelton Costa, Leandro Passos,*
*Clayton Pereira, and Xin-She Yang*

Towards Effective Classification of Imbalanced Data with Convolutional
Neural Networks. . . . . . . . . . . . . . . . . . . . . . . . . . . . . . . . . . . . . . . . . .   150
*Vidwath Raj, Sven Magg, and Stefan Wermter*

On CPU Performance Optimization of Restricted Boltzmann Machine
and Convolutional RBM . . . . . . . . . . . . . . . . . . . . . . . . . . . . . . . . . . . .   163
*Baptiste Wicht, Andreas Fischer, and Jean Hennebert*

Comparing Incremental Learning Strategies for Convolutional Neural
Networks. . . . . . . . . . . . . . . . . . . . . . . . . . . . . . . . . . . . . . . . . . . . . . . .   175
*Vincenzo Lomonaco and Davide Maltoni*

Approximation of Graph Edit Distance by Means of a Utility Matrix. . . . . . .   185
*Kaspar Riesen, Andreas Fischer, and Horst Bunke*

**Applications**

Time Series Classification in Reservoir- and Model-Space: A Comparison . . .   197
*Witali Aswolinskiy, René Felix Reinhart, and Jochen Steil*

Objectness Scoring and Detection Proposals in Forward-Looking Sonar
Images with Convolutional Neural Networks . . . . . . . . . . . . . . . . . . . . . . .   209
*Matias Valdenegro-Toro*

Background Categorization for Automatic Animal Detection in Aerial
Videos Using Neural Networks. . . . . . . . . . . . . . . . . . . . . . . . . . . . . . . . .   220
*Yunfei Fang, Shengzhi Du, Rishaad Abdoola, and Karim Djouani*

Predictive Segmentation Using Multichannel Neural Networks in Arabic
OCR System . . . . . . . . . . . . . . . . . . . . . . . . . . . . . . . . . . . . . . . . . . . . . .   233
*Mohamed A. Radwan, Mahmoud I. Khalil, and Hazem M. Abbas*

Quad-Tree Based Image Segmentation and Feature Extraction to Recognize
Online Handwritten Bangla Characters. . . . . . . . . . . . . . . . . . . . . . . . . . .   246
*Shibaprasad Sen, Mridul Mitra, Shubham Chowdhury, Ram Sarkar,*
*and Kaushik Roy*

A Hybrid Recurrent Neural Network/Dynamic Probabilistic Graphical
Model Predictor of the Disulfide Bonding State of Cysteines from the
Primary Structure of Proteins . . . . . . . . . . . . . . . . . . . . . . . . . . . . . . . . .   257
*Marco Bongini, Vincenzo Laveglia, and Edmondo Trentin*

Using Radial Basis Function Neural Networks for Continuous and Discrete
Pain Estimation from Bio-physiological Signals . . . . . . . . . . . . . . . . . . .    269
  *Mohammadreza Amirian, Markus Kächele, and Friedhelm Schwenker*

Active Learning for Speech Event Detection in HCI . . . . . . . . . . . . . . . .    285
  *Patrick Thiam, Sascha Meudt, Friedhelm Schwenker, and Günther Palm*

Emotion Recognition in Speech with Deep Learning Architectures . . . . . . . .    298
  *Mehmet Erdal, Markus Kächele, and Friedhelm Schwenker*

On Gestures and Postural Behavior as a Modality in Ensemble Methods . . . .    312
  *Heinke Hihn, Sascha Meudt, and Friedhelm Schwenker*

Machine Learning Driven Heart Rate Detection with Camera
Photoplethysmography in Time Domain. . . . . . . . . . . . . . . . . . . . . . . .    324
  *Viktor Kessler, Markus Kächele, Sascha Meudt, Friedhelm Schwenker,
  and Günther Palm*

**Author Index** . . . . . . . . . . . . . . . . . . . . . . . . . . . . . . . . . . . . . . .    335

# Invited Papers

# Learning Sequential Data with the Help of Linear Systems

Luca Pasa and Alessandro Sperduti[✉]

Department of Mathematics, University of Padova, Padua, Italy
{lpasa,sperduti}@math.unipd.it

**Abstract.** The aim of the paper is to show that linear dynamical systems can be quite useful when dealing with sequence learning tasks. According to the complexity of the problem to face, linear dynamical systems may directly contribute to provide a good solution at a reduced computational cost, or indirectly provide support at a pre-training stage for nonlinear models. We present and discuss several approaches, both linear and nonlinear, where linear dynamical systems play an important role. These approaches are empirically assessed on two nontrivial datasets of sequences on a prediction task. Experimental results show that indeed linear dynamical systems can either directly provide a satisfactory solution, as well as they may be crucial for the success of more sophisticated nonlinear approaches.

**Keywords:** Linear dynamical systems · Autoencoders · Learning in sequential domains

## 1 Introduction

With the diffusion of cheap sensors, sensor-equipped devices (e.g., drones), and sensor networks (such as *Internet of Things* [1]), as well as the development of inexpensive human-machine interaction interfaces, the ability to quickly and effectively process sequential data is becoming more and more important. Many are the tasks that may benefit from advancement in this field, ranging from monitoring and classification of human behaviour to prediction of future events. Most of the above tasks require pattern recognition and machine learning capabilities.

Many are the approaches that have been proposed in the past to learn in sequential domains (e.g., [2]). A special mention goes to recent advancements involving Deep Learning [3–5]. Deep Learning is based on very non-linear systems, which reach quite good classification/prediction performances, very often at the expenses of a very high computational burden. Actually, it is common practice, when facing learning in a sequential, or more in general structured, domain to readily resort to non-linear systems. Not always, however, the task really requires a non-linear system. So the risk is to run into difficult and computational expensive training procedures to eventually get a solution that improves

© Springer International Publishing AG 2016
F. Schwenker et al. (Eds.): ANNPR 2016, LNAI 9896, pp. 3–17, 2016.
DOI: 10.1007/978-3-319-46182-3_1

of an epsilon (if not at all) the performances that can be reached by a simple
linear dynamical system involving simpler training procedures and a much lower
computational effort.

The aim of this paper is to open a discussion about the role that linear
dynamical systems may have in learning in sequential domains. On one hand,
we like to point out that a linear dynamical system (LDS) is able, in many
cases, to already provide good performances at a relatively low computational
cost. On the other hand, when a linear dynamical system is not enough to
provide a reasonable solution, we show how to resort to it to design quite effective
pre-training techniques for non-linear dynamical systems, such as Echo State
Networks (ESNs) [6] and simple Recurrent Neural Networks (RNNs) [7].

Specifically, here we consider the task of predicting the next event into a
sequence of events. Two datasets involving polyphonic music and quite long
sequences are used as practical exemplification of this task. We start by introducing a simple state space LDS. Three different approaches to train the LDS are
then considered. The first one is based on random projections and it is particularly efficient from a computational point of view. The second, computationally
more demanding approach, projects the input sequences onto an approximation
of their spanned sub-space obtained via a linear autoencoder naturally associated
to the LDS. For both approaches the output weights are obtained by computing
the pseudo-inverse of the hidden states matrix. Finally, we consider a refinement
via stochastic gradient descent of the solution obtained by the autoencoder-based
training scheme. Of course, this last approach requires additional computational
resources.

We then move to the introduction of non-linearities. From this point of view,
ESNs can be considered a natural extension of the first linear approach, since
non-linear random projections are used to define a coding of input sequences,
and pseudo-inverse exploited to estimate output weights. In addition, these are
the less computationally demanding models in the non-linear models arena. The
second considered family of non-linear models is given by simple RNNs, which
computationally are significantly more demanding. Here we experimentally show
that, at least for the addressed prediction task and the considered datasets,
the introduction of pre-training approaches involving linear systems leads to
quite large improvements in prediction performances. Specifically, we review pre-
training via linear autoencoder previously introduced in [8], and an alternative
based on Hidden Markov Models (HMMs) [9].

Finally, it is worth to notice that linear autoencoders have been exploited in a
recent theoretical work [10] to provide equivalence results between feed-forward
networks, that have simultaneous access to all items composing a sequence, and
single-layer RNNs which access information one step at a time.

## 2   Computational Models

In this section, we introduce the addressed learning task as well as the studied linear and non-linear models. Moreover, we present the different training
approaches for these models that we have experimentally assessed.

## 2.1 Learning Task

In this paper, we focus on a prediction task over sequences that can be formalized as described in the following. We would like to learn a function $\mathcal{F}(\cdot)$ from multivariate bounded length input sequences to desired output values. Specifically, given a training set $\mathcal{T} = \{(\mathbf{s}^q, \mathbf{d}^q) | q = 1, \ldots, N, \ \mathbf{s}^q \equiv (\mathbf{x}_1^q, \mathbf{x}_2^q, \ldots, \mathbf{x}_{l_q}^q), \mathbf{d}^q \equiv (\mathbf{d}_1^q, \mathbf{d}_2^q, \ldots, \mathbf{d}_{l_q}^q), \ \mathbf{x}_t^q \in \mathbb{R}^n, \ \mathbf{d}_t^q \in \mathbb{R}^s\}$, we wish to learn a function $\mathcal{F}(\cdot)$ such that $\forall q, t \ \mathcal{F}(\mathbf{s}^q[1, t]) = \mathbf{d}_t^q$, where $\mathbf{s}^q[1, t] \equiv (\mathbf{x}_1^q, \mathbf{x}_2^q, \ldots, \mathbf{x}_t^q)$. Experimental assessment has been performed in the special case in which $\mathbf{d}_k^q = \mathbf{x}_{k+1}^q$. Different learning approaches have been considered for both linear and non-linear dynamical systems, as described in the following.

## 2.2 Linear and Non-linear Models

The linear model we use is a discrete-state dynamical system defined as:

$$\mathbf{h}_t = \mathbf{A}\,\mathbf{x}_t + \mathbf{B}\,\mathbf{h}_{t-1}, \tag{1}$$

$$\mathbf{o}_t = \mathbf{C}\,\mathbf{h}_t, \tag{2}$$

where $\mathbf{h}_t \in \mathbb{R}^m$ is the state of the system at time $t$, $\mathbf{A} \in \mathbb{R}^{m \times n}$, $\mathbf{B} \in \mathbb{R}^{m \times m}$, $\mathbf{C}^{s \times m}$ are respectively the input matrix, the state matrix and the output matrix. In addition, we assume $\mathbf{h}_0 = \mathbf{0}$, i.e. the null vector.

Associated with this dynamical system, we consider a linear autoencoder obtained by substituting Eq. (2) with

$$\begin{bmatrix} \mathbf{x}_t \\ \mathbf{h}_{t-1} \end{bmatrix} = \mathbf{C}\,\mathbf{h}_t, \tag{3}$$

where $\mathbf{C} \in \mathbb{R}^{(n+m) \times m}$, and $m$ takes the smallest value satisfying Eqs. (1) and (3). Specifically, the smallest value of $m$ can be found as proposed in [11], i.e. by factorisation of the state matrix $\mathbf{H}$ collecting as rows the state vectors of the linear system described by Eq. (1). For the sake of presentation, let illustrate such factorisation for a single sequence $\mathbf{s} \equiv (\mathbf{x}_1, \mathbf{x}_2, \ldots, \mathbf{x}_l)$

$$\underbrace{\begin{bmatrix} \mathbf{h}_1^\mathsf{T} \\ \mathbf{h}_2^\mathsf{T} \\ \mathbf{h}_3^\mathsf{T} \\ \vdots \\ \mathbf{h}_l^\mathsf{T} \end{bmatrix}}_{\mathbf{H}} = \underbrace{\begin{bmatrix} \mathbf{x}_1^\mathsf{T} & \mathbf{0} & \mathbf{0} & \cdots & \mathbf{0} \\ \mathbf{x}_2^\mathsf{T} & \mathbf{x}_1^\mathsf{T} & \mathbf{0} & \cdots & \mathbf{0} \\ \mathbf{x}_3^\mathsf{T} & \mathbf{x}_2^\mathsf{T} & \mathbf{x}_1^\mathsf{T} & \cdots & \mathbf{0} \\ \vdots & \vdots & \vdots & \vdots & \vdots \\ \mathbf{x}_l^\mathsf{T} & \mathbf{x}_{l-1}^\mathsf{T} & \cdots & \mathbf{x}_2^\mathsf{T} & \mathbf{x}_1^\mathsf{T} \end{bmatrix}}_{\Xi} \underbrace{\begin{bmatrix} \mathbf{A}^\mathsf{T} \\ \mathbf{A}^\mathsf{T}\mathbf{B}^\mathsf{T} \\ \mathbf{A}^\mathsf{T}\mathbf{B}^{2^\mathsf{T}} \\ \vdots \\ \mathbf{A}^\mathsf{T}\mathbf{B}^{l-1^\mathsf{T}} \end{bmatrix}}_{\Omega}$$

where, $\Xi \in \mathbb{R}^{l \times n \cdot l}$ is the data matrix which collects the (inverted) subsequences $\mathbf{s}[1, i], \ \forall i = 1, \ldots, l$, as rows, and $\Omega$ is the parameter matrix of the dynamical system.

The smallest value of $m$ preserving all information about $\mathbf{s}$ is obtained by thin svd decomposition of $\Xi = \mathbf{V}\mathbf{\Lambda}\mathbf{U}^\mathsf{T}$. In fact, by imposing that $\mathbf{U}^\mathsf{T}\Omega = \mathbf{I}$, i.e.

the identity matrix, the resulting state space preserves all information and it is of the smallest possibile dimension $m = rank(\Xi)$, i.e. the number of nonzero singular values of $\Lambda$. Matrices $\mathbf{A}$ and $\mathbf{B}$ satisfying such condition can be obtained by exploiting the fact that $\mathbf{U}\mathbf{U}^\mathsf{T} = \mathbf{I}$, which entails $\Omega = \mathbf{U}$. This equality can be met using matrices

$$\mathbf{P}_{n,n\cdot l} \equiv \begin{bmatrix} \mathbf{I}_{n\times n} \\ \mathbf{0}_{n(l-1)\times n} \end{bmatrix}, \text{ and } \mathbf{R}_{n,n\cdot l} \equiv \begin{bmatrix} \mathbf{0}_{n\times n(l-1)} & \mathbf{0}_{n\times n} \\ \mathbf{I}_{n(l-1)\times n(l-1)} & \mathbf{0}_{n(l-1)\times n} \end{bmatrix},$$

to define $\mathbf{A} \equiv \mathbf{U}^\mathsf{T}\mathbf{P}_{n,n\cdot l}$ and $\mathbf{B} \equiv \mathbf{U}^\mathsf{T}\mathbf{R}_{n,n\cdot l}\mathbf{U}$. Moreover, since $\mathbf{H} = \mathbf{V}\Lambda$, the original data $\Xi$ can be fully reconstructed by computing $\mathbf{H}\mathbf{U}^\mathsf{T} = \mathbf{V}\Lambda\mathbf{U}^\mathsf{T} = \Xi$, which can be achieved by running the dynamical system

$$\begin{bmatrix} \mathbf{x}_t \\ \mathbf{h}_{t-1} \end{bmatrix} = \begin{bmatrix} \mathbf{A}^\mathsf{T} \\ \mathbf{B}^\mathsf{T} \end{bmatrix} \mathbf{h}_t$$

starting from $\mathbf{h}_l$, i.e. $\mathbf{C} = \begin{bmatrix} \mathbf{A}^\mathsf{T} \\ \mathbf{B}^\mathsf{T} \end{bmatrix}$. The very same result can be obtained when considering a set of sequences. It is enough to stack all the data matrices corresponding to sequences in the set, and padding with zeros when needed. For example, given the set $\mathcal{S} = \{\mathbf{s}_1, \mathbf{s}_2\}$ with $length(\mathbf{s}_1) = 3$ and $length(\mathbf{s}_2) = 2$, the data matrix corresponding to the set is defined as $\Xi_\mathcal{S} = \begin{bmatrix} \Xi_{\mathbf{s}_1} \\ \Xi_{\mathbf{s}_2}\mathbf{0}_{n\times n} \end{bmatrix}$.

It's crucial to notice that the above method works exactly only in the case where $m$ is equal to the rank of $\Xi$. Due to this fact, applying this method in real world scenarios could be difficult due to the fact that the rank of matrix $\Xi$ is very large. For this reason, in [8] a procedure for approximate computation of the truncated thin svd decomposition of $\Xi$ to a preset value of $m \ll rank(\Xi)$ is proposed.

The considered non-linear model is obtained by adding nonlinear functions $f()$ and $g()$ to Eqs. (1) and (2), i.e.:

$$\tilde{\mathbf{h}}_t = f(\mathbf{A}\,\mathbf{x}_t + \mathbf{B}\,\tilde{\mathbf{h}}_{t-1}), \tag{4}$$

$$\tilde{\mathbf{o}}_t = g(\mathbf{C}\,\tilde{\mathbf{h}}_t). \tag{5}$$

We have considered two specific instances of the above model: *(i)* when using it as an ESN, $f()$ as been instantiated to $tanh()$ and $g()$ to the identity function; *(ii)* when using it as a simple RNN, both $f()$ and $g()$ have been instantiated to $tanh()$.

## 2.3   Training Approaches

For the linear model described by Eqs. (1) and (2), we consider three different training approaches:

$\mathcal{L}_1$: Randomly generate matrices $\mathbf{A}$ and $\mathbf{B}$; compute the corresponding state matrix $\mathbf{H}$ for the full set of training sequences; define $\mathbf{C} = \mathbf{D}\mathbf{H}^{\mathsf{T}^+}$, where

$\mathbf{D} = [\mathbf{d}_1^1, \mathbf{d}_2^1, \ldots, \mathbf{d}_{l_N}^N]$ is the matrix collecting all the target vectors, and $\mathbf{H}^{\mathsf{T}+}$ is the pseudo-inverse of $\mathbf{H}^{\mathsf{T}}$. The use of pseudo-inverse leads to the minimization of the Mean Squared Error between target and actual output. This approach corresponds to a ESN-like training for linear systems.

$\mathcal{L}_2$: Compute $\mathbf{A}$ and $\mathbf{B}$ according to the procedure proposed in [8] for the linear auto-encoder; compute the corresponding state matrix $\mathbf{H}$ for the full set of training sequences; define $\mathbf{C} = \mathbf{D}\mathbf{H}^{\mathsf{T}+}$.

$\mathcal{L}_3$: Perform stochastic gradient descent (SGD) with respect to the regularized error function

$$E_T = \frac{1}{NL} \sum_{q=1}^{N} \sum_{j=1}^{l_q} (\mathbf{d}_j^q - \mathbf{o}_j^q)^2 + R_1 + R_2,$$

where $L = \sum_{q=1}^{N} l_q$, and

$$R_1 = |\sum_{i}^{m} \sum_{j}^{n} \mathbf{A}_{ij}| + |\sum_{i}^{m} \sum_{j}^{m} \mathbf{B}_{ij}| + |\sum_{i}^{s} \sum_{j}^{m} \mathbf{C}_{ij}|,$$

$$R_2 = \sum_{i}^{m} \sum_{j}^{n} \mathbf{A}_{ij}^2 + \sum_{i}^{m} \sum_{j}^{m} \mathbf{B}_{ij}^2 + \sum_{i}^{s} \sum_{j}^{m} \mathbf{C}_{ij}^2,$$

with starting point given by the result of approach $\mathcal{L}_2$ (pre-training).

Concerning approach $\mathcal{L}_1$, a relevant issue is how to generate matrices $\mathbf{A}$ and $\mathbf{B}$. This issue has already be addressed in ESN. Indeed, in order to avoid problems in computing the system state and ensure good results, a set of rules to follow for random matrix initialization has been proposed. This set of rules is called Echo State Property [12], and in particular they prescribe to ensure that the random initialized matrices have *spectral radius* $\rho$ less than or equal to 1. Unfortunately computing the spectral radius of large matrices is computationally demanding, so we use a much faster approach where we require $\mathbf{A}$ and $\mathbf{B}$ to have norm $\|\cdot\|$ (either L1-norm or L2-norm) less than or equal to 1. Since for any matrix $\mathbf{M}$, $\rho(\mathbf{M}) \leq \|\mathbf{M}\|$, in this way the Echo State Property is preserved.

For the non-linear model described by Eqs. (4) and (5), we consider four different training approaches:

$\mathcal{N}_1$: Adopt the Echo State Network training procedure that randomly initializes matrices $\mathbf{A}$ and $\mathbf{B}$ according to the Echo State Property and only trains the output weights using pseudo-inverse of the hidden representations, i.e. compute $\mathbf{C} = \mathbf{D}\tilde{\mathbf{H}}^+$, where $\tilde{\mathbf{H}} = [\tilde{\mathbf{h}}_1^1, \tilde{\mathbf{h}}_2^1, \ldots, \tilde{\mathbf{h}}_{l_N}^N]$ is the matrix that collects all the hidden representations obtained by running the system, with random matrices $\mathbf{A}$ and $\mathbf{B}$, over all sequences in the training set.

$\mathcal{N}_2$: Perform SGD with respect to the regularized error function $E_T$, with standard random initialization for matrices $\mathbf{A}$, $\mathbf{B}$, and $\mathbf{C}$. This corresponds to a standard RNN training procedure.

$\mathcal{N}_3$: Perform SGD with respect to the regularized error function $E_{\mathcal{T}}$, initializing matrices **A** and **B** according to the procedure proposed in [8] for the linear auto-encoder, and **C** with $\mathbf{D}\tilde{\mathbf{H}}^+$, where $\tilde{\mathbf{H}}$ is obtained by first running Eq. (4) with initialized matrices **A** and **B**. This approach corresponds to the pre-training one proposed in [8].

$\mathcal{N}_4$: Perform SGD with respect to the regularized error function $E_{\mathcal{T}}$, starting from matrices **A**, **B**, and **C** obtained by the following pre-training approach: *(i)* train a linear HMM over the training set $\mathcal{T}$; *(ii)* using the obtained HMM, generate $N_{pt}$ sequences that will constitute the pre-training dataset $\mathcal{T}_{pr}$; *(iii)* perform SGD with respect to the regularized error function $E_{\mathcal{T}_{pr}}$ using $\mathcal{T}_{pr}$ and standard random initialization for matrices **A**, **B**, and **C**. This approach corresponds to the pre-training one proposed in [9].

## 3   Experimental Assessment

In this section we are going to compare the capability of the different systems. The task chosen for testing the various models is the prediction of polyphonic music sequences. The learning task consists in predicting the notes played at time $t$ given the sequence of notes played till time $t - 1$. This task turns out to be really interesting because of the nature of the data. Indeed the music sequences are complex, and follow a multi-modal complex distribution that makes difficult to perform the training on them. Moreover prediction of these sequences requires a good capability by the network in managing long-term temporal dependencies. This task has been already tested on different models [8,9,13]. The models and associated training approaches have been tested on two different datasets:

- Nottingham dataset, that contains over 1000 Folk tunes stored in a special text format. These tunes have a simple structure and the set of notes and chords used is quite small.
- Piano-midi.de dataset that contains classic music songs that present a complex structure and that use a wide range of different cords and notes. The songs are longer than the tunes contained in the Nottingham dataset.

We have decided to use these two datasets because they contain very different type of data. Moreover, they allow to stress the models in order to understand the strengths and weaknesses of them. The sequences contained in the dataset are the midi representation of music songs. Each song in the dataset is represented by using a sequence of binary vectors. Each binary vector is composed of 88 values that represent each single note in the piano-roll that span the whole range from A0 to C8. The $i$-th bit of $j$-th vector is set to one only if the $i$-th note is played at $j$-th time step of the considered song. The average number of bits set to 1 (maximum polyphony) is 15 (average 3.9). The music sequences contained in the datasets have variable length. In particular, in Piano-midi.de many sequences are composed of thousands of time steps. Statistics on the dataset (largest sequence length, number of sequences, etc.) are given in [9].

As performance measure we adopted the accuracy measure used in [13] and described in [14]. Each dataset is split in training set, validation set, and test set.

We first discuss the experimental results obtained for approaches where unsupervised linear or non-linear projections are used to define the current state, i.e. there is no supervised learning of the hidden state mapping, while supervision is used for the hidden to output mapping via pseudo-inverse. These approaches are: LDS-$\mathcal{L}_1$, LDS-$\mathcal{L}_2$, and RNN-$\mathcal{N}_1$. Because of the unsupervised projections, a larger state space is supposedly needed for these approaches to get good performances. Subsequently, we present experimental results obtained for SGD-based approaches, i.e. LDS-$\mathcal{L}_3$, RNN-$\mathcal{N}_2$, RNN-$\mathcal{L}_3$, and RNN-$\mathcal{N}_4$, where a smaller state space is required.

## 3.1 Results of Approaches Using Unsupervised Projections

As discussed before, approaches that use random or unsupervised projections, in principle, require larger state spaces in order to get good performances. A profitable size for the state space also depends on the complexity of the dataset. We have explored this issue by performing experimental tests using 500, 1000, 1500 hidden units for the Nottingham dataset, and 500, 1000, 2000 hidden units for the Piano-midi.de dataset. The use of 2000 units takes into account the higher complexity of the Piano-midi.de dataset. The considered approaches are LDS-$\mathcal{L}_1$, LDS-$\mathcal{L}_2$, and RNN-$\mathcal{N}_1$. Experimental results for the Notthingam dataset are shown in Fig. 1, while the results for the Piano-midi.de dataset are shown in Fig. 2.

In general, it seems that random projection-based approaches (i.e. LDS-$\mathcal{L}_1$ and RNN-$\mathcal{N}_1$) are insensitive to the size of the state space, while this seems not to be the case for LDS-$\mathcal{L}_2$. In fact, the autoencoder-based training approach seems to be sensitive to the state space size. This fact can be explained by considering the nature of the training technique. Indeed, the autoncoder-based method exploits the SVD decomposition in order to extract the most relevant information from input and previous states. What happens by increasing the state space size is that those features that have lower variance will be added to the state representation. This allows to collect more relevant information in the state. In other words, by using this approach with increasing size of state space, it very likely that some new relevant features enter the state representation. Therefore, since the size of state space is $\ll rank(\Xi)$, the approach will improve its performance (at least on the training set). For both datasets it seems that a further increase of the state space size would keep improving performances. Finally, it can be noticed that on the Piano-midi.de dataset the RNN-$\mathcal{N}_1$ approach seems to show some overfitting with the increase of state space size.

## 3.2 Results of Approaches Using Supervision and Pre-training

Here we present the results obtained for approaches exploiting supervision and pre-training, i.e. LDS-$\mathcal{L}_3$, RNN-$\mathcal{L}_2$, and RNN-$\mathcal{N}_3$, and RNN-$\mathcal{N}_4$. Actually, all

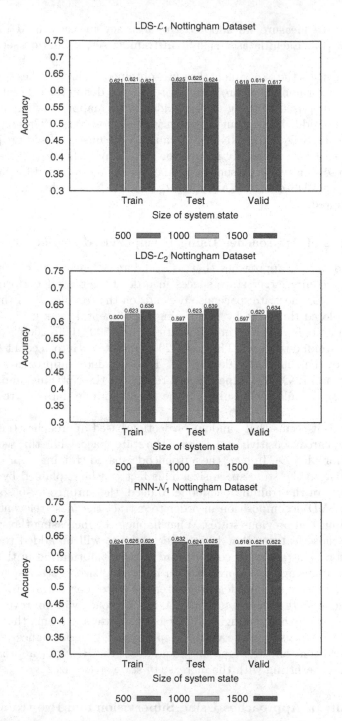

**Fig. 1.** Results achieved for the Nottingham dataset. Each chart shows the accuracy achieved by training LDS-$\mathcal{L}_1$, LDS-$\mathcal{L}_2$, and RNN-$\mathcal{N}_1$ with different state space sizes.

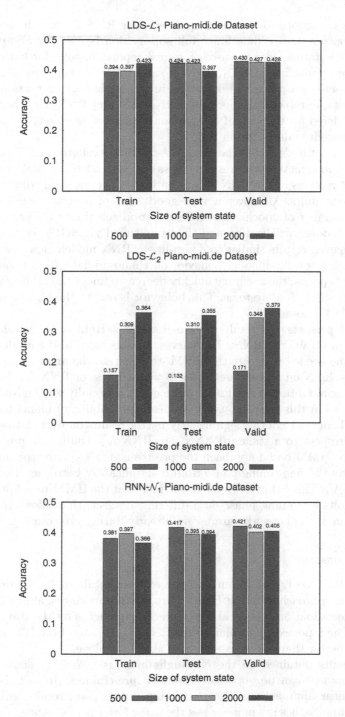

**Fig. 2.** Results achieved by LDS Piano-midi.de dataset. Each chart shows the accuracy achieved by training LDS-$\mathcal{L}_1$, LDS-$\mathcal{L}_2$, and RNN-$\mathcal{N}_1$ with different state space sizes.

approaches use a form of pre-training except for RNN-$\mathcal{L}_2$, which uses random inizialization for the weights. Since full supervision is used for all approaches, much smaller state spaces are used. Figure 3 reports results, for both datasets, obtained for approaches LDS-$\mathcal{L}_3$, RNN-$\mathcal{L}_2$, and RNN-$\mathcal{N}_3$ using the same settings and model selection procedure described in [8], while Fig. 4 reports results, for both datasets, obtained for approaches RNN-$\mathcal{N}_4$, and RNN-$\mathcal{N}_2$, where the last one is considered for the sake of comparison, using the same settings and model selection procedure described in [9].

From Fig. 3 it is clear that the use of pre-training exploiting the linear autoencoder allows to achieve better results in less epochs. Indeed, in both datasets the pre-trained versions of the RNN obtain an higher accuracy regardless the number of hidden units. Moreover a very good level of accuracy can be reached in a lower number of epochs. The LDS-$\mathcal{L}_3$ approach (i.e., LDS pre-trained via linear autoncoder initialization and then fine-tuned via SGD) on Notthingham dataset achieved results similar to a non-linear RNN model. However, the same approach has a totally different behavior on Piano-midi.de. In that case, after a few training epochs, the accuracy quickly decreases under the accuracy achieved by randomly-initialized systems. This behavior is due to the high complexity of sequences in Piano-midi.de.

Figure 4 presents the results obtained when a Hidden Markov Model [15] is used to pre-train the RNN. The curves, in this case, also consider the pre-training time needed to create the HMM, to generate the artificial dataset, and to train the RNN on this dataset. This is why curves for RNN-$\mathcal{N}_4$ do not start from the origin. Obtained results are quite good, especially for the Notthingham dataset. Even in this case the use of a linear (probabilistic) model to pre-train the network allows to obtain significantly better results on both datasets, in less time with respect to a standard RNN, i.e. RNN-$\mathcal{N}_2$. Unlike the previous case, by using the HMM-based approach the pre-trained RNN, after the pre-training phase, starts the fine-tuning phase from a significantly better level of accuracy than RNN-$\mathcal{N}_2$. This behavior is due to the fact that the HMM-based pre-training phase exploits a training phase on a different dataset, that allows to start the optimization of the network parameters already during pre-training.

### 3.3    Discussion

In this section, we try to summarize the experimental results obtained for all the different approaches. In Fig. 5 we report the performances, after model selection via the validation sets, of all the studied approaches on the two considered datasets. The reported performances for RNN-$\mathcal{N}_2$ is taken from [13], since those values are better than the ones we where able to achieve.

The results obtained on the Nottingham dataset seem to show that better performances can be obtained thanks to pre-training. In fact, both linear and non-linear approaches with pre-training return good results, while all the remaining approaches get more or less the same lower performance, regardless of the linearity or non-linearity of the considered model. Among approaches that exploit pre-training, non-linear models get a slighter better performance. It must

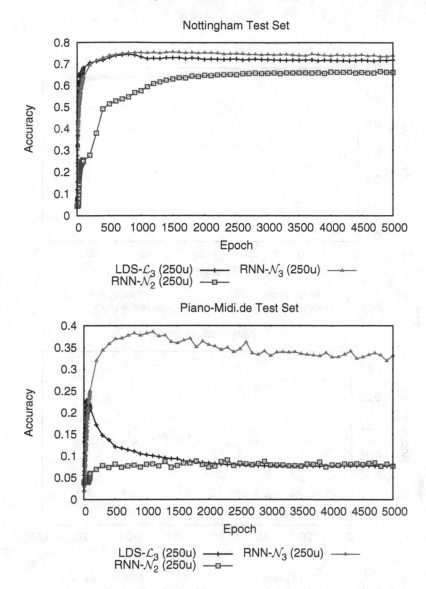

**Fig. 3.** Test curves on the two datasets by models LDS-$\mathcal{L}_3$, RNN-$\mathcal{N}_2$ and RNN-$\mathcal{N}_3$. Curves are sampled at each epoch till epoch 100, and at steps of 100 epochs afterwards.

however be noticed that the computationally less demanding linear model with pre-training already returns a quite good performance.

For the Piano-midi.de, the situation seems to be quite different, since the best performers are based on random projections, independently from the linearity or non-linearity of the model. Pre-training based approaches seem, for this dataset, to be less effective. This may be due to the fact that the Piano-midi.de owns a

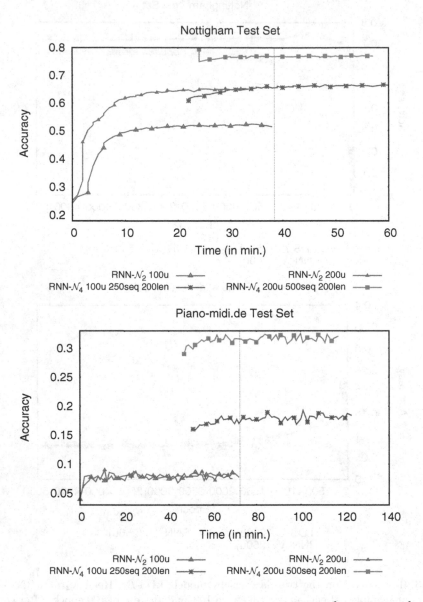

**Fig. 4.** Test curves obtained for the two datasets by using RNN-$\mathcal{N}_2$ and RNN-$\mathcal{N}_4$ with 250 hidden units. Each curve starts at the time when the pre-training ends. The labels associated with RNN-$\mathcal{N}_4$ curves are composed by three identifiers $n_1$u $n_2$ n3, where $n_1$ is the number of used hidden units, $n_2$ is the number of sequences generated by the HMM with 10 states, and $n_3$ is the length of such sequences. Curves that refer to RNN-$\mathcal{N}_2$ are identified by the number of used hidden units. The dotted vertical line is used to mark the end of training of RNN-$\mathcal{N}_2$ after 5000 epochs.

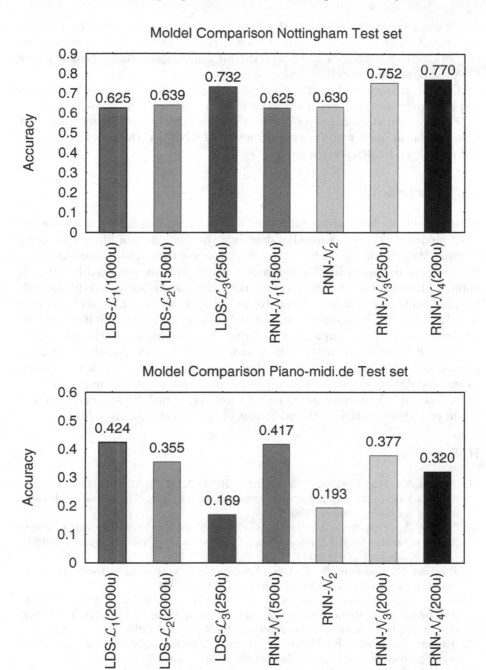

**Fig. 5.** Comparison of the performances obtained by all the studied approaches. For each model the charts report the accuracy achieved on test set. The number reported in brackets is the number of hidden units selected by model selection using the validation set. The results for RNN-$\mathcal{N}_2$ are taken from [13].

higher complexity, and it is likely that far larger state spaces are needed to reach better performances. In fact, it can be noted that approaches based on random projections, either linear or non-linear, exploit a state space that is tenfold larger than the ones used by SGD-based approaches.

Overall, it is clear that the adoption of a pre-training approach allows to systematically improve over the standard RNN approach, i.e. RNN-$\mathcal{N}_2$.

From a computational point of view linear approaches are far more efficient than non-linear ones, with the only exception of RNN-$\mathcal{N}_1$, that does not require training of the hidden state mapping.

## 4   Conclusion

In this paper, we have addressed a issue that is often disregarded when considering prediction tasks in sequential domains: the usefulness of linear dynamical systems. We empirically studied the performances of three linear approaches and four non-linear approaches for sequence learning on two significantly complex datasets. Experimental results seem to show that directly or indirectly (as basis for pre-training approaches) linear dynamical systems may play an important role. In fact, when used directly they may by themselves return state-of-the-art performance (see for example LDS-$\mathcal{L}_3$ for the Nottingham dataset) while requiring a much lower computational effort with respect to their non-linear counterpart. Moreover, even when linear models do not perform well, it is always possible to successfully exploit them within pre-training approaches for non-linear systems. Thus, it is important, when facing a prediction task for sequences, to take in full consideration the contribution that linear models can give.

## References

1. Ashton, K.: That internet of things thing. RFiD J. **22**(7), 97–114 (2009)
2. Sun, R., Giles, C.L. (eds.): Sequence Learning - Paradigms, Algorithms, and Applications. Springer, London (2001)
3. Graves, A., Mohamed, A., Hinton, G.: Speech recognition with deep recurrent neural networks. In: IEEE International Conference on Acoustics, Speech and Signal Processing, pp. 6645–6649. IEEE (2013)
4. Pascanu, R., Gulcehre, C., Cho, K., Bengio, Y.: How to construct deep recurrent neural networks (2013). arXiv preprint arXiv:1312.6026
5. Gregor, K., Danihelka, I., Graves, A., Rezende, D.J., Wierstra, D.: Draw: A recurrent neural network for image generation (2015). arXiv preprint arXiv:1502.04623
6. Jaeger, H.: Echo state network. Scholarpedia **2**(9), 2330 (2007)
7. Jerome, T., Connor, R., Martin, D.: Recurrent neural networks and robust time series prediction. IEEE Trans. Neural Netw. **5**(2), 240–254 (1994)
8. Pasa, L., Sperduti, A.: Pre-training of recurrent neural networks via linear autoencoders. In: Advances in Neural Information Processing Systems, pp. 3572–3580 (2014)
9. Pasa, L., Testolin, A., Sperduti, A.: Neural networks for sequential data: a pre-training approach based on hidden Markov models. Neurocomputing **169**, 323–333 (2015)

10. Sperduti, A.: Equivalence results between feedforward and recurrent neural networks for sequences. In: Proceedings of the 24th International Conference on Artificial Intelligence, pp. 3827–3833. AAAI Press (2015)
11. Sperduti, A.: Exact solutions for recursive principal components analysis of sequences and trees. In: Kollias, S.D., Stafylopatis, A., Duch, W., Oja, E. (eds.) ICANN 2006. LNCS, vol. 4131, pp. 349–356. Springer, Heidelberg (2006)
12. Herbert, J.: The echo state approach to analysing and training recurrent neural networks-with an erratum note. Technical report, German National Research Center for Information Technology GMD, Bonn, Germany, 148:34 (2001)
13. Boulanger-Lewandowski, N., Bengio, Y., Vincent, P.: Modeling temporal dependencies in high-dimensional sequences: application to polyphonic music generation and transcription. In: ICML (2012)
14. Bay, M., Ehmann, AF., Downie, J.S.: Evaluation of multiple-F0 estimation and tracking systems. In: ISMIR, pp. 315–320 (2009)
15. Rabiner, L.R.: A tutorial on hidden Markov models and selected applications in speech recognition. Proc. IEEE $77$(2), 257–286 (1989)

# A Spiking Neural Network for Personalised Modelling of Electrogastrography (EGG)

Vivienne Breen[1](✉), Nikola Kasabov[1], Peng Du[2], and Stefan Calder[2]

[1] Knowledge Engineering and Discovery Research Institute,
Auckland University of Technology, Auckland, New Zealand
{vbreen,nkasabov}@aut.ac.nz
[2] Auckland Bioengineering Institute, University of Auckland,
Auckland, New Zealand
{peng.du,scal044}@auckland.ac.nz

**Abstract.** EGG records the resultant body surface potential of gastric slow waves (electrical activity); while slow waves regulate contractions of gastric muscles, it is the electrical activity we are recording, not movement (like ECG records the cardiac electrical activity, but not the contractions of the heart, even the two are essentially related).

**Keywords:** Spiking neural network · Personalised modelling · Electrogastrography · Functional dyspepsia · Slow wave dysrhythmia

## 1 Introduction

Spiking neural networks have been used successfully to model spatiotemporal data, especially neurological data. The present study utilizes the spiking neural network setup within NeuCube [1] developed at the Knowledge Engineering and Development Research Institute (KEDRI), at Auckland University of Technology. NeuCube has been used to model not just brain data but a number of other types of temporal based data, for example stroke prediction [2], ecological data [3], and seismic data which is currently under investigation by KEDRI researchers. The power within this system lies in the customization available in many aspects of its execution, from the neuron locations, the spike encoding method, the combination of supervised and unsupervised learning, to the visualization of interactions within the network and predictive results. The full capabilities of this type of infrastructure within NeuCube are yet to be realized, with continual development and improvement from the researchers within KEDRI.

## 2 Background

Digestion is facilitated by the motility of the stomach, which in turn is governed by an electrophysiological event called slow waves. Dysrhythmias of slow waves have been associated with a number of digestive diseases, including gastroparesis, unexplained nausea and vomiting, and functional dyspepsia [4]. Electrogastrography (EGG) is a non-invasive method of recording the resultant body surface potential of gastric slow

© Springer International Publishing AG 2016
F. Schwenker et al. (Eds.): ANNPR 2016, LNAI 9896, pp. 18–25, 2016.
DOI: 10.1007/978-3-319-46182-3_2

wave propagation. EGG has the ability to convey information about gastric slow wave activity in terms of frequency, amplitude and propagation, therefore it holds potential as a routinely deployable tool to aid the diagnosis of gastric functions.

The present study investigates the spatiotemporal data recorded by EGG as it relates specifically to functional dyspepsia. Functional dyspepsia is the condition where the gut does not function as it should but this is not caused by biological or bacterial factors. An efficient way of measuring what is occurring in the gut in a non-invasive manner is to utilize EGG, where a patient wears mesh of sensors on around their torso that records the resultant surface electrical potentials due to gastric slow waves. The patient was asked to stay still and the movement recorded at 100 Hz. EGG readings are able to reflect a number of different types of functional dyspepsia as it relates to the dysrhythmic slow waves, which in turn cause digestive difficulties. EGG is less invasive than the alternative of endoscopic investigation where a scope is introduced into the stomach via the mouth. Although this is an effective method of investigation for visual inspection and biopsy taking, it is generally deployed for a limited time and does not perform functional measurements on the stomach.

Personalised modelling utilizes a different focus than that of global or local modelling, for a description of the differences see [5]. This approach not only allows for the prediction of outcomes and model development for an individual, but functions well in situations where low numbers of samples are present where more traditional modelling and statistical methods are not applicable. This type of modelling also allows for more accurate modelling of subgroups within a large population space.

## 3  Investigation

EGG recordings were acquired using a mesh of sensors on the torso. The sensors detect the far-field potential of the sequential activation of gastric slow waves generated by the stomach as it contracts to digest food. With this in mind, retaining the spatiotemporal relationships is vitally important. Each trial was sampled at 100 Hz and a total of 400 time units were recorded. This covered two repeated cycles of the gastric functionality being tracked. There were 7 samples available for each of the 4 classes (28 samples in total). The samples represented the baseline data: normal (N), and three common cases of slow wave dysrhythmias: re-entry (R), conduction block (C) and ectopic pacemaker (E).

To determine the best way of encoding the original signal into spikes for the SNN three different methods were tested. Firstly thresholding the difference between two consecutive variables (TR), secondly Ben's Spiking Algorithm (BSA), and thirdly Moving Window (MW). Each of these methods captures different components of the input signals in a different way and it was not known beforehand which would best capture the idiosyncrasies of the input data.

As NeuCube has the functionality to allow user specification of node location this facility was used to determine the configuration of both input and computational nodes within the cube itself. As Fig. 1 shows the input nodes are arranged as they are spatially on a person retaining their relationship and allowing the visualization of input node activation sequences which directly reflect the actual movement of the subjects'

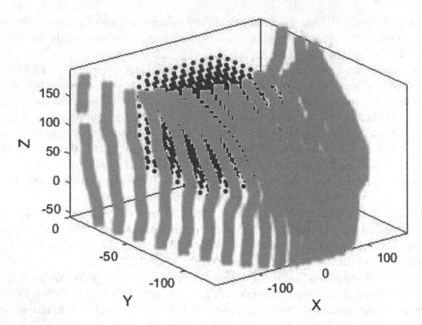

**Fig. 1.** NeuCube node layout. Yellow are input, blue are computational nodes (Color figure online)

muscles. The computational portion is represented as a cube of nodes located away from the input nodes to determine the type of input pattern received. The visualization of the connections between the calculation nodes and their activation levels gives a representation of the nature of the computation required.

## 4  Results and Discussion

The classification results of each sample computed individually are grouped and reported in Table 1. With 7 samples of each type being tested the result of 81.71 % for Ectopic pacemaker under BSA spike encoding, represents only one sample having been misidentified.

As this is a "proof of concept" investigation several combinations of neural structure, encoding method and class combination were tried. The best classification ac-curacy is listed in Table 1 which exceeded that if any of the two class tests. In the two class testing each dysrhythmia was tested against the normal data. Some dys-rhythmias, for example re-entry, returned a 100 % accuracy level for each of the spike encoding methods. Other dysfunctions, for example ectopic pacemaker, re-turned mixed results for each encoding method which were much lower than the four class situation. The worst of these was a result of 57.14 % under TR encoding. Comparing the various spike encoding methods over all testing scenarios the moving window (MW) method gave the most consistent and accurate classification results. Viewing the spike encoding versus the original signal (Fig. 2) for this method shows that it

**Table 1.** Testing all 4 classes in one model - modified configuration

| Condition | TR | BSA | MW |
| --- | --- | --- | --- |
| Normal | 100 | 100 | 100 |
| Re-entry | 71.43 | 71.43 | 100 |
| Ectopic pacemaker | 100 | 85.71 | 100 |
| Conduction block | 100 | 85.71 | 100 |
| Overall | 92.86 | 85.71 | 100 |

represents the physical activity under investigation very closely. Figure 2 represents a single input node of the 851 total input nodes. Each node reveals a slightly different pattern of spikes relating to the placement of the sensor.

NeuCube has the capability to retain the activation sequence for the input nodes. Replaying this allows the user to watch repeatedly the sequence of node activation which represents the physical activity of the patient; to stop at any point, replay sections, and concentrate on individual points of interest whilst the patient need only be recorded once. This allows the reviewer, be it physician or trainee doctor, to look more deeply into what is happening at any point in time and to follow any sequences of interest. This is especially useful when dealing with irregular irregularities, as dys-rhythmias may occur infrequently and often at unpredictable intervals [6].

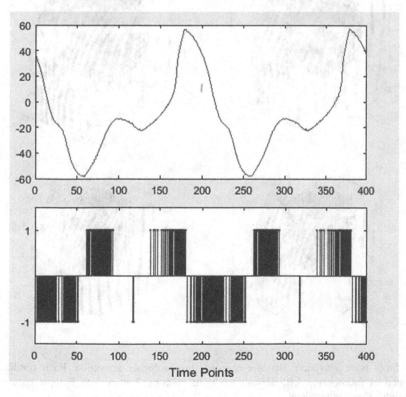

**Fig. 2.** Moving window spike encoding representation.

From the computational perspective the final version of node layout and size (see Fig. 1) resulted in a reduction in the number of nodes needed, the computational time, and memory overhead. The first version had over 4000 nodes which included the input nodes; whereas this final version contains the 851 input nodes and 1000 computational nodes. Spatially separating the input nodes from the computational ones was done in an attempt to increase accuracy. Retesting each of the previous scenarios of classification, both 2 class and 4 class setups, resulted in each class re-turning better results, or at the minimum remaining the previous accuracy. The best overall result (Table 1) is the only one where a full 100 % accuracy was achieved for all dysrhythmias and exceeded that of the two class versions. Analysis of the computational nodes reveals a very active matrix where a large number of connections are present between nodes (Fig. 4). Node activation is seen as light and dark shading of the node as in the input node activation of Fig. 3, and connections by joining lines. Connection direction is represented by a blue line if the connection is exhibitive of the next node and red if it is inhibitive of the node. The strength of this connection is seen in the strength of each line.

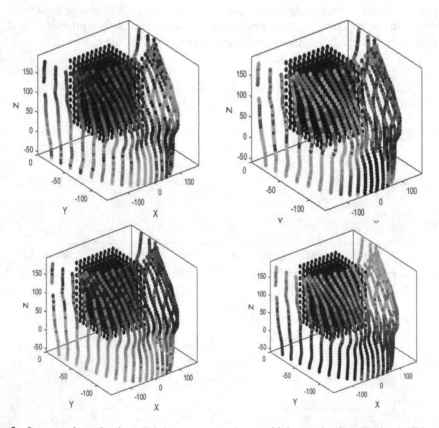

**Fig. 3.** Input node activation. Brighter neurons show a higher activation. Each condition is represented as follows; top left, normal; top right, re-entry; bottom left, Ectopic pacemaker; bottom right, Conduction block.

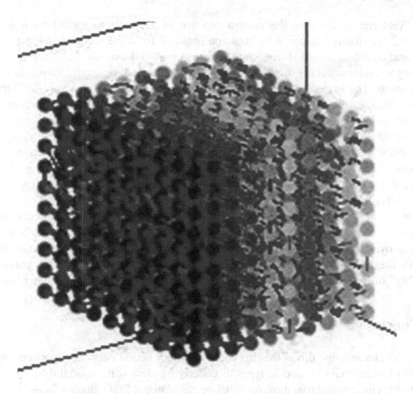

**Fig. 4.** Connections in the computational nodes (Color figure online)

The current parameter settings available in NeuCube have not been optimized at this point in time. The results so far indicate that this method of analysis has great potential for further development. Parameter optimization is left to a later stage as the next version of the input sensor mesh is very likely to contain fewer sensors which will necessitate alterations to the structure of the current design and parameter settings. These early results reveal both the computational and visual representational sides of NeuCube as being integral to the development of diagnostic aids and increased knowledge of the individual patients.

## 4.1 Implications of This Research

To our best knowledge this is the first study in which neural network algorithms were applied to relate EGG to specific types of underlying slow wave spatial dysrhythmias. Prior to the current analysis, most previous neural network EGG studies focused on the temporal variations of slow waves, which might have missed certain spatial dysrhythmias that could occur at the same frequency as the normal activity [6]. By analyzing EGG over sustained dysrhythmias, the algorithms were able to detect exact intervals when irregularities occurred, which would be useful as slow wave dysrhythmias have often been reported to be dynamic, as demonstrated by the re-entry case. Furthermore, because the current practice of EGG is to record from a few channels, only limited information could be obtained from the recording. The present study analyzed data recorded from multiple sensors placed over the torso surface, thus ensuring a greater accuracy in pinpointing the type of spatial dysrhythmia occurring in the stomach. The increased amount of EGG data could also offer a way for further analysis after each recording session for greater details of gastric slow waves and functions, in order to minimize patient's time in hospital. Once trained and validated, the algorithms should allow clinicians a relatively quick and accurate way of analyzing EGG data with greater insight, and ability to monitor gastric functions before and after treatment. This technology has the potential to also be used as a teaching aid for clinicians increasing their understanding of the various dysrhythmias, especially irregular dysrhythmias.

The ability of NeuCube to encase two, or possibly more, distinct neural node structures bodes well for future investigation into multi-structured neural nets within the same overall network. The use of the different structures resulted in a vast increase in both the classification accuracy and repeatability of the results. The improved "stability" of both network and result indicates that further investigation is warranted.

## 5   Conclusion

To our best knowledge this is the first study in which neural network algorithms were applied to relate EGG to specific types of underlying slow wave spatial dysrhythmias. Prior to the current analysis, most previous neural network EGG studies focused on the temporal variations of slow waves, which might have missed certain spatial dysrhythmias that could occur at the same frequency as the normal activity [6]. By analyzing EGG over sustained dysrhythmias, the algorithms were able to detect exact intervals when irregularities occurred, which would be useful as slow wave dysrhythmias have often been reported to be dynamic, as demonstrated by the re-entry case. Furthermore, because the current practice of EGG is to record from a few channels, only limited information could be obtained from the recording. The present study analyzed data recorded from multiple sensors placed over the torso surface, thus ensuring a greater accuracy in pinpointing the type of spatial dysrhythmia occurring in the stomach. The increased amount of EGG data could also offer a way for further analysis after each recording session for greater details of gastric slow waves and functions, in order to minimize patient's time in hospital. Once trained and validated, the algorithms should allow clinicians a relatively quick and accurate way of analyzing EGG data with greater insight, and ability to monitor gastric functions before and after treatment. This technology has the potential to also be used as a teaching aid for clinicians increasing their understanding of the various dysrhythmias, especially irregular dysrhythmias.

The ability of NeuCube to encase two, or possibly more, distinct neural node structures bodes well for future investigation into multi-structured neural nets within the same overall network. The use of the different structures resulted in a vast increase in both the classification accuracy and repeatability of the results. The improved "stability" of both network and result indicates that further investigation is warranted.

# References

1. Kasabov, N.: NeuCube: a spiking neural network architecture for mapping, learning and understanding of spatio-temporal brain data. Neural Netw. **52**, 62–76 (2014)
2. Kasabov, N.: Evolving spiking neural networks for personalised modelling, classification and prediction of spatio-temporal patterns with a case study on stroke. Neurocomputing **134**, 11 (2014)
3. Kasabov, N., et al.: Evolving spatio-temporal data machines based on the NeuCube neuromorphic framework: design methodology and selected applications. Neural Netw. **78**, 1–14 (2016). Special Issue on "Neural Network Learning in Big Data"
4. O'Grady, G., et al.: Recent progress in gastric arrhythmia: pathophysiology, clinical significance and future horizons. Clin. Exp. Pharmacol. Physiol. **41**(10), 854–862 (2014)
5. Kasabov, N.: Global, local and personalised modeling and pattern discovery in bioinformatics: an integrated approach. Pattern Recogn. Lett. **28**, 673–685 (2016)
6. Du P., O'Grady, G., Paskaranandavadivel, N., Tang, S-j., Abell, T., Cheng, L.K.: Simultaneous anterior and posterior serosal mapping of gastric slow-wave dysrhythmias induced by vasopressin. Exp. Physiol. (2016). doi:10.1113/EP085697

# Learning Algorithms and Architectures

# Improving Generalization Abilities of Maximal Average Margin Classifiers

Shigeo Abe$^{(\boxtimes)}$

Kobe University, Rokkodai, Nada, Kobe, Japan
abe@kobe-u.ac.jp
http://www2.kobe-u.ac.jp/~abe

**Abstract.** Maximal average margin classifiers (MAMCs) maximize the average margin without constraints. Although training is fast, the generalization abilities are usually inferior to support vector machines (SVMs). To improve the generalization abilities of MAMCs, in this paper, we propose optimizing slopes and bias terms of separating hyperplanes after the coefficient vectors of the hyperplanes are obtained. The bias term is optimized so that the number of misclassifications is minimized. To optimized the slope, we introduce a weight to the average of mapped training data for one class and optimize the weight by cross-validation. To improve the generalization ability further, we propose equally constrained MAMCs and show that they reduce to least squares SVMs. Using two-class problems, we show that the generalization ability of the unconstrained MAMCs are inferior to those of the constrained MAMCs and SVMs.

## 1 Introduction

Since the introduction of support vector machines (SVMs) [1,2] various variants have been developed to improve the generalization ability. Because SVMs do not assume a specific data distribution, a priori knowledge on the data distribution can improve the generalization ability. The Mahalanobis distance, instead of the Euclidean distance is useful for this purpose. One approach reformulates SVMs so that the margin is measured by the Mahalanobis distance [3–7], and another approach uses Mahalanobis kernels, which calculate the kernel value according to the Mahalanobis distance [8–13].

In SVMs, the minimum margin is maximized. But in AdaBoost [14], the margin distribution, instead of the minimum margin, has been known to be important in improving the generalization ability [15,16].

Several approaches have been proposed to control the margin distribution in SVM-like classifiers [17–22]. In [18], a maximum average margin classifier (MAMC) is proposed, in which instead of maximizing the minimum margin, the margin mean for the training data is maximized without slack variables. In [21,22], in addition to maximizing the margin mean, the margin variance is minimized and the classifier is called large margin distribution machine (LDM).

© Springer International Publishing AG 2016
F. Schwenker et al. (Eds.): ANNPR 2016, LNAI 9896, pp. 29–41, 2016.
DOI: 10.1007/978-3-319-46182-3_3

According to the computer experiments in [21], the generalization ability of MAMCs is inferior to SVMs and LDMs.

In this paper, we clarify why MAMCs perform poorly for some classification problems and propose two methods to improve the generalization ability. Because the MAMC does not include constraints associated with training data, the determined bias term depends only on the difference between the numbers of training data for the two classes. To solve this problem, after the weight vector is obtained by the MAMC, we optimize the bias term so that the classification error is minimized. Then to improve the generalization ability further, we introduce a weight parameter to the average vector of one class and determine the parameter value by cross-validation. This results in optimizing the slope of the separating hyperplane. To improve the generalization ability further, we define the equality-constrained MAMC (EMAMC), which is shown to be equivalent to the least squares (LS) SVM. Using two-class problems, we show that the generalization ability of the unconstrained MAMCs with the optimized bias term and slopes are inferior to that of the EMAMC.

In Sect. 2, we explain the architecture of the MAMC and clarify the problems of MAMC. Then, we propose bias term optimization and slope optimization and develop the EMAMC. In Sect. 3, we compare the generalization abilities of the MAMC with those of the proposed MAMC with optimized bias terms and slopes, the EMAMC, and the SVM.

## 2    Maximum Average Margin Classifiers

### 2.1    Architecture

In the following we explain the maximum average margin classifiers (MAMCs) according to [18].

We consider a classification problem with $M$ training input-output pairs $\{\mathbf{x}_i, y_i\}$ $(i = 1, \ldots, M)$, where $\mathbf{x}_i$ are $m$-dimensional training inputs and belong to Class 1 or 2 and the associated labels are $y_i = 1$ for Class 1 and $-1$ for Class 2. We map the $m$-dimensional input vector $\mathbf{x}$ into the $l$-dimensional feature space using the nonlinear vector function $\boldsymbol{\phi}(\mathbf{x})$. In the feature space, we determine the decision function that separates Class 1 data from Class 2 data:

$$f(\mathbf{x}) = \mathbf{w}^\top \boldsymbol{\phi}(\mathbf{x}) + b = 0, \tag{1}$$

where $\mathbf{w}$ is the $l$-dimensional vector, $\top$ denotes the transpose of a vector (matrix), and $b$ is the bias term.

The margin of $\mathbf{x}_i$, $\delta_i$, which is the distance from the hyperplane, is given by

$$\delta_i = y_i \left( \mathbf{w}^\top \boldsymbol{\phi}(\mathbf{x}_i) + b \right) / \|\mathbf{w}\|. \tag{2}$$

Under the assumption of

$$\mathbf{w}^\top \mathbf{w} = 1, \tag{3}$$

(2) becomes

$$\delta_i = y_i \left( \mathbf{w}^\top \boldsymbol{\phi}(\mathbf{x}_i) + b \right). \tag{4}$$

With $b = 0$, the MAMC, which maximizes the average margin, is defined by

$$\text{maximize} \quad Q(\mathbf{w}) = \frac{1}{M} \sum_{i=1}^{M} y_i \, \mathbf{w}^{\top} \phi(\mathbf{x}_i) \tag{5}$$

$$\text{subject to} \quad \mathbf{w}^{\top} \mathbf{w} = 1. \tag{6}$$

Introducing the Lagrange multiplier $\lambda (> 0)$, we obtain the unconstrained optimization problem:

$$\text{maximize} \quad Q(\mathbf{w}) = \frac{1}{M} \sum_{i=1}^{M} y_i \, \mathbf{w}^{\top} \phi(\mathbf{x}_i) - \frac{\lambda}{2} (\mathbf{w}^{\top} \mathbf{w} - 1). \tag{7}$$

Taking the derivative of $Q$ with respect to $\mathbf{w}$, we obtain the optimal $\mathbf{w}$:

$$\lambda \mathbf{w} = \frac{1}{M} \sum_{i=1}^{M} y_i \, \phi(\mathbf{x}_i). \tag{8}$$

In [18], $\lambda$ is determined using (6) and (8), but $\lambda$ can take on any positive value because that does not change the decision boundary. Therefore, in the following we set $\lambda = 1$.

In calculating the decision function given by (1), we use kernels $K(\mathbf{x}, \mathbf{x}') = \phi^{\top}(\mathbf{x}) \, \phi(\mathbf{x})$ to avoid treating the variables in the feature space explicitly.

The resulting decision function $f(\mathbf{x})$ with $b = 0$ is given by

$$f(\mathbf{x}) = \frac{1}{M} \sum_{i=1}^{M} y_i \, K(\mathbf{x}, \mathbf{x}_i). \tag{9}$$

Among several kernels, radial basis function (RBF) kernels are widely used and thus in the following study we use RBF kernels:

$$K(\mathbf{x}, \mathbf{x}') = \exp(-\gamma \|\mathbf{x} - \mathbf{x}'\|^2 / m), \tag{10}$$

where $m$ is the number of inputs for normalization and $\gamma$ is to control a spread of a radius.

## 2.2   Problems with MAMCs

The MAMC is derived without a bias term, i.e., $b = 0$. To include the bias term we change (7) to

$$\text{maximize} \quad Q(\mathbf{w}, b) = \frac{1}{M} \sum_{i=1}^{M} y_i \, (\mathbf{w}^{\top} \phi(\mathbf{x}_i) + b) - \frac{1}{2} (\mathbf{w}^{\top} \mathbf{w} + b^2). \tag{11}$$

Here, we replace $\lambda$ with 1 and delete the constant term. Then, $Q$ is maximized when

$$\mathbf{w} = \frac{1}{M} \sum_{i=1}^{M} y_i \, \phi(\mathbf{x}_i), \tag{12}$$

$$b = \frac{1}{M} \sum_{i=1}^{M} y_i. \tag{13}$$

From (13), $b$ is determined by the deference of the numbers of the data belonging to Classes 1 and 2, not by the distributions of the data belonging to the two classes. And if the numbers are the same, $b = 0$, irrespective of $\mathbf{x}_i$ $(i = 1, \ldots, M)$. This occurs because the coefficient of $b$ becomes zero in (11); the value of $b$ does not affect optimality of the solution.

This means that the constraints are lacking for determining the optimal value of $b$. Similar to SVMs, the addition of inequality or equality constraints for the training data may solve the problem, which will be discussed later.

## 2.3  Bias Term Optimization

In this section we propose two-stage training; in the first stage, we determine the coefficient vector $\mathbf{w}$ by (12), and in the second stage, we optimize the value of $b$ by

$$\text{minimize} \quad E_{\mathrm{R}} = \sum_{i=1}^{M} I(\xi_i) \tag{14}$$

$$\text{subject to} \quad y_i (\mathbf{w}^{\top} \boldsymbol{\phi}(\mathbf{x}_i) + b) \geq \rho - \xi_i$$
$$\rho > 0, \quad \xi_i \geq 0, \tag{15}$$

where $E_{\mathrm{R}}$ is the number of misclassifications, $\rho$ is a positive constant, $\xi_i$ $(\geq 0)$ is a slack variable, $I(\xi_i) = 0$ for $\xi_i = 0$ and $I(\xi_i) = 1$ for $\xi_i > 0$. If there are multiple $b$ values that minimize (14), we break the tie by

$$\text{minimize} \quad E_{\mathrm{S}} = \sum_{i=1}^{M} \xi_i, \tag{16}$$

where $E_{\mathrm{S}}$ is the sum of slack variables.

First we consider the case where the classification problem is separable in the feature space. Suppose that

$$\max_{\substack{j=1,\ldots,M \\ y_j = -1}} \mathbf{w}^{\top} \boldsymbol{\phi}(\mathbf{x}_j) < \min_{\substack{i=1,\ldots,M \\ y_i = 1}} \mathbf{w}^{\top} \boldsymbol{\phi}(\mathbf{x}_i) < 0 \tag{17}$$

is satisfied, where training data belonging to Class 2 are correctly classified but some of the training data belonging to Class 1 are misclassified. Because of the first inequality in (17), by setting a proper value to $b$, all the training data are correctly classified.

Let

$$j = \arg_j \max_{\substack{j=1,\ldots,M \\ y_j = -1}} \mathbf{w}^{\top} \boldsymbol{\phi}(\mathbf{x}_j), \quad i = \arg_i \min_{\substack{i=1,\ldots,M \\ y_i = 1}} \mathbf{w}^{\top} \boldsymbol{\phi}(\mathbf{x}_i). \tag{18}$$

Then, from (15), to make $\mathbf{x}_i$ and $\mathbf{x}_j$ be correctly classified with margin $\rho$,

$$\mathbf{w}^\top \phi(\mathbf{x}_i) + b = \rho, \quad -(\mathbf{w}^\top \phi(\mathbf{x}_j) + b) = \rho \tag{19}$$

must be satisfied. Thus,

$$b = -\frac{1}{2}(\mathbf{w}^\top \phi(\mathbf{x}_i) + \mathbf{w}^\top \phi(\mathbf{x}_j)), \quad \rho = \frac{1}{2}(\mathbf{w}^\top \phi(\mathbf{x}_i) - \mathbf{w}^\top \phi(\mathbf{x}_j)). \tag{20}$$

The above equations are also valid when

$$0 < \max_{\substack{j=1,\ldots,M \\ y_j=-1}} \mathbf{w}^\top \phi(\mathbf{x}_j) < \min_{\substack{i=1,\ldots,M \\ y_i=1}} \mathbf{w}^\top \phi(\mathbf{x}_i), \tag{21}$$

where some of the training data for Class 2 are misclassified.

It is clear that (20) satisfies $E_{\mathrm{R}} = E_{\mathrm{S}} = 0$ and that $\rho$ is the maximum.

Now consider the inseparable case. Let the misclassified training data for Class 1 be $\mathbf{x}_{i_k}$ $(k = 1, \ldots, p)$ and

$$\mathbf{w}^\top \phi(\mathbf{x}_{i_1}) \leq \mathbf{w}^\top \phi(\mathbf{x}_{i_2}) \leq \cdots \leq \mathbf{w}^\top \phi(\mathbf{x}_{i_p}) \leq 0. \tag{22}$$

Likewise, let the misclassified training data for Class 2 be $\mathbf{x}_{j_k}$ $(k = 1, \ldots, n)$ and

$$0 \leq \mathbf{w}^\top \phi(\mathbf{x}_{j_1}) \leq \mathbf{w}^\top \phi(\mathbf{x}_{j_2}) \leq \cdots \leq \mathbf{w}^\top \phi(\mathbf{x}_{j_n}). \tag{23}$$

Similar to the separable case, it is clear that the optimal $b$ occurs at (20) with $i = i_k$ $(k \in \{1, \ldots, p\})$ and $j$ being given by

$$j = \arg_j \max_{\substack{\mathbf{w}^\top \phi(\mathbf{x}_j) < \mathbf{w}^\top \phi(\mathbf{x}_{i_k}) \\ y_j=-1, j=1,\ldots,M}} \mathbf{w}^\top \phi(\mathbf{x}_j) \tag{24}$$

or with $j = j_k$ $(k \in \{1, \ldots, n\})$ and $i$ being given by

$$i = \arg_i \max_{\substack{\mathbf{w}^\top \phi(\mathbf{x}_{j_k}) < \mathbf{w}^\top \phi(\mathbf{x}_i) \\ y_i=1, i=1,\ldots,M}} \mathbf{w}^\top \phi(\mathbf{x}_i). \tag{25}$$

Let $E_{\mathrm{R}}(i, j)$ and $E_{\mathrm{S}}(i, j)$ denote that $E_{\mathrm{R}}$ and $E_{\mathrm{S}}$ are evaluated with $b$ determined using $\mathbf{x}_i$ and $\mathbf{x}_j$ by (20), where $i = i_k$ $(k \in \{1, \ldots, p\})$ and $j$ is given by (24) or $j = j_k$ $(k \in \{1, \ldots, n\})$ and $i$ is given by (25). For each pair of $i$ and $j$, we calculate $E_{\mathrm{R}}(i, j)$ and select the value of $b$ that minimizes $E_{\mathrm{R}}(i, j)$. If there are multiple pairs of $i$ and $j$, we select the value of $b$ that minimizes $E_{\mathrm{S}}(i, j)$.

## 2.4 Extension of MAMCs

**Characteristics of Solutions.** Rewriting (8) with $\lambda = 1$,

$$\mathbf{w} = \frac{M_+}{M}\bar{\phi}_+ - \frac{M_-}{M}\bar{\phi}_- \tag{26}$$

where

$$\bar{\phi}_+ = \frac{1}{M_+} \sum_{\substack{i=1 \\ y_i=1}}^{M} \phi(\mathbf{x}_i), \quad \bar{\phi}_- = \frac{1}{M_-} \sum_{\substack{i=1 \\ y_i=-1}}^{M} \phi(\mathbf{x}_i), \tag{27}$$

and $\bar{\phi}_+$ and $\bar{\phi}_-$ are the averages of the mapped training data belonging to Classes 1 and 2, respectively, and $M_+$ and $M_-$ are the numbers of training data belonging to Classes 1 and 2, respectively.

If $M_+ = M_-$, $\mathbf{w}$ is the vector which is from $\bar{\phi}_-/2$ to $\bar{\phi}_+/2$. Therefore the decision function is orthogonal to the vector. If $M_+ \neq M_-$, the decision function is orthogonal to $\bar{\phi}_+ - (M_-/M_+)\,\bar{\phi}_-$.

**Slope Optimization.** To control the decision function, we introduce a positive hyperparameter $C_m$ as follows:

$$\mathbf{w} = \frac{M_+}{M}\bar{\phi}_+ - \frac{C_m M_-}{M}\bar{\phi}_-, \tag{28}$$

where $C_m$ works to lengthen or shorten the length of vector $\bar{\phi}_-$ according to whether $C_m > 1$ or $0 < C_m < 1$. Therefore, by changing the value of $C_m$, the slope of the decision function is changed.

Then the decision function becomes

$$f(\mathbf{x}) = \frac{1}{M} \sum_{i=1, y_i=1}^{M} K(\mathbf{x}, \mathbf{x}_i) - \frac{C_m}{M} \sum_{i=1, y_i=-1}^{M} K(\mathbf{x}, \mathbf{x}_i) + b. \tag{29}$$

We determine the value of $C_m$ by cross-validation.

In $k$-fold cross-validation, we divide the training data set into $k$ almost-equal-size subsets and train the classifier using the $k - 1$ subsets and test the trained classifier using the remaining subset. We iterate this procedure $k$ times for different combinations and calculate the classification error.

Calculation of the classification error for a given $C_m$ value is as follows:

1. Calculate (29) with $b = 0$ using the $k - 1$ subsets.
2. Calculate the bias term using the method discussed in Sect. 2.3.
3. Calculate the classification error for the remaining subset using the decision function generated in Steps 1 and 2.

Repeat the above procedure for the $k$ different combinations and calculate the classification error for the decision function.

For a given set of $C_m$ values, we calculate the classification errors and select the value of $C_m$ with the minimum classification error.

## 2.5   Equality-Constrained MAMCs

To improve the generalization ability of MAMCs further, we consider equality-constrained MAMCs (EMAMCs) as follows:

$$\text{maximize } Q(\mathbf{w}, b) = -\frac{1}{2}\mathbf{w}^\top \mathbf{w} + \frac{C_a}{M}\sum_{i=1}^{M} y_i \left(\mathbf{w}^\top \boldsymbol{\phi}(\mathbf{x}_i) + b\right) - \frac{C}{2}\sum_{i=1}^{M}\xi_i^2 \quad (30)$$

$$\text{subject to } y_i \left(\mathbf{w}^\top \boldsymbol{\phi}(\mathbf{x}_i) + b\right) = 1 - \xi_i \quad \text{for} \quad i = 1,\ldots, M, \quad (31)$$

where $C_a$ and $C$ are parameters to control the trade-off between the generalization ability and the classification error for the training data, and $\xi_i$ are the slack variables for $\mathbf{x}_i$.

We solve (30) and (31) in the empirical feature space [2] and define

$$\boldsymbol{\phi}(\mathbf{x}) = (K(\mathbf{x}, \mathbf{x}_1), \ldots, K(\mathbf{x}, \mathbf{x}_M))^\top. \quad (32)$$

Solving (31) for $\xi_i$ and substituting it into (30), we obtain the unconstrained optimization problem:

$$\text{maximize }\quad Q(\mathbf{w}, b) = -\frac{1}{2}\mathbf{w}^\top \mathbf{w} + \frac{C_a}{M}\sum_{i=1}^{M} y_i \left(\mathbf{w}^\top \boldsymbol{\phi}(\mathbf{x}_i) + b\right)$$

$$-\frac{C}{2}\sum_{i=1}^{M}(1 - y_i \left(\mathbf{w}^\top \boldsymbol{\phi}(\mathbf{x}_i) + b\right))^2. \quad (33)$$

If we delete the second term (the average margin) in the above equation, the optimization problem result in the least squares (LS) SVM defined in the empirical feature space [2].

Taking the partial derivative of (33) with respect to $\mathbf{w}$ and $b$ and setting the results to zero, we obtain the optimality conditions:

$$\left(1 + C\sum_{i=1}^{M}\boldsymbol{\phi}(\mathbf{x}_i)\,\boldsymbol{\phi}^\top(\mathbf{x}_i)\right)\mathbf{w} + C\sum_{i=1}^{M} y_i\,\boldsymbol{\phi}(\mathbf{x}_i)\,b = \left(\frac{C_a}{M} + C\right)\sum_{i=1}^{M} y_i\,\boldsymbol{\phi}(\mathbf{x}_i), \quad (34)$$

$$C\sum_{i=1}^{M}\boldsymbol{\phi}(\mathbf{x}_i)\,\mathbf{w} + C\,M\,b = \left(\frac{C_a}{M} + C\right)\sum_{i=1}^{M} y_i. \quad (35)$$

The above optimality conditions can be solved for $\mathbf{w}$ and $b$ by matrix inversion. The coefficient $(C_a/M + C)$ can be deleted because it is a scaling factor and does not change the decision boundary. Then, because $C_a$ is not included in the left-hand sides of (34) and (35), the value of $C_a$ does not influence the location of the decision boundary. This means that the second term in (33) can be safely deleted.

In addition, if we delete the $\mathbf{w}^\top \mathbf{w}$ term from (33), all the terms in the left-hand sides of (34) and (35) include $C$, thus $C$ can be deleted; $C$ does not work to control the trade-off. Therefore, the $\mathbf{w}^\top \mathbf{w}$ term is essential.

Accordingly, dividing (34) and (35) by $C$ and deleting the constant term $(C_a/C\,M + 1)$ from the right-hand sides of (34) and (35), we obtain

$$\left(\frac{1}{C} + \sum_{i=1}^{M} \phi(\mathbf{x}_i)\,\phi^{\top}(\mathbf{x}_i)\right)\mathbf{w} + \sum_{i=1}^{M} y_i\,\phi(\mathbf{x}_i)\,b = \sum_{i=1}^{M} y_i\,\phi(\mathbf{x}_i), \qquad (36)$$

$$\sum_{i=1}^{M} \phi(\mathbf{x}_i)\,\mathbf{w} + M\,b = \sum_{i=1}^{M} y_i. \qquad (37)$$

The above formulation is exactly the same as the LS SVM defined in the empirical feature space. Therefore, the EMAMC results in the LS SVM.

## 3    Performance Evaluation

### 3.1    Experimental Conditions

We compared the proposed MAMC including the EMAMC (LS SVM) with the plain MAMC and the L1 SVM using two-class data sets [23]. The L1 SVM we used is as follows:

$$\text{maximize} \quad Q(\boldsymbol{\alpha}) = \sum_{i=1}^{M} \alpha_i - \frac{1}{2}\sum_{i,j=1}^{M} \alpha_i\alpha_j\,y_i\,y_j K(\mathbf{x}_i,\mathbf{x}_j) \qquad (38)$$

$$\text{subject to} \quad \sum_{i=1}^{M} y_i\,\alpha_i = 0, \quad 0 \le \alpha_i \le C \quad \text{for} \quad i = 1,...,M, \qquad (39)$$

where $\alpha_i$ are Lagrange multipliers associated with $\mathbf{x}_i$ and $C\,(> 0)$ is a margin parameter that controls the trade-off between the classification error of the training data and the generalization ability.

Table 1 lists the numbers of inputs, training data, test data, and data set pairs of two class problems. Each data set pair consists of the training data set and the test data set. We trained classifiers using the training data set and evaluated the performance using the test data set. Then we calculated the average accuracy and the standard deviation for all the data set pairs. We determined the parameter values by fivefold cross-validation. We selected the $\gamma$ value of the RBF kernels from $\{0.01, 0.1, 0.5, 1, 5, 10, 15, 20, 50, 100, 200, 300, 400, 500, 600, 700\}$. For the $C_m$, we selected from $\{0.05, 0.1, 0.2, \ldots, 0.9, 1.0, 1.1111, \ldots, 20\}$. For the EMAMC and L1 SVM, we selected the $\gamma$ value from 0.01 to 200 and the $C$ value, from $\{0.1, 1, 10, 50, 100, 500, 1000, 2000\}$. We trained the L1 SVM using SMO-NM [24], which fuses SMO (Sequential minimal optimization) and NM (Newton's method).

**Table 1.** Benchmark data sets for two-class problems

| Data | Inputs | Train | Test | Sets |
|------|--------|-------|------|------|
| Banana | 2 | 400 | 4,900 | 100 |
| Breast cancer | 9 | 200 | 77 | 100 |
| Diabetes | 8 | 468 | 300 | 100 |
| Flare-solar | 9 | 666 | 400 | 100 |
| German | 20 | 700 | 300 | 100 |
| Heart | 13 | 170 | 100 | 100 |
| Image | 18 | 1,300 | 1,010 | 20 |
| Ringnorm | 20 | 400 | 7,000 | 100 |
| Splice | 60 | 1,000 | 2,175 | 20 |
| Thyroid | 5 | 140 | 75 | 100 |
| Titanic | 3 | 150 | 2,051 | 100 |
| Twonorm | 20 | 400 | 7,000 | 100 |
| Waveform | 21 | 400 | 4,600 | 100 |

## 3.2   Results

Table 2 shows the average accuracies and their standard deviations of the six classifiers with RBF kernels. In the table, MAMC is given by (12) and (13) and the $\gamma$ value is optimized by cross-validation. In MAMC$_b$, the bias term is optimized as discussed in Sect. 2.3. In MAMC$_{bs}$, after $\gamma$ value is optimized with $C_m = 1$, the $C_m$ value is optimized. We call this strategy line search in contrast to grid search. In MAMC$_{bsg}$, the $\gamma$ and $C_m$ values are optimized by grid search.

Among the six classifiers including the L1 SVM the best average accuracy is shown in bold and the worst average accuracy is underlined. The "Average" row shows the average accuracy of the 13 average accuracies and the two numerals in the parentheses show the numbers of the best and worst accuracies in the order. We performed Welch's t test with the confidence intervals of 95 %. The "W/T/L" row shows the results; W, T, and L denote the numbers that the MAMC$_{bs}$ shows statistically better than, the same as, and worse than the remaining five classifiers, respectively.

From the "Average" row, the EMAMC performed best in the average accuracy and the L1 SVM the second best. The difference between MAMC$_{bsg}$ and MAMC$_{bs}$ is small. The MAMC is the worst. From the "W/T/L" row, the accuracies of the MAMC$_{bs}$ and the MAMC$_{bsg}$ are statistically comparable and the accuracy of the MAMC$_{bs}$ is slightly better than that of the MAMC$_b$ but always better than that of the MAMC. The accuracy of the MAMC$_{bs}$ is worse than that of the EMAMC and L1 SVM.

In Sect. 2.2, we clarified that the bias term is not optimized by the original MAMC formulation. This is exemplified by the experiments. By optimizing the bias term as proposed in Sect. 2.3, the accuracy is improved drastically. The effect

Table 2. Accuracy comparison for the two-class problems

| Data | MAMC$_{bs}$ | MAMC$_{bsg}$ | MAMC$_b$ | MAMC | EMAMC | L1 SVM |
|---|---|---|---|---|---|---|
| Banana | 88.46±0.85 | 88.49±0.86 | 88.51±0.73 | 59.69±9.17 | 89.13±0.63 | **89.17±0.72** |
| B. cancer | 74.14±4.33 | 74.00±4.43 | **74.27±4.36** | 71.19±4.53 | 73.57±4.55 | 73.03±4.51 |
| Diabetes | 73.03±2.26 | 72.19±2.49 | 72.66±2.32 | 65.22±2.15 | **76.67±1.76** | 76.29±1.73 |
| Flare-solar | 66.10±2.00 | 65.64±2.05 | 66.35±2.03 | 59.48±5.83 | 66.25±2.00 | **66.99±2.12** |
| German | 75.53±2.18 | 75.93±2.21 | 69.51±2.28 | 70.18±1.95 | **76.27±2.04** | 75.95±2.24 |
| Heart | 81.00±3.58 | 80.99±3.24 | 81.32±3.38 | 58.87±8.84 | 82.70±3.70 | **82.82±3.37** |
| Image | 93.59±1.22 | 94.11±1.24 | 92.90±1.12 | 56.81±1.10 | 96.97±0.74 | **97.16±0.41** |
| Ringnorm | **98.27±0.27** | 98.25±0.30 | **98.27±0.25** | 70.26±21.99 | 98.04±0.43 | 98.14±0.35 |
| Splice | 84.43±1.11 | 85.17±0.88 | 84.08±0.99 | 54.57±8.76 | **89.07±0.59** | 88.89±0.91 |
| Thyroid | 95.15±2.24 | 95.41±2.30 | 95.11±2.17 | 70.25±4.36 | **95.43±2.35** | 95.35±2.44 |
| Titanic | 77.68±0.84 | **77.70±1.03** | 77.64±0.84 | 67.69±0.30 | 77.69±0.82 | 77.39±0.74 |
| Twonorm | 97.28±0.31 | 97.21±0.40 | 97.35±0.27 | 79.25±16.20 | **97.41±0.23** | 97.38±0.26 |
| Waveform | 88.93±1.53 | 89.23±1.29 | 88.84±1.64 | 67.07±0.19 | **90.20±0.50** | 89.76±0.66 |
| Average | 84.12 (1/0) | 84.18 (2/0) | 83.60 (2/1) | 65.43 (0/11) | 85.34 (6/0) | 85.26 (4/0) |
| W/T/L | — | 1/11/1 | 1/12/0 | 13/0/0 | 1/4/8 | 2/3/8 |

of slope optimization to the accuracy is small. However, by the bias term and slope optimization, the generalization ability is still below that of EMAMC or L1 SVM. This indicates that the equality or inequality constraints are essential in realizing the high generalization ability.

We measured the average CPU time per data set including time for model selection by fivefold cross-validation, training a classifier, and classifying the test data by the trained classifier. We used a personal computer with 3.4 GHz CPU and 16 GB memory. Table 3 shows the results. From the table the MAMC is the fastest and the MAMC$_{bs}$ and MAMC$_b$ are comparable to the MAMC. Comparing the MAMC$_{bs}$ and the MAMC$_{bsg}$, the MAMC$_{bsg}$ requires more time because of the grid search. Because the classification performance is comparable, line search seems to be sufficient. The EMAMC, L1 SVM and MAMC$_{bsg}$ are in the slowest group.

## 3.3  Discussions

The advantage of the MAMC is its simplicity: The coefficient vector of the decision hyperplane is calculated by addition or subtraction of kernel values. The inferior generalization ability of the original MAMC is mitigated by bias and slope optimization, but the improvement is still not sufficient compared to the EMAMC and L1 SVM. Therefore, the introduction of the equality or inequality constraints are essential. But it leads to the LS SVM or L1 SVM and the simplicity of the MAMC is completely lost.

**Table 3.** Training time comparison for the two-class problems (in seconds)

| Data | $MAMC_{bs}$ | $MAMC_{bsg}$ | $MAMC_b$ | MAMC | EMAMC | L1 SVM |
|---|---|---|---|---|---|---|
| Banana | 1.10 | 9.74 | **0.59** | **0.59** | <u>22.40</u> | 4.92 |
| B. cancer | 0.31 | 2.77 | **0.13** | 0.14 | 2.95 | <u>7.08</u> |
| Diabetes | 1.51 | 14.39 | 0.78 | **0.73** | <u>35.86</u> | 22.96 |
| Flare-solar | 3.24 | 29.89 | 1.61 | **1.51** | 111.43 | <u>218.67</u> |
| German | 4.42 | 40.03 | 2.07 | **2.04** | 148.65 | <u>776.53</u> |
| Heart | 0.24 | 2.13 | 0.12 | **0.11** | <u>2.05</u> | 1.75 |
| Image | 15.04 | 144.43 | 7.62 | **7.18** | <u>2290.87</u> | 56.7 |
| Ringnorm | 1.60 | 12.91 | 0.99 | **0.96** | <u>23.27</u> | 12.57 |
| Splice | 13.34 | 126.51 | 7.16 | **6.87** | <u>887.16</u> | 30.71 |
| Thyroid | 0.15 | <u>1.35</u> | **0.07** | **0.07** | 1.16 | 0.33 |
| Titanic | 0.17 | 1.31 | **0.09** | **0.09** | 1.32 | <u>21.25</u> |
| Twonorm | 1.67 | 12.92 | 0.99 | **0.92** | <u>22.51</u> | 10.46 |
| Waveform | 1.51 | 12.84 | **0.88** | **0.88** | 22.44 | <u>35.61</u> |
| B/W | 0/0 | 0/1 | 5/0 | 12/0 | 0/7 | 0/5 |

# 4   Conclusions

We discussed two ways to improve the generalization ability of the maximum average margin classifier (MAMC). One is to optimize the bias term after calculating the weight vector, and the other is to optimize the slope of the decision function by introducing the weight parameter to the average vector of one class. The parameter value is determined by cross validation. To improve the generalization ability further, we introduced the EMAMC, which is the equality constrained MAMC, but this is shown to be equivalent to the LS SVM defined in the empirical feature space.

According to the experiments for two-class problems, we show that the generalization ability is improved by the bias term and slope optimization. However, the obtained generalization ability is inferior to the EMAMC and L1 SVM. Therefore, the unconstrained MAMC is not so powerful as the EMAMC and L1 SVM.

**Acknowledgment.** This work was supported by JSPS KAKENHI Grant Number 25420438.

# References

1. Vapnik, V.N.: Statistical Learning Theory. Wiley, New York (1998)
2. Abe, S.: Support Vector Machines for Pattern Classification, 2nd edn. Springer, London (2010)

40     S. Abe

3. Lanckriet, G.R.G., El Ghaoui, L., Bhattacharyya, C., Jordan, M.I.: A robust minimax approach to classification. J. Mach. Learn. Res. **3**, 555–582 (2002)
4. Huang, K., Yang, H., King, I., Lyu, M.R.: Learning large margin classifiers locally and globally. In: Proceedings of the Twenty-First International Conference on Machine Learning (ICML 2004), pp. 1–8 (2006)
5. Yeung, D.S., Wang, D., Ng, W.W.Y., Tsang, E.C.C., Wang, X.: Structured large margin machines: sensitive to data distributions. Mach. Learn. **68**(2), 171–200 (2007)
6. Xue, H., Chen, S., Yang, Q.: Structural regularized support vector machine: a framework for structural large margin classifier. IEEE Trans. Neural Netw. **22**(4), 573–587 (2011)
7. Peng, X., Xu, D.: Twin Mahalanobis distance-based support vector machines for pattern recognition. Inf. Sci. **200**, 22–37 (2012)
8. Abe, S.: Training of support vector machines with Mahalanobis kernels. In: Duch, W., Kacprzyk, J., Oja, E., Zadrożny, S. (eds.) ICANN 2005. LNCS, vol. 3697, pp. 571–576. Springer, Heidelberg (2005)
9. Wang, D., Yeung, D.S., Tsang, E.C.C.: Weighted Mahalanobis distance kernels for support vector machines. IEEE Trans. Neural Netw. **18**(5), 1453–1462 (2007)
10. Shen, C., Kim, J., Wang, L.: Scalable large-margin Mahalanobis distance metric learning. IEEE Trans. Neural Netw. **21**(9), 1524–1530 (2010)
11. Liang, X., Ni, Z.: Hyperellipsoidal statistical classifications in a reproducing kernel Hilbert space. IEEE Trans. Neural Netw. **22**(6), 968–975 (2011)
12. Fauvel, M., Chanussot, J., Benediktsson, J.A., Villa, A.: Parsimonious Mahalanobis kernel for the classification of high dimensional data. Pattern Recogn. **46**(3), 845–854 (2013)
13. Reitmaier, T., Sick, B.: The responsibility weighted Mahalanobis kernel for semi-supervised training of support vector machines for classification. Inf. Sci. **323**, 179–198 (2015)
14. Freund, Y., Schapire, R.E.: A decision-theoretic generalization of on-line learning and an application to boosting. J. Comput. Syst. Sci. **55**(1), 119–139 (1997)
15. Reyzin, L., Schapire, R.E.: How boosting the margin can also boost classifier complexity. In: Proceedings of the 23rd International Conference on Machine learning, pp. 753–760. ACM (2006)
16. Gao, W., Zhou, Z.-H.: On the doubt about margin explanation of boosting. Artif. Intell. **203**, 1–18 (2013)
17. Garg, A., Roth, D.: Margin distribution and learning. In: Proceedings of the Twentieth International Conference (ICML) on Machine Learning, Washington, DC, USA, pp. 210–217 (2003)
18. Pelckmans, K., Suykens, J., Moor, B.D.: A risk minimization principle for a class of parzen estimators. In: Platt, J.C., Koller, D., Singer, Y., Roweis, S.T. (eds.) Advances in Neural Information Processing Systems, vol. 20, pp. 1137–1144. Curran Associates Inc., New York (2008)
19. Aiolli, F., Da San Martino, G., Sperduti, A.: A kernel method for the optimization of the margin distribution. In: Kůrková, V., Neruda, R., Koutník, J. (eds.) ICANN 2008, Part I. LNCS, vol. 5163, pp. 305–314. Springer, Heidelberg (2008)
20. Zhang, L., Zhou, W.-D.: Density-induced margin support vector machines. Pattern Recogn. **44**(7), 1448–1460 (2011)
21. Zhou, Z.-H., Zhang, T.: Large margin distribution machine. In: Twentieth ACM SIGKDD Conference on Knowledge Discovery and Data Mining, pp. 313–322 (2014)

22. Zhou, Z.-H.: Large margin distribution learning. In: El Gayar, N., Schwenker, F., Suen, C. (eds.) ANNPR 2014. LNCS, vol. 8774, pp. 1–11. Springer, Heidelberg (2014)
23. Rätsch, G., Onoda, T., Müller, K.-R.: Soft margins for AdaBoost. Mach. Learn. **42**(3), 287–320 (2001)
24. Abe, S.: Fusing sequential minimal optimization and Newton's method for support vector training. Int. J. Mach. Learn. Cybern. **7**, 345–364 (2016). doi:10.1007/s13042-014-0265-x

# Finding Small Sets of Random Fourier Features for Shift-Invariant Kernel Approximation

Frank-M. Schleif[1,2,3]([✉]), Ata Kaban[3], and Peter Tino[3]

[1] School of Computer Science, University of Applied Sciences Würzburg-Schweinfurt,
97074 Würzburg, Germany
schleify@cs.bham.ac.uk
[2] Computational Intelligence Group, University of Applied Sciences Mittweida,
Technikumplatz 17, 09648 Mittweida, Germany
[3] School of Computer Science, University of Birmingham, Edgbaston B15 2TT, UK

**Abstract.** Kernel based learning is very popular in machine learning, but many classical methods have at least quadratic runtime complexity. Random fourier features are very effective to approximate *shift-invariant* kernels by an explicit kernel expansion. This permits to use efficient linear models with much lower runtime complexity. As one key approach to kernelize algorithms with linear models they are successfully used in different methods. However, the number of features needed to approximate the kernel is in general still quite large with substantial memory and runtime costs. Here, we propose a simple test to identify a small set of random fourier features with linear costs, substantially reducing the number of generated features for low rank kernel matrices, while widely keeping the same representation accuracy. We also provide generalization bounds for the proposed approach.

## 1 Introduction

Kernel based learning methods are very popular in various machine learning tasks like regression, classification or clustering [1–5]. The operations, used to calculate the respective models, typically evaluate the full kernel matrix, leading to quadratic or even cubic complexity. As a consequence, the approximation of positive semi-definite (psd) kernels has raised wide interest [6,7]. Most approaches focus on approximating the kernel by the (clustered) Nyström approximation or specific variations of the singular value decompositions [8,9]. A recent approach effectively combining multiple strategies was presented in [6]. Random fourier features (RFF) have been introduced in [10] to the field of kernel based learning. The aim is to approximate shift invariant kernels by mapping the input data into a randomized feature space and then apply existing fast *linear* methods [11–13]. This is of special interest if the number of samples N is very large and the obtained kernel matrix $K \in \mathbb{R}^{N \times N}$ leads to high storage and calculation costs. The features are constructed so that the inner products of the transformed data are approximately equal to those in the feature space of a user specified shift-invariant kernel:

© Springer International Publishing AG 2016
F. Schwenker et al. (Eds.): ANNPR 2016, LNAI 9896, pp. 42–54, 2016.
DOI: 10.1007/978-3-319-46182-3_4

$$k(\mathbf{x}, \mathbf{y}) = \langle \phi(\mathbf{x}), \phi(\mathbf{y}) \rangle \approx z(\mathbf{x})'z(\mathbf{y}) \tag{1}$$

With $\phi : \mathcal{X} \mapsto \mathcal{H}$ being a non-linear mapping of patterns from the original input space $\mathcal{X}$ to a high-dimensional space $\mathcal{H}$. The mapping function is in general not given in an explicit form. Unlike the kernel lifting by using $\phi(\cdot)$, $z$ is a (comparable) low dimensional feature vector. In [14] it was empirically shown that for data with a large eigenvalue gap random fourier features are less efficient than a standard Nyström approximation. However, the authors used only a rather small data independent set of fourier features. Here, we propose a selection strategy which not only reduces the number of necessary random fourier features but also helps to select a reasonable set of features, which provides a good approximation of the original kernel function. In this line our focus is less on best possible approximation accuracy but rather on saving memory and obtaining compact representations to address the usage of random fourier features in low resource environments. In [10] it is shown how random fourier feature vectors z can be constructed for various shift invariant kernels $k(\mathbf{x} - \mathbf{y})$, e.g. the RBF kernel upto an error using only $D = \mathcal{O}(d\epsilon^{-2} \log \frac{1}{\epsilon^2})$ dimensions, where $d$ is the input dimension of the original input data. Assuming that $d = 10$ and $\epsilon = 0.01$ one gets $D \approx 100.000$.

However, in [10] it is empirically shown that the approximation is already reasonable enough for smaller $D \approx 500\text{--}5000$. While very efficient in general there is not yet a reasonable strategy how to choose an appropriate number $D$ nor which features have to be generated in a more systematic way. If we assume that the images of training inputs in the feature space given (implicitly) by the kernel lie in an intrinsically low dimensional space one can expect that a much smaller number of features should be sufficient to describe the data. A reasonable strategy to test for a reliable number $D$ of random fourier features is to compare the approximated kernel using Eq. (1) with the true kernel $K$ based on the original data using some appropriate measure. This however is in general not possible or very costly for larger N because one would need to generate two $N \times N$ kernel matrices. We suggest to use a constructive approach, generating as many features as necessary to obtain a low reconstruction error between the two kernel matrices. Our approach is very generic as we do not focus on dedicated cost functions used in (semi-)supervised classification nor clustering or embedding measures, such that the constructed feature set provides a reasonable approximation of the shift-invariant kernel matrix. Standard feature reduction techniques for high dimensional data sets like random projection [15,16], unsupervised feature selection techniques based on statistical measures [17] or supervised approaches [18,19] are not suitable because they start from a high-dimensional feature space or are specific to the underlying cost-function. In [1] random fourier features were used in combination with a singular value decomposition to reduce the number of features after generating the random fourier features, again with in general rather large initial D. To avoid high costs in the construction procedure we employ the Nyström approximation at different points to evaluate the accuracy of the constructed random fourier feature set using the Frobenius norm. We assume that the considered kernel is in fact intrinsically low dimensional. The paper is organized as follows, first we review the main theory for random Fourier features, subsequently we detail our approach of finding small sets of random fourier features, still sufficiently accurate to approximate the kernel matrix using the Frobenius norm. In a subsection we review the Nyström approximation and derive a linear time calculation of the Frobenius norm for the difference of two Nyström approximated matrices. Later we derive error bounds for the presented approach and show the efficiency of our method on various standard datasets employing two state of the art linear time classifiers for vectorial input data.

## 2  Random Fourier Features

Random fourier features as introduced in [10], project the vectorial data points onto a randomly chosen line, and then pass the resulting scalar through a sinusoid. The random lines are drawn from a distribution so as to guarantee that the inner product of two transformed points approximates the desired shift-invariant kernel. The motivation for this approach is given by Bochners theorem:

**Definition 1.** *A continuous kernel* $k(\mathbf{x}, \mathbf{y}) = k(\mathbf{x} - \mathbf{y})$ *on* $\mathbb{R}^d$ *is positive definite if and only if* $k(\mathbf{x} - \mathbf{y})$ *is the Fourier transform of a non-negative measure.*

---

**Algorithm 1.** RFF selection by optimizing an approximated Frobenius norm

---
1: **Input:** $X, NyK, n, D_{max}, iter, \epsilon$
2: **Output:** $H_{\text{final}}$
3: **function** RFF SELECTION($(X, NyK, n, D_{max}, iter, \epsilon)$)
4:     try=0,#$F = 0$, $E_{\text{old}} = $ realmax
5:     **while** #$F < D_{max}$ **do**
6:         get n random fourier features $H$
7:         construct $N\hat{y}K$ based on $H$
8:         calculate $E_{\text{new}} = \|N\hat{y}K - NyK\|_F$ Eq. (7)
9:         **if** $E_{\text{old}} > E_{\text{new}}$ & $E_{\text{old}} - E_{\text{new}} \geq \epsilon$ **then**
10:             add $H$ to $H_{\text{final}}$
11:             #F := #F + n
12:             update $E_{\text{old}}$
13:             try := 0
14:         **else**
15:             **if** try > iterMax **then**
16:                 break
17:             **else**
18:                 try := try +1
19:             **end if**
20:         **end if**
21:     **end while**
22: **return** $H_{\text{final}}$
23: **end function**

---

If the kernel $k(\mathbf{x} - \mathbf{y})$ is properly scaled, Bochners theorem guarantees that its Fourier transform $p(\omega)$ is a proper probability distribution. The idea in [10] is to approximate the kernel as

$$k(\mathbf{x} - \mathbf{y}) = \int_{\mathbb{R}^d} p(\omega)e^{j\omega(\mathbf{x} - \mathbf{y})}d\omega$$

with some extra normalizations and simplifications one can sample the features for $k$ using the mapping $z_\omega(\mathbf{x}) = [\cos(\mathbf{x})\sin(\mathbf{x})]$. In [10] the authors also give a proof for the uniform convergence of Fourier features to the kernel $k(\mathbf{x} - \mathbf{y})$. A detailed derivation can be found in [10].

To generate the random fourier features one eventually needs a psd kernel matrix $k(\mathbf{x},\mathbf{y}) = k(\mathbf{x} - \mathbf{y})$ and a random feature map $z(\mathbf{x}) : \mathbb{R}^d \to \mathbb{R}^{2D}$ s.t. $z(\mathbf{x})^\top z(\mathbf{y}) \approx k(\mathbf{x}-\mathbf{y})$. One draws $D$ i.i.d. samples $\{\omega_1,\ldots,\omega_D\} \in \mathbb{R}^d$ from $p(\omega)$ and generates $z(\mathbf{x}) = \sqrt{1/D}[\cos(\omega_1^\top \mathbf{x})\ldots\cos(\omega_D^\top \mathbf{x})\ \sin(\omega_1^\top \mathbf{x})\ldots\sin(\omega_D^\top \mathbf{x})]^\top$.

## 3   Finding Small Sets of Random Fourier Features

To incrementally add fourier features to the approximation of kernel k we use the Frobenius norm to calculate the difference between the two kernels. For real valued data the Frobenius norm of two squared matrices is simply the sum of the squared difference between the individual kernel entries:

$$\left\| \hat{k} - k \right\|_F = \sqrt{\sum_i^N \sum_j^N (\hat{k}(\mathbf{x}_i - \mathbf{y}_j) - k(\mathbf{x}_i - \mathbf{y}_j))^2} \tag{2}$$

This has $\mathcal{O}(N^2)$ costs in memory and runtime and we would need to generate the full kernel $\hat{k}$ and $k$. To avoid these costs we use the Nyström approximation for kernel matrices [20] to approximate both kernels by using only $\mathcal{O}(N)$ coefficients and provide a formulation for calculating the Frobenius norm of the difference of two Nyström approximated matrices. The Nyström approximation of the original kernel matrix NyK (detailed in the next sub-section) can be done once prior to calculations of the random fourier features. Subsequently the approximated kernel is constructed by iteratively adding those n random fourier features which significantly improve the Frobenius error in Eq. (2) with $\epsilon = 1e^{-3}$. This iterative procedure is continued until either no further significant improvement was found for a number $iterMax = 5$ of random selections or an upper limit of features $D_{max}$ is obtained. The detailed procedure is given in Algorithm 1.

## 4   Nyström Approximated Matrix Processing

The Nyström approximation technique has been proposed in the context of kernel methods in [20]. One well known way to approximate a $N \times N$ Gram matrix, is to use a low-rank approximation. This can be done by computing the eigendecomposition of the kernel matrix $K = U\Lambda U^T$, where $U$ is a matrix, whose columns are orthonormal eigenvectors, and $\Lambda$ is a diagonal matrix consisting of eigenvalues $\Lambda_{11} \geq \Lambda_{22} \geq \ldots \geq 0$, and keeping only the $m$ eigenspaces which correspond to the $m$ largest eigenvalues of the matrix. The approximation is $\tilde{K} \approx U_{(N,m)}\Lambda_{(m,m)}U_{(m,N)}$, where the indices refer to the size of the corresponding submatrix restricted to the larges $m$ eigenvalues. The Nyström method approximates a kernel in a similar way, without computing the eigende-composition of the whole matrix, which is an $O(N^3)$ operation.

By the Mercer theorem, kernels $k(\mathbf{x}, \mathbf{x}')$ can be expanded by orthonormal eigenfunctions $\varphi_i$ and non negative eigenvalues $\lambda_i$ in the form

$$k(\mathbf{x}, \mathbf{x}') = \sum_{i=1}^{\infty} \lambda_i \varphi_i(\mathbf{x}) \varphi_i(\mathbf{x}').$$

The eigenfunctions and eigenvalues of a kernel are defined as solutions of the integral equation

$$\int k(\mathbf{x}', \mathbf{x}) \varphi_i(\mathbf{x}) p(\mathbf{x}) d\mathbf{x} = \lambda_i \varphi_i(\mathbf{x}'),$$

where $p(\mathbf{x})$ is a probability density over the input space. This integral can be approximated based on the Nyström technique by an i.i.d. sample $\{\mathbf{x}_k\}_{k=1}^m$ from $p(\mathbf{x})$:

$$\frac{1}{m} \sum_{k=1}^{m} k(\mathbf{x}', \mathbf{x}_k) \varphi_i(\mathbf{x}_k) \approx \lambda_i \varphi_i(\mathbf{x}'). \tag{3}$$

Using this approximation we denote with $K^{(m)}$ the corresponding $m \times m$ Gram sub-matrix and get the corresponding matrix eigenproblem equation as:

$$\frac{1}{m} K^{(m)} U^{(m)} = U^{(m)} \Lambda^{(m)}$$

with $U^{(m)} \in \mathbb{R}^{m \times m}$ is column orthonormal and $\Lambda^{(m)}$ is a diagonal matrix.

Now we can derive the approximations for the eigenfunctions and eigenvalues of the kernel $k$

$$\lambda_i \approx \frac{\lambda_i^{(m)} \cdot N}{m}, \quad \varphi_i(\mathbf{x}') \approx \frac{\sqrt{m/N}}{\lambda_i^{(m)}} \mathbf{k}_x'^{\top} \mathbf{u}_i^{(m)}, \tag{4}$$

where $\mathbf{u}_i^{(m)}$ is the $i$th column of $U^{(m)}$. Thus, we can approximate $\varphi_i$ at an arbitrary point $\mathbf{x}'$ as long as we know the vector $\mathbf{k}_x' = (k(\mathbf{x}_1, \mathbf{x}'), ..., k(\mathbf{x}_m, \mathbf{x}'))$. For a given $N \times N$ Gram matrix $K$ one may randomly choose $m$ rows and respective columns. The corresponding indices are called landmarks, and should be chosen such that the data distribution is sufficiently covered. Strategies how to chose the landmarks have recently been addressed in [8,21] and [22,23]. We denote these rows by $K_{(m,N)}$. Using the formulas Eq. (4) we can reconstruct the original kernal matrix,

$$\tilde{K} = \sum_{i=1}^{m} 1/\lambda_i^{(m)} \cdot K_{(m,N)}^T (\mathbf{u}_i^{(m)})^T (\mathbf{u}_i^{(m)}) K_{(m,N)},$$

where $\lambda_i^{(m)}$ and $\mathbf{u}_i^{(m)}$ correspond to the $m \times m$ eigenproblem (3). Thus we get the approximation,

$$\tilde{K} = K_{(N,m)} K_{(m,m)}^{-} K_{(m,N)}. \tag{5}$$

This approximation is exact, if $K_{(m,m)}$ has the same rank as $K$.

**Nyström Approximation Based Frobenius Norm.** Instead of the Frobenius norm definition given in Eq. (2) we will use an equivalent formulation based on the trace of the matrix:

$$\left\|\hat{k} - k\right\|_F = \sqrt{\sum_{i=1}^{N} \ddot{k}(\mathbf{x}_i, \mathbf{y}_i)} \tag{6}$$

$\ddot{k}(\mathbf{x}_i, \mathbf{y}_i)$ is given by the $(i,j)$'th entry of the matrix $\ddot{K}$ defined as $\ddot{K} = (\hat{K} - K) \cdot (\hat{K} - K)^\top$ (in matrix notation). This formulation is useful because we can obtain the diagonal elements of a Nyström approximated matrix very easy.

We approximate $\hat{k}(\mathbf{x}_i, \mathbf{y}_j)$ and $k(\mathbf{x}_i, \mathbf{y}_j)$ using the Nyström approximation and obtain matrices $\hat{K}_{(nm)}, \hat{K}_{(nm)}^{-1}$ and $K_{(nm)}, K_{(nm)}^{-1}$ as defined before. With some basic algebraic operations one gets the following equation for the Frobenius norm of the difference of two Nyström approximated matrices (in matrix notation). Let $C = \hat{K}_{(nm)} \otimes K_{(nm)}$ and $W = C_{(m,m)}^{-1}$. Further we introduce matrices $\hat{C}$ with entries $\hat{C}_{[i,j]} = C_{[i,j]}^2, \hat{W} = \hat{C}_{(m,m)}^{-1}$ and $C'$ with entries $C'_{[i,j]} = K_{[i,j]}^2, W' = K_{(m,m)}^{-1}$. Then the approximated Frobenius norm can be derived as:

$$\left\|\hat{k} - k\right\|_F = \sqrt{\sum(\sum(\hat{C}\cdot\hat{W}))\cdot\hat{C}^\top + (\sum(C'\cdot W'))\cdot C'^{,\top} - 2\cdot(\sum(C\cdot W))\cdot C^\top} \tag{7}$$

This operation can be done with linear costs.

**Complexity and Error Analysis.** The Nyström approximation of $k$ can be calculated once prior to random fourier feature selection, with costs of $\mathcal{O}(N \times m + m^3)$, to obtain the two submatrices of the Nyström approximation. If we assume that $m \ll N$ this is summarized by $\mathcal{O}(N \times m)$. The Nyström approximation of $\hat{k}$ needs to be calculated in each enhancing step of the feature construction. If we add one feature per iteration the costs of $m^3$ can be avoided by use of the matrix inversion lemma. If we assume that in each step 1 feature is added. If we restrict the number of added features by $D_{max}$ extra costs of $\mathcal{O}(D_{max} \times N \times m)$ are present to calculate the Nyström approximation of $k$. If we assume that $D_{max} \ll N$ this is again reduced to $\mathcal{O}(N \times m)$.

The calculation of the Frobenius norm can be done in linear time using Eq. (7). Hence, we finally have costs of $\mathcal{O}(N \times m)$ for generating the random fourier features. The number of effectively chosen random fourier features is in general much smaller then $D_{max}$. In the following we use $m = 50$ and $D_{max} = 5000$ and report crossvalidation accuracies using the approximated $\hat{k}$ in comparison to $k$ for different classification tasks.

As mentioned before and shown from the runtime analysis the approach is reasonable only if the number of landmarks $m$ is low with respect to $N$, or the intrinsic dimensionality of the datasets is low, respectively. Taking e.g. an RBF kernel, the $\sigma$ parameter controls the width of the Gaussian. If the RBF kernel is employed in a kernel classifier we observe that for very small $\sigma$

a Nearest Neighbor approach is approximated and the intrinsic dimensionality of the data or number of non-vanishing eigenvalues gets large. In these cases the RBF representation can not be approximated without a corresponding high loss in the prediction accuracy of the model and our approach can not be used. In the proposed procedure we observe two approximation errors, namely the error introduced by the random fourier feature approximation and the error introduced by the Nyström approximation. We have:

$$\left\| \hat{K} - K \right\|_F = \sqrt{\sum_{i=1}^{N} \sum_{j=1}^{N} |\hat{K}(\mathbf{x}_i, \mathbf{x}_j) - K(\mathbf{x}_i, \mathbf{x}_j)|^2} \tag{8}$$

where $\hat{K}$ is the Nyström approximated kernel matrix of the kernel matrix obtained from the random fourier features of the training data. By the triangle inequality we get

$$\left\| \hat{K} - K \right\|_F \leq \left\| \varphi^\top \varphi - K \right\|_F + \left\| \varphi^\top \varphi - \tilde{K} \right\|_F \tag{9}$$

where $\tilde{K}$ is the Nyström approximated kernel matrix of the linear kernel matrix on the random fourier features and $\varphi^\top \varphi$ is given as

$$\varphi(\mathbf{x}_i)^\top \varphi(\mathbf{x}_j) = \frac{\alpha}{D} \sum_{l=1}^{D} \cos(\omega_l^\top (\mathbf{x}_i - \mathbf{x}_j)) \tag{10}$$

where $D$ is the number of random fourier features $\omega_l \sim N(0, I_d), \alpha > 0$ and $\mathbf{x}_i, \mathbf{x}_j$ are training points with RFF feature values stored in $\varphi$. In the following we derive and combine the bounds for both approximation schemes. The Frobenius error of the approximated kernel using the random fourier features is given as

$$\left\| \varphi^\top \varphi - K \right\|_F = \sqrt{\sum_{i=1}^{N} \sum_{j=1}^{N} |\varphi(\mathbf{x}_i)^\top \varphi(\mathbf{x}_j) - K(\mathbf{x}_i, \mathbf{x}_j)|^2} \tag{11}$$

with $K$ as the kernel matrix of the training points $\{\mathbf{x}_1, \ldots, \mathbf{x}_N\}$. For a fixed pair of points $(\mathbf{x}_i, \mathbf{x}_j)$ we have:

$$Pr\{|\varphi(\mathbf{x}_i)^\top \varphi(\mathbf{x}_j) - \underbrace{E[\varphi(\mathbf{x}_i)^\top \varphi(\mathbf{x}_j)]}_{K(\mathbf{x}_i, \mathbf{x}_j)}| > t\} =$$

$$= Pr\{|\frac{\alpha}{D} \sum_{l=1}^{D} \cos(\omega_l^\top (\mathbf{x}_i - \mathbf{x}_j)) - K(\mathbf{x}_i, \mathbf{x}_j)| > t\} \tag{12}$$

$$\leq 2 \exp\left\{ \frac{-t^2 D}{2\alpha^2} \right\} \tag{13}$$

the last inequation follows by the Höffding inequality because the terms $\cos(\omega_l^\top (\mathbf{x}_i, \mathbf{x}_j))$ are independent w.r.t. $\{\omega_1, \ldots, \omega_D\}$ and are bounded $\in [-1, 1]$.

The above condition can be generalized asymptotically to all pairs from $\{\mathbf{x}_1, \ldots, \mathbf{x}_N\}$. Hence the following holds simultaneously for all pairs:

$$Pr\{\exists (i,j) : |\varphi(\mathbf{x}_i)^\top \varphi(\mathbf{x}_j) - K(\mathbf{x}_i, \mathbf{x}_j)| > t\} \leq 2N^2 \exp\left\{\frac{-t^2 D}{2\alpha^2}\right\} \tag{14}$$

by union bound. Hence we obtain

$$\left\|\varphi^\top \varphi - K\right\|_F \leq \sqrt{N^2 t^2} = N \cdot t \tag{15}$$

with probability of at least $1 - 2N^2 \exp\left\{\frac{-t^2 D}{2\alpha^2}\right\}$. To get the failure probability to an arbitrary small $\delta$:

$$2N^2 \exp\left\{\frac{-t^2 D}{2\alpha^2}\right\} = \delta$$

$$\frac{t^2 D}{2\alpha^2} = \log \frac{2N^2}{\delta} = \log \frac{2}{\delta} + 2 \log N$$

$$t = \alpha \sqrt{\frac{2}{D}\left(\log \frac{2}{\delta} + 2 \log N\right)}$$

We get

$$\left\|\varphi^\top \varphi - K\right\|_F \leq N \cdot \alpha \sqrt{\frac{2}{D}\left(\log \frac{2}{\delta} + 2 \log N\right)} = \tilde{O}\left(\frac{N}{\sqrt{D}}\right) \tag{16}$$

We can ensure that $\left\|\varphi^\top \varphi - K\right\|_F \leq \epsilon$ by choosing $D$ large enough:

$$N \cdot \alpha \sqrt{\frac{2}{D}\left(\log \frac{2}{\delta} + 2 \log N\right)} \leq \epsilon$$

$$\frac{2}{D}\left(\log \frac{2}{\delta} + 2 \log N\right) \leq \frac{\epsilon^2}{N^2 \alpha^2} \rightarrow \frac{2}{D} \leq \frac{\epsilon^2}{\left(\log \frac{2}{\delta} + 2 \log N\right) N^2 \alpha^2}$$

$$D \geq \frac{2N^2 \alpha^2 \left(\log \frac{2}{\delta} + 2 \log N\right)}{\epsilon^2} = \tilde{O}\left(\frac{N}{\epsilon}\right)$$

For the second approximation error we bound the error of inner product of the random fourier feature vectors obtained from the training data with respect to a Nyström approximation of the kernel based on the random fourier features. This is just a classical Nyström approximation of a kernel matrix. Hence we can use bounds already provided in [24]. According to Theorem 2 given in [24] the following inequality holds with probability of at least $1 - \delta$:

$$\left\|\varphi^\top \varphi - \hat{K}\right\|_F \leq \left\|\varphi^\top \varphi - K_k\right\|_F + \left[\frac{64D}{m}\right]^{\frac{1}{4}} N K_{max} \left[1 + \sqrt{\frac{N-m}{n-1/2}\frac{1}{\beta(m,N)}\log \frac{1}{\delta} d_{max}^K / K_{max}^{\frac{1}{2}}}\right]^{\frac{1}{2}} \tag{17}$$

with $\beta(m,n) = 1 - \frac{1}{2\max m, N-m}$ and $K_k$ the best $k$ approximation of $K$ and $K_{max}$ the maximal diagonal entry of $K$ and $d^K_{max}$ the maximum Euclidean distance defined over $K$. Which maybe summarized in accordance to [22] as $\left\| \varphi^\top \varphi - K_k \right\|_F + [\frac{D}{m}]^{\frac{1}{4}} N \left\| K \right\|_2$. Combining both bounds we get

$$
\begin{aligned}
\left\| \hat{K} - K \right\|_F &\leq \left\| \varphi^\top \varphi - K \right\|_F + \left\| \varphi^\top \varphi - \tilde{K} \right\|_F \\
&\leq \left\| \varphi^\top \varphi - K_k \right\|_F + \\
&\quad N \cdot \left( \alpha \sqrt{\frac{2}{D} \left( \log \frac{2}{\delta} + 2 \log N \right)} + [\frac{D}{m}]^{\frac{1}{4}} \left\| K \right\|_2 \right)
\end{aligned}
\tag{18}
$$

We see that both approximation terms increase as $\tilde{\mathcal{O}}(N)$ - that is, up to log factors the kernel approximation error increases linearly with the number of training points $N$. This was expected since the gram matrix $K$ has size $N \times N$. We may also notice a tradeoff for the value of $D$: The random Fourier feature approximation bound tightens as $k$ increases whereas the Nyström approximation loosens. (One could possibly use the value of $D$ that minimizes the approximation error bound, although we have not tried this in the experiments.) The approximation error bound presented here is uniform only over the training points - which was much simpler to achieve than a bound that holds uniformly over the whole input domain - as in the original paper [10]. Nevertheless we can still expect it to be informative since for kernel based learning there exist generalization bounds whose complexity term only depends on the gram matrix constructed from the training set (e.g. the Rademacher complexity for kernel based linear classification works out as the trace of the gram matrix).

## 5   Experiments

We evaluate the approach on multiple public datasets most of them already used in the paper [10][1]. For the Nyström approximation step we use 50 landmarks. The checkerboard data (checker) is a two class problem consisting of 9000 2d samples organized like a checkerboard on a $3 \times 3$ grid. The data are separable with low error using an rbf kernel. The coil-20 dataset (coil) consists of 1440 image files in 16384 dimensions from the coil database categorized in 20 classes. The spam database consists of 4601 samples in 57 dimensions in two classes. The adult dataset consists of 30162 samples in 44 dimensions given in 2 classes[2]. The code-rna dataset with 59535 samples and 8 dimensions from http://www.csie.ntu.edu.tw/~cjlin/libsvmtools/datasets/. Two further used datasets are the famous USPS data with 11000 samples in 256 dimensions and the MNIST data with 60000 samples and 256 dimensions. Both datasets are organized in 10 classes

---

[1] We skip the KDDCUP data which is very simple as already reported in [10]. Further for some of the datasets the original configuration was not exactly reconstructable e.g. Adult data such that we could not directly copy results from.

[2] Preprocessed as reported in http://ssdi.di.fct.unl.pt/nmm/scripts/mdatasets/.

**Table 1.** Test set accuracy (% ± std) of the various benchmark datasets for constructing small sets of random fourier features. The second row of each dataset contains the mean cputime of a single cycle in the crossvalidation.

| | $D$ | $D^*$ | SVM+Full-RFF | SVM+Small-RFF | LS+Full-RFF | LS+Small-RFF |
|---|---|---|---|---|---|---|
| coil | 5000 | 175 | 98.75 ± 1.46 | 97.22 ± 2.15 | 99.38 ± 0.51 | 96.67 ± 1.49 |
| | | | 5 s | 0.1 s | 7 s | 1 s |
| spam | 5000 | 185 | 92.98 ± 0.76 | 89.74 ± 1.18 | 92.87 ± 1.09 | 89.13 ± 1.71 |
| | | | 1 s | 0.1 s | 1 s | 0.1 s |
| checker | 5000 | 20 | 100.00 ± 0 | 99.96 ± 0.06 | 100.00 ± 0.00 | 99.09 ± 0.29 |
| | | | 1.5 s | 0.02 s | 2 s | 0.02 s |
| usps | 5000 | 820 | 97.96 ± 0.32 | 96.85 ± 0.49 | 98.11 ± 0.42 | 96.25 ± 0.76 |
| | | | 19 s | 3 s | 28.15 s | 4.33 s |
| adult | 500 | 105 | 82.44 ± 0.50 | 82.20 ± 0.58 | 82.48 ± 0.44 | 82.08 ± 0.76 |
| | | | 1 s | 0.35 s | 0.65 s | 0.14 s |
| code-rna | 500 | 20 | 95.31 ± 0.24 | 91.40 ± 0.50 | 94.65 ± 0.37 | 91.27 ± 0.39 |
| | | | 1.4 s | 0.21 s | 1.4 s | 0.1 s |
| mnist | 5000 | 235 | 96.30 ± 0.30 | 89.89 ± 0.41 | 95.12 ± 0.20 | 87.12 ± 0.43 |
| | | | 3.4 min | 14 s | 1.5 min | 6 s |
| cover | 1000 | 90 | 83.44 ± 0.20 | 75.85 ± 0.18 | 80.56 ± 0.19 | 76.18 ± 0.18 |
| | | | 19 min | 48 s | 3 min | 17 s |
| forest | 1000 | 150 | 83.15 ± 0.08 | 76.18 ± 0.18 | 79.94 ± 0.11 | 74.31 ± 0.21 |
| | | | 2.8 min | 2 s | 1.9 min | 6 s |

originating from a character recognition problem. Finally the covertype data with 495141 entries (classes 1 and 2) and 54 dimensions and the Forest data with 522.000 samples and 54 dimensions both taken from the UCI database were analyzed. All datasets have been z-transformed. For the $\sigma$ parameter of the rbf kernel we use values reported before elsewhere. To evaluate the classification performance we follow [10] and use a least squares regression (LS) model as well as the liblinear, which is a high performance linear Support Vector Machine[3]. The parameter C of the Liblinear-SVM was fixed to 1 as suggested by the liblinear authors. Multiclass problems have been approached in LS using a one vs rest scheme. In Table 1 we report 10-fold crossvalidation results and the minimal number of features $D^*$ as obtained by the proposed strategy. The maximal number of random fourier features D per dataset is in general 5000 as suggested in [10] with exceptions for the larger datasets to keep memory consumption tractable only 500–1000 features where chosen.

For the coil data we see that the identified small RFF-model contains 29−times less features than the full model while loosing ≈2 % discrimination accuracy on the test set. For the spam database we observe a similar result with

[3] http://www.csie.ntu.edu.tw/~cjlin/liblinear/.

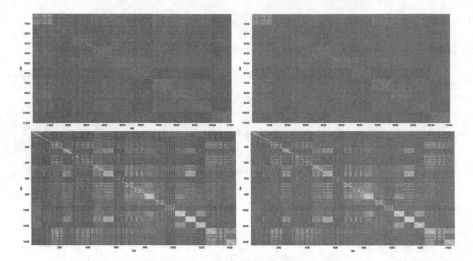

**Fig. 1.** Top left: reconstruction of the USPS radial basis function kernel with 5000 random fourier features, right: reconstruction of the USPS radial basis function kernel with the identified random fourier features. Bottom-left: reconstruction of the coil radial basis function kernel with 5000 random fourier features and right with the random fourier features as obtained by the proposed approach.

27−times less features and a small decay in the accuracy of 3 % for SVM. At the simulated checkerboard data we have almost the same accuracy in the reduced set while the number of features is reduced by a factor of 250. For the USPS data we have ≈6 times less features with almost the same prediction accuracy, slightly reduced by 1–2 %. The Adult dataset keeps almost the same accuracy while having 5-times less features similar observations can be made for the code-dna data. For MNIST the accuracy drops by 7–8 % with 21 times less features. Finally the cover data are represented by 28 times less features with a similar good accuracy like the full model and the forest data could be represented with 22 times less features with a slight decay on 7 % in the accuracy. For USPS and MNIST we found that the number of remaining features is still a bit high which can be potentially attributed to a more complex eigenvalue structure of these datasets such that the proposed test was less efficient. The other datasets have basically almost the same accuracy on a drastically reduced feature set. For the coil and the usps data the kernel reconstructions are exemplarily shown in Fig. 1.

## 6    Conclusions

In this paper we proposed a test for selecting a small set of random fourier features such that the approximated shift invariant kernel is close to the original one with respect to the Frobenius norm. In general we found that the proposed approach is efficient to reduce the number of features, already during the construction, by in general a magnitude or more with low costs with respect to N.

The approach is especially applicable if the approximated kernel is of low rank and N is large. Thereby the proposed selection procedure is efficient to obtain small random fourier features sets with high representation accuracy. The effect of sometimes reduced accuracy for random fourier features as observed in [14] could not be confirmed as long as the RFF set is either large enough or appropriately chosen by the proposed method. The proposed approach saves runtime and memory costs during training but is also very valuable if memory is constrained under test conditions e.g. within an embedded system environment. The obtained transformation matrix P has $d \times D$ coefficients which is most often small enough to be of use also under system conditions with limited resources. The original data needs to be transformed into the random fourier feature space using P by a simple matrix multiplication and can subsequently be fed into a linear classifier. The obtained models are in general very efficient as seen above. The small $D*$ also avoids the need to sparsify the linear models by using ridge regression (instead of simple LS) or sparse linear SVM models like the support feature machine [18], such that efficient high performance implementations of linear classifiers can be directly used. In future work we will analyze the effect of our approach on tensor sketching [25] which was used to approximate polynomial kernels.

**Acknowledgment.** Marie Curie Intra-European Fellowship (IEF): FP7-PEOPLE-2012-IEF (FP7-327791-ProMoS) is greatly acknowledged.

# References

1. Chitta, R., Jin, R., Jain, A.K.: Efficient kernel clustering using random Fourier features. In: 12th IEEE International Conference on Data Mining, ICDM, pp. 161–170. IEEE (2012)
2. Villmann, T., Haase, S., Kaden, M.: Kernelized vector quantization in gradient-descent learning. Neurocomputing **147**, 83–95 (2015)
3. Schleif, F.-M., Villmann, T., Hammer, B., Schneider, P.: Efficient kernelized prototype-based classification. J. Neural Syst. **21**(6), 443–457 (2011)
4. Hofmann, D., Schleif, F.-M., Hammer, B.: Learning interpretable kernelized prototype-based models. Neurocomputing **131**, 43–51 (2014)
5. Schleif, F.-M., Zhu, X., Gisbrecht, A., Hammer, B.: Fast approximated relational and kernel clustering. In: Proceedings of ICPR 2012, pp. 1229–1232. IEEE (2012)
6. Si, S., Hsieh, C.-J., Dhillon, I.S.: Memory efficient kernel approximation. In: Proceedings of the 31th International Conference on Machine Learning, ICML, volume 32 of JMLR Proceedings, pp. 701–709. JMLR.org (2014)
7. Cortes, C., Mohri, M., Talwalkar, A.: On the impact of kernel approximation on learning accuracy. In: Proceedings of the 13th International Conference on Artificial Intelligence and Statistics, AISTATS, volume 9 of JMLR Proceedings, pp. 113–120. JMLR.org (2010)
8. Zhang, K., Kwok, J.T.: Clustered Nyström method for large scale manifold learning and dimension reduction. IEEE Trans. Neural Netw. **21**(10), 1576–1587 (2010)
9. Gisbrecht, A., Schleif, F.-M.: Metric and non-metric proximity transformations at linear costs. Neurocomputing **167**, 643–657 (2015)

10. Rahimi, A., Recht, B.: Random features for large-scale kernel machines. In: Proceedings of the 21st Annual Conference on Neural Information Processing Systems, NIPS 2007. Curran Associates, Inc. (2007)
11. Agarwal, A., Kakade, S.M., Karampatziakis, N., Song, L., Valiant, G.: Least squares revisited: scalable approaches for multi-class prediction. In: Proceedings of the 31th International Conference on Machine Learning, ICML, volume 32 of JMLR Proceedings, pp. 541–549. JMLR.org (2014)
12. Bunte, K., Kaden, M., Schleif, F.-M.: Low-rank kernel space representations in prototype learning. WSOM 2016. AISC, vol. 428, pp. 341–353. Springer, Switzerland (2016)
13. Schleif, F.-M., Hammer, B., Villmann, T.: Margin based active learning for LVQ networks. Neurocomputing **70**(7–9), 1215–1224 (2007)
14. Yang, T., Li, Y.-F., Mahdavi, M., Jin, R., Zhou, Z.-H., Nystroem method vs random Fourier features: a theoretical and empirical comparison. In: Proceedings of the 26st Annual Conference on Neural Information Processing Systems, NIPS 2012, pp. 485–493 (2012)
15. Durrant, R.J., Kabán, A.: Random projections as regularizers: learning a linear discriminant from fewer observations than dimensions. Mach. Learn. **99**(2), 257–286 (2015). doi:10.1007/s10994-014-5466-8
16. Freund, Y., Dasgupta, S., Kabra, M., Verma, N.: Learning the structure of manifolds using random projections. In: Proceedings of the 21st Annual Conference on Neural Information Processing Systems, NIPS 2007. Curran Associates, Inc. (2007)
17. Vergara, J.R., Estévez, P.A.: A review of feature selection methods based on mutual information. Neural Comput. Appl. **24**(1), 175–186 (2014)
18. Klement, S., Anders, S., Martinetz, T.: The support feature machine: classification with the least number of features and application to neuroimaging data. Neural Comput. **25**(6), 1548–1584 (2013)
19. Schleif, F.-M., Villmann, T., Zhu, X.: High dimensional matrix relevance learning. In: Proceedings of IEEE Internation Conference on Data Mining Workshop (ICDMW), pp. 661–667 (2014)
20. Williams, C.K.I., Seeger, M.: Using the Nyström method to speed up kernel machines. In: Proceedings of the 13th Annual Conference on Neural Information Processing Systems, NIPS 2000, pp. 682–688 (2000)
21. Zhang, K., Tsang, I.W., Kwok, J.T.: Improved Nystrom low-rank approximation and error analysis. In: Proceedings of the 25th International Conference on Machine Learning, ICML 2008, pp. 1232–1239. ACM, New York (2008)
22. Gittens, A., Mahoney, M.W.: Revisiting the Nystrom method for improved large-scale machine learning. In: Proceedings of the 30th International Conference on Machine Learning, ICML 2013, volume 28 of JMLR Proceedings, pp. 567–575. JMLR.org (2013)
23. De Brabanter, K., De Brabanter, J., Suykens, J.A.K., De Moor, B.: Optimized fixed-size kernel models for large data sets. Comput. Stat. Data Anal. **54**(6), 1484–1504 (2010)
24. Kumar, S., Mohri, M., Talwalkar, A.: Sampling methods for the Nyström method. J. Mach. Learn. Res. **13**, 981–1006 (2012)
25. Pham, N., Pagh, R.: Fast and scalable polynomial kernels via explicit feature maps. In: Proceedings of the 19th ACM SIGKDD International Conference on Knowledge Discovery and Data Mining, KDD 2013, pp. 239–247. ACM (2013)

# Incremental Construction of Low-Dimensional Data Representations

Alexander Kuleshov[1] and Alexander Bernstein[1,2(✉)]

[1] Skolkovo Institute of Science and Technology, Moscow, Russia
{A.Kuleshov,a.bernstein}@skoltech.ru
[2] Kharkevich Institute for Information Transmission Problems RAS,
Moscow, Russia

**Abstract.** Various Dimensionality Reduction algorithms transform initial high-dimensional data into their lower-dimensional representations preserving chosen properties of the initial data. Typically, such algorithms use the solution of large-dimensional optimization problems, and the incremental versions are designed for many popular algorithms to reduce their computational complexity. Under manifold assumption about high-dimensional data, advanced manifold learning algorithms should preserve the Data manifold and its differential properties such as tangent spaces, Riemannian tensor, etc. Incremental version of the Grassmann&Stiefel Eigenmaps manifold learning algorithm, which has asymptotically minimal reconstruction error, is proposed in this paper and has significantly smaller computational complexity in contrast to the initial algorithm.

**Keywords:** Machine learning · Dimensionality reduction · Manifold learning · Tangent bundle manifold learning · Incremental learning

## 1 Introduction

The general goal of data analysis is to extract previously unknown information from a given dataset. Many data analysis tasks, such as pattern recognition, classification, clustering, prognosis, and others, deal with real-world data that are presented in high-dimensional spaces, and the 'curse of dimensionality' phenomena are often an obstacle to the use of many methods for solving these tasks.

Fortunately, in many applications, especially in pattern recognition, the real high-dimensional data occupy only a very small part in the high dimensional 'observation space' $R^p$; it means that an intrinsic dimension q of the data is small compared to the dimension p (usually, q << p) [1, 2]. Various dimensionality reduction (feature extraction) algorithms, whose goal is a finding of a low-dimensional parameterization of such high-dimensional data, transform the data into their low-dimensional representations (features) preserving certain chosen subject-driven data properties [3, 4].

The most popular model of high-dimensional data, which occupy a small part of observation space $R^p$, is Manifold model in accordance with which the data lie on or near an unknown Data manifold (DM) of known lower dimensionality q < p embedded in an ambient high-dimensional space $R^p$ (Manifold assumption [5] about high-dimensional data). Typically, this assumption is satisfied for 'real-world' high-dimensional data obtained from 'natural' sources.

© Springer International Publishing AG 2016
F. Schwenker et al. (Eds.): ANNPR 2016, LNAI 9896, pp. 55–67, 2016.
DOI: 10.1007/978-3-319-46182-3_5

Dimensionality reduction under the manifold assumption about processed data are usually referred to as the Manifold learning [6, 7] whose goal is constructing a low-dimensional parameterization of the DM (global low-dimensional coordinates on the DM) from a finite dataset sampled from the DM. This parameterization produces an Embedding mapping from the DM to low-dimensional Feature space that should preserve specific properties of the DM determined by chosen optimized cost function which defines an 'evaluation measure' for the dimensionality reduction and reflects the desired properties of the initial data which should be preserved in their features.

Most manifold learning algorithms include the solution of large-dimensional global optimization problems and, thus, are computationally expensive. The incremental versions of many popular algorithms (Locally Linear Embedding, Isomap, Laplacian Eigenmaps, Local Tangent Space Alignment, Hessian Eigenmaps, etc. [6, 7]), which reduce their computational complexity, were developed [8–17].

The manifold learning algorithms are usually used as a first key step in solution of machine learning tasks: the low-dimensional features are used in reduced learning procedures instead of initial high-dimensional data avoiding the curse of dimensionality [18]: 'dimensionality reduction may be necessary in order to discard redundancy and reduce the computational cost of further operations' [19]. If the low-dimensional features preserve only specific properties of data, then substantial data losses are possible when using the features instead of the initial data. To prevent these losses, the features should preserve as much as possible available information contained in the high-dimensional data [20]; it means the possibility for recovering the initial data from their features with small reconstruction error. Such Manifold reconstruction algorithms result in both the parameterization and recovery of the unknown DM [21].

Mathematically [22], a 'preserving the important information of the DM' means that manifold learning algorithms should 'recover the geometry' of the DM, and 'the information necessary for reconstructing the geometry of the manifold is embodied in its Riemannian metric (tensor)' [23]. Thus, the learning algorithms should accurately recover Riemannian data manifold that is the DM equipped by Riemannian tensor.

Certain requirement to the recovery follows from the necessity of providing a good generalization capability of the manifold reconstruction algorithms and preserving local structure of the DM: the algorithms should preserve a differential structure of the DM providing proximity between tangent spaces to the DM and Recovered data manifold (RDM) [24]. In the Manifold theory [23, 25], the set composed of the manifold points equipped by tangent spaces at these points is called the Tangent bundle of the manifold; thus, a reconstruction of the DM, which ensures accurate reconstruction of its tangent spaces too, is referred to as the Tangent bundle manifold learning.

Earlier proposed geometrically motivated Grassmann&Stiefel Eigenmaps algorithm (GSE) [24, 26] solves the Tangent bundle manifold learning and recovers Riemannian tensor of the DM; thus, it solves the Riemannian manifold recovery problem.

The GSE, like most manifold learning algorithms, includes the solution of large-dimensional global optimization problems and, thus, is computationally expensive.

In this paper, we propose an incremental version of the GSE that reduces the solution of the computationally expensive global optimization problems to the solution of a sequence of local optimization problems solved in explicit form.

The rest of the paper is organized as follows. Section 2 contains strong definition of the Tangent bundle manifold learning and describes main ideas realized in its GSE-solution. The proposed incremental version of the GSE is presented in Sect. 3.

# 2 Tangent Bundle Manifold Learning

## 2.1 Definitions and Assumptions

Consider unknown q-dimensional Data manifold with known intrinsic dimension q

$$\mathbf{M} = \{X = g(y) \in R^p : y \in \mathbf{Y} \subset R^q\}$$

covered by a single chart g and embedded in an ambient p-dimensional space $R^p$, q < p. The chart g is one-to-one mapping from open bounded Coordinate space $\mathbf{Y} \subset R^q$ to the manifold $\mathbf{M} = g(\mathbf{Y})$ with differentiable inverse mapping $h_g(X) = g^{-1}(X)$ whose values $y = h_g(X) \in \mathbf{Y}$ give low-dimensional coordinates (representations, features) of high-dimensional manifold-valued data X.

If the mappings $h_g(X)$ and $g(y)$ are differentiable and $J_g(y)$ is p × q Jacobian matrix of the mapping $g(y)$, than q-dimensional linear space $L(X) = Span(J_g(h_g(X)))$ in $R^p$ is tangent space to the DM $\mathbf{M}$ at the point $X \in \mathbf{M}$; hereinafter, Span(H) is linear space spanned by columns of arbitrary matrix H.

The tangent spaces can be considered as elements of the Grassmann manifold Grass(p, q) consisting of all q-dimensional linear subspaces in $R^p$.

Standard inner product in $R^p$ induces an inner product on the tangent space L(X) that defines Riemannian metric (tensor) $\Delta(X)$ in each manifold point $X \in \mathbf{M}$ smoothly varying from point to point; thus, the DM $\mathbf{M}$ is a Riemannian manifold $(\mathbf{M}, \Delta)$.

Let $\mathbf{X}_n = \{X_1, X_2, \ldots, X_n\}$ be a dataset randomly sampled from the DM $\mathbf{M}$ according to certain (unknown) probability measure whose support coincides with $\mathbf{M}$.

## 2.2 Tangent Bundle Manifold Learning Definition

Conventional manifold learning problem, called usually Manifold embedding problem [6, 7], is to construct a low-dimensional parameterization of the DM from given sample $\mathbf{X}_n$, which produces an Embedding mapping $h : \mathbf{M} \subset R^p \to \mathbf{Y}_h = h(\mathbf{M}) \subset R^q$ from the DM $\mathbf{M}$ to the Feature space (FS) $\mathbf{Y}_h \subset R^q$, q < p, which preserves specific chosen properties of the DM.

Manifold reconstruction algorithm, which provides additionally a possibility of accurate recovery of original vectors X from their low-dimensional features $y = h(X)$, includes a constructing of a Recovering mapping $g(y)$ from the FS $\mathbf{Y}_h$ to the Euclidean space $R^p$ in such a way that the pair (h, g) ensures approximate equalities

$$r_{h,g}(X) \equiv g(h(X)) \approx X \text{ for all points } X \in \mathbf{M}. \tag{1}$$

The mappings (h, g) determine q-dimensional Recovered data manifold

$$\mathbf{M}_{h,g} = r_{h,g}(\mathbf{M}) = \{r_{h,g}(X) \in R^p : X \in \mathbf{M}\} = \{X = g(y) \in R^p : y \in \mathbf{Y}_h \subset R^q\} \quad (2)$$

which is embedded in the ambient space $R^p$, covered by a single chart g, and consists of all recovered values $r_{h,g}(X)$ of manifold points $X \in \mathbf{M}$. Proximities (1) imply manifold proximity $\mathbf{M}_{h,g} \approx \mathbf{M}$ meaning a small Hausdorff distance $d_H(\mathbf{M}_{h,g}, \mathbf{M})$ between the DM $\mathbf{M}$ and RDM $\mathbf{M}_{h,g}$ due inequality $d_H(\mathbf{M}_{h,g}, \mathbf{M}) \leq \sup_{X \in \mathbf{M}} |r_{h,g}(X) - X|$.

Let $G(y) = J_g(y)$ be $p \times q$ Jacobian matrix of the mapping g(y) which determines q-dimensional tangent space $L_{h,g}(X)$ to the RDM $\mathbf{M}_{h,g}$ at the point $r_{h,g}(X) \in \mathbf{M}_{h,g}$:

$$L_{h,g}(X) = Span(G(h(X))) \quad (3)$$

Tangent bundle manifold learning problem is to construct the pair (h, g) of mappings h and g from given sample $\mathbf{X}_n$ ensuring both the proximities (1) and proximities

$$L_{h,g}(X) \approx L(X) \text{ for all points } X \in \mathbf{M}; \quad (4)$$

proximities (4) are defined with use certain chosen metric on the Grass(p, q).

The matrix G(y) determines also metric tensor $\Delta_{h,g}(X) = G^T(h(X)) \times G(h(X))$ on the RMD $\mathbf{M}_{h,g}$ which is $q \times q$ matrix consisting of inner products $\{(G_i(h(X)), G_j(h(X)))\}$ between $i^{th}$ and $j^{th}$ columns $G_i(h(X))$ and $G_j(h(X))$ of the matrix G(h(X)). Thus, the pair (h, g) determines Recovered Riemannian manifold $(\mathbf{M}_{h,g}, \Delta_{h,g})$ that accurately approximates initial Riemannian data manifold $(\mathbf{M}, \Delta)$.

## 2.3   Grassmann&Stiefel Eigenmaps: An Approach

Grassmann&Stiefel Eigenmaps algorithm gives the solution to the Tangent bundle manifold learning problem and consists of three successively performed parts: Tangent manifold learning, Manifold embedding, and Manifold recovery.

**Tangent Manifold Learning Part.** A sample-based family $\mathbf{H}$ consisting of $p \times q$ matrices H(X) smoothly depending on $X \in \mathbf{M}$ is constructed to meet relations

$$L_H(X) \equiv Span(H(X)) \approx L(X) \text{ for all } X \in \mathbf{M} \quad (5)$$

in certain chosen metric on the Grassmann manifold. In next steps, the mappings h and g will be built in such a way that both the equalities (1) and

$$G(h(X)) \approx H(X) \text{ for all points } X \in \mathbf{M} \quad (6)$$

are fulfilled. Hence, linear space $L_H(X)$ (5) approximates the tangent space $L_{h,g}(X)$ (3) to the RDM $\mathbf{M}_{h,g}$ at the point $r_{h,g}(X)$.

**Manifold Embedding Part.** Given the family $\mathbf{H}$ already constructed, the embedding mapping y = h(X) is constructed as follows. The Taylor series expansions

$$g(h(X')) - g(h(X)) \approx G(h(X)) \times (h(X') - h(X)) \tag{7}$$

of the mapping g at near points $h(X')$, $h(X) \in \mathbf{Y}_h$, under the desired approximate equalities (1), (6) for the mappings h and g to be specified further, imply equalities:

$$X' - X \approx H(X) \times (h(X') - h(X)) \tag{8}$$

for near points X, $X' \in \mathbf{M}$. These equations considered further as regression equations allow constructing the embedding mapping h and the FS $\mathbf{Y}_h = h(\mathbf{M})$.

**Manifold Reconstruction Step.** Given the family $\mathbf{H}$ and mapping $h(X)$ already constructed, the expansion (7), under the desired proximities (1) and (6), implies relation

$$g(y) \approx X + H(X) \times (y - h(X)) \tag{9}$$

for near points y, $h(X) \in \mathbf{Y}_h$ which is used for constructing the mapping g.

## 2.4    Grassmann&Stiefel Eigenmaps: Some Details

Details of the GSE are presented below. The numbers $\{\varepsilon_i > 0\}$ denote the algorithms parameters whose values are chosen depending on the sample size n ($\varepsilon_i = \varepsilon_{i,n}$) and tend to zero as $n \rightarrow \infty$ with rate $O(n^{-1/(q+2)})$.

**Step S1: Neighborhoods (Construction and Description).** The necessary preliminary calculations are performed at first step S1.

*Euclidean Kernel.* Introduce Euclidean kernel $K_E(X, X') = I\{|X' - X| < \varepsilon_1\}$ on the DM at points X, $X' \in \mathbf{M}$, here $I\{\cdot\}$ is indicator function.

*Grassmann Kernel.* An applying the Principal Component Analysis (PCA) [27] to the points from the set $U_n(X, \varepsilon_1) = \{X' \in \mathbf{X}_n: |X' - X| < \varepsilon_1\} \cup \{X\}$, results in $p \times q$ orthogonal matrix $Q_{PCA}(X)$ whose columns are PCA principal eigenvectors corresponding to the q largest PCA eigenvalues. These matrices determine q-dimensional linear spaces $L_{PCA}(X) = \mathrm{Span}(Q_{PCA}(X))$ in $R^p$, which, under certain conditions, approximate the tangent spaces $L(X)$:

$$L_{PCA}(X) \approx L(X). \tag{10}$$

In what follows, we assume that sample size n is large enough to ensure a positive value of the $q^{th}$ PCA-eigenvalue in sample points and provide proximities (10). To provide trade-off between 'statistical error' (depending on number $n(X)$ of sample points in set $U_n(X, \varepsilon_1)$) and 'curvature error' (caused by deviation of the manifold-valued points from the 'assumed in the PCA' linear space) in (10), ball radius $\varepsilon_1$ should tend to 0 as $n \rightarrow \infty$ with rate $O(n^{-1/(q+2)})$, providing, with high probability, the order $O(n^{-1/(q+2)})$ for the error in (10) [28, 29]; here 'an event occurs with high probability' means that its probability exceeds the value $(1 - C_\alpha/n^\alpha)$ for any n and $\alpha > 0$, and the constant $C_\alpha$ depends only on $\alpha$.

Grassmann kernel $K_G(X, X')$ on the DM at points X, $X' \in M$ is defined as

$$K_G(X, X') = I\{d_{BC}(L_{PCA}(X), L_{PCA}(X')) < \varepsilon_2\} \times K_{BC}(L_{PCA}(X), L_{PCA}(X'))$$

with use Binet-Cauchy kernel $K_{BC}(L_{PCA}(X), L_{PCA}(X')) = \text{Det}^2[S(X, X')]$ and Binet-Cauchy metric $d_{BC}(L_{PCA}(X), L_{PCA}(X')) = \{1 - \text{Det}^2[S(X, X')]\}^{1/2}$ on the Grassmann manifold Grass(p, q) [30, 31], here $S(X, X') = Q_{PCA}^T(X) \times Q_{PCA}(X')$.

Orthogonal $p \times p$ matrix $\pi_{PCA}(X) = Q_{PCA}(X) \times Q_{PCA}^T(X)$ is projector onto linear space $L_{PCA}(X)$ which approximates projection matrix $\pi(X)$ onto the tangent space $L(X)$.

*Aggregate Kernel.* Introduce the kernel $K(X, X') = K_E(X, X') \times K_G(X, X')$, which reflects not only geometrical nearness between points X and X' but also nearness between the linear spaces $L_{PCA}(X)$ and $L_{PCA}(X')$ (and, thus (10), nearness between the tangent spaces $L(X)$ and $L(X')$), as a product of the Euclidean and Grassmann kernels.

**Step S2: Tangent Manifold Learning.** The matrices H(X) will be constructed to meet the equalities $L_H(X) = L_{PCA}(X)$ for all points $X \in M$ that implies a representation

$$H(X) = Q_{PCA}(X) \times v(X), \tag{11}$$

in which $q \times q$ matrices v(x) should provide a smooth depending H(X) on point X.

At first, the $p \times q$ matrices $\{H_i = Q_{PCA}(X_i) \times v_i\}$ are constructed to minimize a form

$$\Delta_{H,n} = \frac{1}{2}\sum_{i,j=1}^n K(X_i, X_j) \times \|H_i - H_j\|_F^2 \tag{12}$$

over $q \times q$ matrices $v_1, v_2, \ldots, v_n$, under normalizing constraint

$$\sum_{i=1}^n K(X_i) \times (H_i^T \times H_i) = \sum_{i=1}^n K(X_i) \times (v_i^T \times v_i) = K \times I_q \tag{13}$$

used to avoid a degenerate solution; here $K(X) = \sum_{j=1}^n K(X, X_j)$ and $K = \sum_{i=1}^n K(X_i)$.

The quadratic form (12) and the constraint (13) take the forms $(K - \text{Tr}(V^T \times \Phi \times V))$ and $V^T \times F \times V = K \times I_q$, respectively, here V is $(nq) \times q$ matrix whose transpose consists of the consecutively written transposed $q \times q$ matrices $v_1, v_2, \ldots, v_n$, $\Phi = \|\Phi_{ij}\|$ and $F = \|F_{ij}\|$ are $nq \times nq$ matrices consisting, respectively, of $q \times q$ matrices

$$\{\Phi_{ij} = K(X_i, X_j) \times S(X_i, X_j)\} \text{ and } \{F_{ij} = \delta_{ij} \times K(X_i) \times I_q\}.$$

Thus, a minimization (12), (13) is reduced to the generalized eigenvector problem

$$\mathbf{\Phi} \times \mathbf{V} = \lambda \times \mathbf{F} \times \mathbf{V}, \tag{14}$$

and (nq) $\times$ q matrix $\mathbf{V}$, whose columns $V_1$, $V_2$, ..., $V_q \in R^{nq}$ are orthonormal eigenvectors corresponding to the q largest eigenvalues in the problem (14), determines the required q $\times$ q matrices $v_1$, $v_2$, ..., $v_n$.

The value H(X) (11) at arbitrary point $X \in \mathbf{M}$ is chosen to minimize a form

$$d_{H,n}(H) = \sum\nolimits_{j=1}^{n} K(X, X_j) \times \|Q_{PCA}(X) \times v(X) - H_j\|_F^2 \tag{15}$$

over v(X) under condition Span(H) = $L_{PCA}(X)$, whose solution is

$$H(X) = Q_{PCA}(X) \times v(X) = Q_{PCA}(X) \times \frac{1}{K(X)} \sum\nolimits_{j=1}^{n} K(X, X_j) \times S(X, X_j) \times v_j. \tag{16}$$

It follows from above formulas that the q $\times$ p matrix

$$G_h(X) = H^-(X) \times \pi_{PCA}(X) = v^{-1}(X) \times Q_{PCA}^T(X)$$

estimates Jacobian matrix $J_h(X)$ of Embedding mapping h(X) constructed afterward, here $H^-(X)$ is q $\times$ p pseudoinverse Moore-Penrose matrix of p $\times$ q matrix H(X) [32].

**Step S3: Manifold Embedding.** Embedding mapping h(X) with already known (estimated) Jacobian $G_h(X)$ is constructed to meet equalities (8) written for all pairs of near points X, X' $\in \mathbf{M}$ which can be considered as regression equations.

At first, the vector set $\{h_1, h_2, ..., h_n\} \subset R^q$ is computed as a standard least squares solution in this regression problem by minimizing the residual

$$\Delta_{h,n} = \sum\nolimits_{i,j=1}^{n} K(X_i, X_j) \times |X_j - X_i - H_i \times (h_j - h_i)|^2 \tag{17}$$

over the vectors $h_1$, $h_2$, ..., $h_n$ under normalizing condition $h_1 + h_2 + ... + h_n = \mathbf{0}$.

Then, considering the obtained vectors $\{h_j\}$ as preliminary values of the mapping h(X) at sample points, choose the value

$$h(X) = \frac{1}{K(X)} \sum\nolimits_{i=1}^{n} K(X, X_i) \times \{h_i + G_h(X) \times (X - X_i)\} \tag{18}$$

for arbitrary point X $\in \mathbf{M}$ as a result of minimizing over h the residual

$$d_{h,n}(h) = \sum\nolimits_{j=1}^{n} K(X, X_j) \times |X_j - X - H(X) \times (h_j - h)|^2. \tag{19}$$

The mapping (18) determines Feature sample $Y_{h,n} = \{y_{h,i} = h(X_i), i = 1, 2, ..., n\}$.

**Step S4: Manifold Recovery.** A kernel on the FS $Y_h$ and, then, the recovering mapping g(y) and its Jacobian matrix G(y) are constructed in this step.

*Kernel on the Feature Space.* It follows from (8) that proximities

$$|X-X_i| \approx d(y,\ y_{h,i}) = \{(y-y_{h,i})^T \times [H^T(X_i) \times H(X_i)] \times (y-y_{h,i})\}^{1/2}$$

hold true for near points $y = h(X)$ and $y_{h,i} \in Y_{h,n}$. Let $u_E(y, \varepsilon_1) = \{y_{h,i}: d(y, y_{h,i}) < \varepsilon_1\}$ be a neighborhood of the feature $y = h(X)$ consisting of sample features which are images of the sample points from $U_n(X, \varepsilon_1)$.

An applying the PCA to the set $h^{-1}(u_E(y, \varepsilon_1)) = \{X_i: y_{h,i} \in u_E(y, \varepsilon_1)\}$ results in the linear space $L_{PCA*}(y) \in \text{Grass}(p, q)$ which meets proximity $L_{PCA*}(h(X)) \approx L_{PCA}(X)$.

Introduce feature kernel $k(y, y_{h,i}) = I\{y_{h,i} \in u_E(y, \varepsilon_1)\} \times K_G(L_{PCA*}(y), L_{PCA*}(y_{h,i}))$ that meets equalities $k(h(X), h(X')) \approx K(X, X')$ for near points $X \in M$ and $X' \in X_n$.

*Constructing the Recovering Mapping and its Jacobian.* The matrix $G(y)$, which should meet both the conditions (6) and constraint $\text{Span}(G(y)) = L_{PCA*}(y)$, is chosen by minimizing quadratic form $\sum_{j=1}^n k(y, y_{h,j}) \times \|G(y) - H_j\|_F^2$ over G, that results in

$$G(y) = \pi^*(y) \times \frac{1}{k(y)} \sum_{j=1}^n k(y, y_{h,j}) \times H_j, \qquad (20)$$

here $\pi^*(y)$ is the projector onto the linear space $L_{PCA*}(y)$ and $k(y) = \sum_{j=1}^n k(y, y_{h,j})$.

Based on expansions (9) written for features $y_{h,j} \in u_E(y, \varepsilon_1)$, g(y) is chosen by minimizing quadratic form $\sum_{j=1}^n k(y, y_{h,j}) \times |X_j - g(y) - G(y) \times (y_{h,j} - y)|^2$ over g, thus

$$g(y) = \frac{1}{k(y)} \sum_{j=1}^n k(y, y_{h,j}) \times \{X_j + G(y) \times (y - y_{h,j})\}. \qquad (21)$$

The constructed mappings (18), (21) allow recovering the DM **M** and its tangent spaces L(X) by the formulas (2) and (4).

## 2.5    Grassmann&Stiefel Eigenmaps: Some Properties

Under asymptotic $n \to \infty$, when $\varepsilon_1 = O(n^{-1/(q+2)})$, relation $d_H(M_{h,g}, M) = O(n^{-2/(q+2)})$ hold true uniformly in points $X \in M$ with high probability [33]. This rate coincides with the asymptotically minimax lower bound for the Hausdorff distance $d_H(M_{h,g}, M)$ [34]; thus, the RDM $M_{h,g}$ estimates the DM **M** with optimal rate of convergence.

The main computational complexity of the GSE-algorithm is in the second and third steps, in which global high-dimensional optimization problems are solved.

First problem is generalized eigenvector problem (14) with $nq \times nq$ matrices **F** and **Φ**. This problem is solved usually with use the Singular value decomposition (SVD) [32] whose computational complexity is $O(n^3)$ [35].

Second problem is regression problem (17) for nq-dimensional estimated vector. This problem is reduced to the solution of the linear least-square normal equations with $nq \times nq$ matrix whose computational complexity is $O(n^3)$ also [32].

Thus, the GSE has total computational complexity $O(n^3)$ and is computationally expensive under large sample size n.

# 3  Incremental Grassmann&Stiefel Eigenmaps

The incremental version of the GSE divides the most computationally expensive generalized eigenvector and regression problems into n local optimization procedures, each time k solved in explicit form for one new variable (matrix $H_k$ and feature $h_k$) only, k = 1, 2, ..., n.

The proposed incremental version includes an additional preliminary step $S1^+$ performed after the Step S1, in which a weighted undirected sample graph $\Gamma(X_n)$ consisting of the sample points $\{X_i\}$ as nodes is constructed and the shortest ways between arbitrary node chosen as an origin of the graph and all the other nodes are calculated.

The second and third steps S2 and S3 are replaced by common incremental step S2–3 in which the matrices $\{H_k\}$ and features $\{h_k\}$ are computed sequentially at the graph nodes, moving along the shortest paths starting from the chosen origin of the graph. Step S4 in the GSE remains unchanged in the incremental version.

## 3.1  Step $S1^+$: Sample Graph

Introduce a weighted undirected sample graph $\Gamma(X_n)$ consisting of the sample points $\{X_i\}$ as nodes. The edges in $\Gamma(X_n)$ connect the nodes $X_i$ and $X_j$ if and only when $K(X_i, X_j) > 0$; the lengths of such edge $(X_i, X_j)$ equal to $|X_i - X_j|/K(X_i, X_j)$.

Choose arbitrary node $X_{(1)} \in \Gamma(X_n)$ as an origin of the graph. Using the Dijksra algorithm [36], compute the shortest paths between the chosen node and all the other nodes $X_{(2)}, X_{(3)}, ..., X_{(n)}$ writing in ascending order of the lengths of the shortest paths from the origin $X_{(1)}$. Denote $\Gamma_k$ a subgraph consisting of the nodes $\{X_{(1)}, X_{(2)}, ..., X_{(k)}\}$ and connected them edges.

*Note.* The origin $X_{(1)}$ can be chosen as a node with minimal eccentricity; an eccentricity of some node equals to maximum of lengths of the shortest paths between the node under consideration and all the other nodes. But a calculation of the shortest ways between all nodes in the graph $\Gamma(X_n)$, which should be computed for this construction, require n-fold applying of the Dijksra algorithm.

## 3.2  Step S2–3: Incremental Tangent Manifold Learning and Manifold Embedding

Incremental version computes sequentially the matrices H(X) and h(X) at the points $X_{(1)}, X_{(2)}, ..., X_{(n)}$, starting from matrix $H_{(1)}$ and $h_{(1)}$ (initialization). Thus, step S2–3 consists of n substeps $\{S2–3_k, k = 1, 2, ..., n\}$ in which initialization substep is

**Initialization substep S2–3$_1$.** Put $v_{(1)} = I_q$ and $h_{(1)} = 0$; thus, $H(X_{(1)}) = Q_{PCA}(X_{(1)})$.

At the k-th substep S2–3$_k$, $k > 1$, when the matrices $H_{(j)}$, $j < k$, have already computed, quadratic form $\Delta_{H,k}$, similar to the form (12) but written only for the points $X_i$, $X_j \in \Gamma_k$, is minimized over single unknown matrix $H_{(k)} = Q_{PCA}(X_{(k)}) \times v_{(k)}$. This problem, in turn, is reduced to a minimization over $v_{(k)}$ of the form $d_{H,k}(H_{(k)})$, similar to the form $d_{H,n}(H_{(k)})$ (15) but written only for points $X_j \in \Gamma_{k-1}$. Its solution $v_{(k)}$, which is similar to the solution (16), is written in explicit form.

Let $\Delta_{h,k}$ be a quadratic form, similar to the form $\Delta_{h,n}$ (17) but written only for points $X_i$, $X_j \in \Gamma_k$. The value $h_{(k)}$, under the already computed values $h_{(j)}$, $j < k$, is calculated by minimizing the quadratic form $\Delta_{h,k}$ over single vector $h_{(k)}$. This problem, in turn, is reduced to a minimization over $h_{(k)}$ the form $d_{h,k}(h_{(k)})$, similar to the form $d_{h,n}(h_{(k)})$ (19) but written only for points $X_j \in \Gamma_{k-1}$; its solution, similar to the solution (18), is written in explicit form also.

Thus, the substeps S2–3$_k$, $k = 1, 2, \ldots, n$, are:

**Typical substep S2–3$_k$, $1 < k \le n$.** Given $\{(H_{(j)}, h_{(j)}), j < k\}$ already obtained, put

$$
H_{(k)} = Q_{PCA}(X_{(k)}) \times v_{(k)} = Q_{PCA}(X_{(k)}) \times \frac{\sum_{j<k} K(X_{(k)}, X_{(j)}) \times S(X_{(k)}, X_{(j)}) \times v_{(j)}}{\sum_{j<k} K(X_{(k)}, X_{(j)})},
$$
(22)

$$
h_{(k)} = \frac{\sum_{j<k} K(X_{(k)}, X_{(j)}) \times \left\{ h_{(j)} + v_{(k)}^{-1} \times Q_{PCA}^T(X_{(k)}) \times (X_{(k)} - X_{(j)}) \right\}}{\sum_{j<k} K(X_{(k)}, X_{(j)})}.
$$
(23)

Given $\{(H_{(k)}, h_{(k)}), k = 1, 2, \ldots, n\}$, the value $H(X) = Q_{PCA}(X) \times v(X)$ and $h(X)$ at arbitrary point $X \in \mathbf{M}$ are calculated with use formulas (16) and (18), respectively.

### 3.3   Incremental GSE: Properties

**Computational Complexity.** Incremental GSE works mainly with sample data lying in a neighborhood of some point X contained in $\varepsilon_1$-ball $U_n(X, \varepsilon_1)$ centered at X. The number $n(X)$ of sample points fallen into this ball, under $\varepsilon_1 = \varepsilon_{1,n} = O(n^{-1/(q+2)})$, with high probability equals to $n \times O(n^{-q/(q+2)}) = O(n^{2/(q+2)})$ uniformly on $X \in \mathbf{M}$ [37].

The sample graph $\Gamma(\mathbf{X}_n)$ consists of $V = n$ nodes and E edges connecting the graph nodes $\{X_k\}$. Each node $X_k$ is connected with no more than $n(X_k)$ other nodes, thus $E < 0.5 \times n \times \max_k n(X_k) = O(n^{(q+4)/(q+2)})$ and, hence, $\Gamma(\mathbf{X}_n)$ is sparse graph.

The running time of the Dijksra algorithm (Step S1$^+$), which computes the shortest paths in the sparse connected graph $\Gamma(\mathbf{X}_n)$, is $O(E \times \ln V) = O(n^{(q+4)/(q+2)} \times \ln n)$ in the worst case; the Fibonacci heap improves this rate to $O(E + V \times \ln V) = O(n^{(q+4)/(q+2)})$ [38].

The running time of k-th Step S2–3$_k$ (formulas (22) and (23)) is proportional to $n(X_k)$; thus total running time of the Step S2–3 is $n \times O(n^{-q/(q+2)}) = O(n^{(q+4)/(q+2)})$.

Therefore, the running time of the incremental version of the GSE is $O(n^{(q+4)/(q+2)})$, in contrast to the running time $O(n^3)$ of the initial GSE.

**Accuracy.** It follows from (18), (21) that $X - r_{h,g}(X) \approx \left(\pi^T_{PCA}(X) \times e(X)\right) \times |\delta(X)|$, in which $\delta(X) = X - \frac{1}{K(X)} \sum_{i=1}^n K(X, X_i) \times X_i$ and $e(X) = \delta(X)/|\delta(X)|$. The first and second multipliers are majorized by the PCA-error in (10) and $\varepsilon_{1,n}$, respectively, each of them has rate $O(n^{-1/(q+2)})$. Thus, reconstruction error $(X - r_{h,g}(X))$ in the incremental GSE has the same asymptotically optimal rate $O(n^{-2/(q+2)})$ as in the original GSE.

# 4 Conclusion

The incremental version of the Grassmann&Stiefel Eigenmaps algorithm, which constructs the low-dimensional representations of high-dimensional data with asymptotically minimal reconstruction error, is proposed. This version has the same optimal convergence rate $O(n^{-2/(q+2)})$ of the reconstruction error and a significantly smaller computational complexity on the sample size n: running time $O(n^{(q+4)/(q+2)})$ of the incremental version in contrast to $O(n^3)$ of the original algorithm.

**Acknowledgments.** This work is partially supported by the Russian Foundation for Basic Research, research project 16-29-09649 ofi-m.

# References

1. Donoho, D.L.: High-Dimensional Data Analysis: The Curses and Blessings of Dimensionality. Lecture at the "Mathematical Challenges of the 21st Century" Conference of the AMS, Los Angeles (2000). http://www-stat.stanford.edu/donoho/Lectures/AMS2000/AMS2000.html
2. Verleysen, M.: Learning high-dimensional data. In: Ablameyko, S., Goras, L., Gori, M., Piuri, V. (eds.) Limitations and Future Trends in Neural Computation. NATO Science Series, III: Computer and Systems Sciences, vol. 186, pp. 141–162. IOS Press, Netherlands (2003)
3. Bengio, Y., Courville, A., Vincent, P.: Representation Learning: A Review and New Perspectives, pp. 1–64 (2014). arXiv:1206.5538v3[cs.LG]. Accessed 23 Apr 2014
4. Bernstein, A., Kuleshov, A.: Low-dimensional data representation in data analysis. In: El Gayar, N., Schwenker, F., Suen, C. (eds.) ANNPR 2014. LNCS, vol. 8774, pp. 47–58. Springer, Heidelberg (2014)
5. Seung, H.S., Lee, D.D.: The manifold ways of perception. Science **290**(5500), 2268–2269 (2000)
6. Huo, X., Ni, X., Smith, A.K.: Survey of manifold-based learning methods. In: Liao, T.W., Triantaphyllou, E. (eds.) Recent Advances in Data Mining of Enterprise Data, pp. 691–745. World Scientific, Singapore (2007)
7. Ma, Y., Fu, Y. (eds.): Manifold Learning Theory and Applications. CRC Press, London (2011)

8. Law, M.H.C., Jain, A.K.: Nonlinear manifold learning for data stream. In: Berry, M., Dayal, U., Kamath, C., Skillicorn, D. (eds.) Proceedings of the 4th SIAM International Conference on Data Mining, Like Buena Vista, Florida, USA, pp. 33–44 (2004)
9. Law, M.H.C., Jain, A.K.: Incremental nonlinear dimensionality reduction by manifold learning. IEEE Trans. Pattern Anal. Mach. Intell. **28**(3), 377–391 (2006)
10. Gao, X., Liang, J.: An improved incremental nonlinear dimensionality reduction for isometric data embedding. Inf. Process. Lett. **115**(4), 492–501 (2015)
11. Saul, L.K., Roweis, S.T.: Think globally, fit locally: unsupervised learning of low dimensional manifolds. J. Mach. Learn. Res. **4**, 119–155 (2003)
12. Kouropteva, O., Okun, O., Pietikäinen, M.: Incremental locally linear embedding algorithm. In: Kalviainen, H., Parkkinen, J., Kaarna, A. (eds.) SCIA 2005. LNCS, vol. 3540, pp. 521–530. Springer, Heidelberg (2005)
13. Kouropteva, O., Okun, O., Pietikäinen, M.: Incremental locally linear embedding. Pattern Recogn. **38**(10), 1764–1767 (2005)
14. Schuon, S., Đurković, M., Diepold, K., Scheuerle, J., Markward, S.: Truly incremental locally linear embedding. In: Proceedings of the CoTeSys 1st International Workshop on Cognition for Technical Systems, 6–8 October 2008, Munich, Germany, p. 5 (2008)
15. Jia, P., Yin, J., Huang, X., Hu, D.: Incremental Laplacian eigenmaps by preserving adjacent information between data points. Pattern Recogn. Lett. **30**(16), 1457–1463 (2009)
16. Liu, X., Yin, J.-w., Feng, Z., Dong, J.: Incremental manifold learning via tangent space alignment. In: Schwenker, F., Marinai, S. (eds.) ANNPR 2006. LNCS (LNAI), vol. 4087, pp. 107–121. Springer, Heidelberg (2006)
17. Abdel-Mannan, O., Ben Hamza, A., Youssef, A.: Incremental line tangent space alignment algorithm. In: Proceedings of 2007 Canadian Conference on Electrical and Computer Engineering (CCECE 2007), 22–26 April 2007, Vancouver, pp. 1329–1332. IEEE (2007)
18. Kuleshov, A., Bernstein, A.: Manifold learning in data mining tasks. In: Perner, P. (ed.) MLDM 2014. LNCS, vol. 8556, pp. 119–133. Springer, Heidelberg (2014)
19. Lee, J.A., Verleysen, M.: Nonlinear Dimensionality Reduction. Information Science and Statistics. Springer, New York (2007)
20. Lee, J.A., Verleysen, M.: Quality assessment of dimensionality reduction: rank-based criteria. Neurocomputing **72**(7–9), 1431–1443 (2009)
21. Bernstein, A.V., Kuleshov, A.P.: Data-based manifold reconstruction via tangent bundle manifold learning. In: ICML-2014, Topological Methods for Machine Learning Workshop, Beijing, 25 June 2014. http://topology.cs.wisc.edu/KuleshovBernstein.pdf
22. Perrault-Joncas, D., Meilă, M.: Non-linear Dimensionality Reduction: Riemannian Metric Estimation and the Problem of Geometric Recovery, pp. 1–25 (2013). arXiv:1305.7255v1 [stat.ML]. Accessed 30 May 2013
23. Jost, J.: Riemannian Geometry and Geometric Analysis, 6th edn. Springer, Berlin (2011)
24. Bernstein, A.V., Kuleshov, A.P.: Manifold learning: generalizing ability and tangent proximity. Int. J. Softw. Inf. **7**(3), 359–390 (2013)
25. Lee, J.M.: Manifolds and Differential Geometry. Graduate Studies in Mathematics, vol. 107. American Mathematical Society, Providence (2009)
26. Bernstein, A.V., Kuleshov, A.P.: Tangent bundle manifold learning via Grassmann&Stiefel eigenmaps, pp. 1–25, December 2012. arXiv:1212.6031v1[cs.LG]
27. Jollie, T.: Principal Component Analysis. Springer, New York (2002)
28. Singer, A., Wu, H.-T.: Vector diffusion maps and the connection Laplacian. Commun. Pure Appl. Math. **65**(8), 1067–1144 (2012)
29. Tyagi, H., Vural, E., Frossard, P.: Tangent space estimation for smooth embeddings of Riemannian manifold, pp. 1–35 (2013). arXiv:1208.1065v2[stat.CO]. Accessed 17 May 2013

30. Hamm, J., Daniel, L.D.: Grassmann discriminant analysis: a unifying view on subspace-based learning. In: Proceedings of the 25th International Conference on Machine Learning (ICML 2008), pp. 376–383 (2008)
31. Wolf, L., Shashua, A.: Learning over sets using kernel principal angles. J. Mach. Learn. Res. **4**, 913–931 (2003)
32. Golub, G.H., Van Loan, C.F.: Matrix Computation, 3rd edn. Johns Hopkins University Press, Baltimore (1996)
33. Kuleshov, A., Bernstein, A., Yanovich, Y.: Asymptotically optimal method in manifold estimation. In: Abstracts of the XXIX-th European Meeting of Statisticians, 20–25 July 2013, Budapest, Hungary, p. 325 (2013). http://ems2013.eu/conf/upload/BEK086_006.pdf
34. Genovese, C.R., Perone-Pacifico, M., Verdinelli, I., Wasserman, L.: Minimax manifold estimation. J. Mach. Learn. Res. **13**, 1263–1291 (2012)
35. Trefethen, L.N.: Bau III, David: Numerical Linear Algebra. SIAM, Philadelphia (1997)
36. Cormen, T., Leiserson, C., Rivest, R., Stein, C.: Introduction to Algorithms. MIT Press, Cambridge (2001)
37. Yanovich, Y.: Asymptotic properties of local sampling on manifolds. J. Math. Stat. (2016)
38. Fredman, M.L., Tarjan, R.E.: Fibonacci heaps and their uses in improved network optimization algorithms. J. Assoc. Comput. Mach. **34**(3), 596–615 (1987)

# Soft-Constrained Nonparametric Density Estimation with Artificial Neural Networks

Edmondo Trentin$^{(\boxtimes)}$

Dipartimento di Ingegneria dell'Informazione e Scienze Matematiche,
Università degli Studi di Siena, Siena, Italy
trentin@dii.unisi.it

**Abstract.** The estimation of probability density functions (pdf) from
unlabeled data samples is a relevant (and, still open) issue in pattern
recognition and machine learning. Statistical parametric and nonparametric approaches present severe drawbacks. Only a few instances of
neural networks for pdf estimation are found in the literature, due to the
intrinsic difficulty of unsupervised learning under the necessary integral-
equals-one constraint. In turn, also such neural networks do suffer from
serious limitations. The paper introduces a soft-constrained algorithm
for training a multilayer perceptron (MLP) to estimate pdfs empirically. A variant of the Metropolis-Hastings algorithm (exploiting the
very probabilistic nature of the MLP) is used to satisfy numerically the
constraint on the integral of the function learned by the MLP. The preliminary outcomes of a simulation on data drawn from a mixture of
Fisher-Tippett pdfs are reported on, and compared graphically with the
estimates yielded by statistical techniques, showing the viability of the
approach.

**Keywords:** Density estimation · Nonparametric estimation · Unsupervised learning · Constrained learning · Multilayer perceptron

## 1 Introduction

The estimation of probability density functions (pdf) has long been a relevant
issue in pattern classification [3], data compression and model selection [5], coding [2], genomic analysis [4] and bioinformatics in general [15]. There are intrinsic
difficulties to pdf estimation, to the point that Vladimir Vapnik stated that it
is inherently a "hard (...) computational problem" where "one cannot guarantee
a good approximation using a limited number of observations" [14]. Albeit popular, statistical parametric techniques (e.g. maximum-likelihood for Gaussian
mixture models) rely on an arbitrary and unrealistic assumption on the form
of the underlying, unknown distribution. This holds true also for radial basis
functions networks, which realize mixture densities of fixed forms (say, again,
Gaussian). On the other hand, nonparametric techniques like the Parzen Window (PW) and the $k_n$-nearest neighbors ($k_n$-NN) [3] drop this assumption, and
attempt a direct estimation of the pdf from a data sample. Still, they suffer from
several limitations, including the following:

© Springer International Publishing AG 2016
F. Schwenker et al. (Eds.): ANNPR 2016, LNAI 9896, pp. 68–79, 2016.
DOI: 10.1007/978-3-319-46182-3_6

1. in PW and $k_n$-NN there is no actual *learning* (thus, no generalization). The models are "memory based", thence (by definition) they are prone to overfitting;
2. the local nature of the quantities used to express the estimates (either the window functions in PW, or the hyper-spheres used in $k_n$-NN, both centered at the individual locations of the specific data at hand) tend to yield a fragmented, unsmooth model;
3. the whole training set has to be kept always in memory in order to compute the estimate of the pdf over any new input pattern. This may result in a dramatic computational burden (in both space and time) of the estimation process during the test.

In order to overcome these limitations, multilayer perceptrons (MLP) can be regarded as a promising alternative family of nonparametric estimators, at least in principle. Given their "universal approximation" property, they might be a suitable model for any given (continuous) form of the pdf underlying the data. Unfortunately, MLPs have been scarcely exploited for pdf estimation so far. This is mostly due to the difficulty of the present unsupervised learning task and by the requirement that, at least numerically, the integral of the function learned by the MLP over its definition domain equals 1. Some exceptions are represented by the hybrid model presented in [13], where the MLP is trained via maximum-likelihood (ML) in order to estimate the emission pdfs of a hidden Markov model (although the algorithm is strictly HMM-specific, and it gives no guarantee on the integral of the MLP), and by the MLP-based ML density estimator discussed in [7], which faces theoretically the integral-equals-one issue but without offering focused solutions to the resultant numerical integration problems. Besides, the nature of the activation function required in the output layer of such estimator entails pdf estimates that belong to the exponential family [7], which can be seen as a step back to parametric estimation. In [6] the focus is shifted from the direct estimation of the pdf to the (much simpler) estimation of the corresponding cumulative distribution function (cdf). After properly training the MLP model $\phi(.)$ of the cdf, the pdf can be recovered by applying differentiation to $\phi(.)$. Unfortunately, a good approximation of the cdf does not necessarily translate into a similarly good estimate of its derivative. In fact, a small squared error between $\phi(.)$ and the true cdf does not entail that $\phi(.)$ is free from steep fluctuation that imply huge, rapidly changing values of the derivative. Negative values of $\frac{\partial \phi(x)}{\partial x}$ may even occur, violating the definition of pdf, and the integral-equals-one constraint is not accounted for.

In [12] we proposed a simple technique for training MLPs for pdf estimation relying on PW-generated target outputs. Albeit this technique results in much smoother results than the PW, it does not yield pdf models that satisfy the integral-equals-one constraint (actually, the corresponding integral is not even bounded). Besides, the technique is sensitive to the form and bandwidth of the window functions used (albeit less sensitive than the bare PW is). In this paper we propose a more general and sounder algorithm, which relies on an empirical estimate of the unknown pdf which does not involve any specific window functions. The optimization technique enforces the numerical satisfaction of the soft-constraint on the integral of the estimated function, by means of a variant of the Markov chain Monte Carlo method that exploits the very probabilistic nature of the resulting MLP. The algorithm is presented in the next section, while Sect. 3 reports on a graphical comparative simulation involving data drawn from a mixture density, showing the viability of the approach. Preliminary conclusions and on-going research activities are handed out in Sect. 4.

## 2  The Algorithm

Let $\mathcal{T} = \{\mathbf{x}_1, \ldots, \mathbf{x}_n\}$ be an unlabeled dataset of $n$ independent observations collected in the field. The observations are assumed to be represented as random vectors in a $d$-dimensional feature space, say $\mathbb{R}^d$, and to be identically distributed according to the unknown pdf $p(\mathbf{x})$. A MLP having $d$ input units, an output unit, and at least one hidden layer is used to estimate $p(\mathbf{x})$ from $\mathcal{T}$. An activation function $f_i(a_i)$ is associated with the generic unit $i$ in the MLP, where the activation $a_i$ is given by summing over all the weighted contributions to $i$ from the previous layer, say $a_i = \sum_j w_{ij} f_j(a_j)$ where $w_{ij} \in W$ is the connection weight between units $j$ and $i$, as usual, and $W$ is the set of all the parameters in the MLP (i.e., weights and bias values). For the input units only, the activation function reduces to the identity map over current input $\mathbf{x} = (x_1, \ldots, x_d)$, such that $f_i(a_i) = f_i(x_i) = x_i$. While the usual logistic sigmoid is used in the hidden layers, the activation function associated with the output unit shall have a counterdomain that fits the definition of pdf, that is the range $[0, +\infty)$. This may be granted in several different ways, for instance by setting $f_i(a_i) = \exp(a_i)$ (resulting in a pdf model belonging to the broad exponential family). In this paper we resort to a logistic sigmoid with adaptive amplitude $\lambda \in \mathbb{R}^+$, that is $f_i(a_i) = \lambda/(1 + \exp(-a_i))$ as described in [11]. In so doing, the MLP can stretch its output over any required interval $[0, \lambda)$, which is not bounded a priori but is learned (along with the other parameters of the MLP) in order to fit the nature of the specific pdf at hand. Moreover, having sigmoids in both the output and the hidden layers eases the training process via backpropagation from a numerical viewpoint, as well. Other advantages entailed by adaptive amplitudes are pointed out in [11]. Eventually, the MLP realizes a function $\varphi(\mathbf{x}, W)$ of its input $\mathbf{x}$. In the following, without loss of generality for all the present intents and purposes, we assume that the observations of interest are limited to a compact $S \subset \mathbb{R}^d$ (in practice, any data normalization technique may be used in order to ensure satisfaction of this requirement) such that, in turn, $S$ can be treated as the domain of $\varphi(\mathbf{x}, W)$. The goal of the training algorithm is a proper modification of the MLP parameters $W$ given the unlabeled sample $\mathcal{T}$ such that $p(\mathbf{x})$ can be properly estimated relying on $\varphi(\mathbf{x}, W)$. This requires pursuing two purposes: (1) exploiting the information encapsulated in the unlabeled data $\mathcal{T}$ to approximate the unknown pdf; (2) preventing the MLP from developing spurious solutions, i.e. enforcing the constraint $\int_S \varphi(\mathbf{x}, W)d\mathbf{x} = 1$. To these ends, we first let the estimated pdf model be defined as $\tilde{p}(\mathbf{x}, W) = \frac{\varphi(\mathbf{x}, W)}{\int_S \varphi(\mathbf{x}, W)d\mathbf{x}}$ for all $\mathbf{x}$ in $S$. Afterwards, given a generic observation $\hat{\mathbf{x}}$ in $\mathcal{T}$, an empirical estimate of $p(\hat{\mathbf{x}})$

is achieved as follows (see [3]). Consider a $d$-ball $B(\hat{\mathbf{x}}, \mathcal{T})$ centered in $\hat{\mathbf{x}}$ and having the minimum radius $r(\hat{\mathbf{x}}, \mathcal{T})$ such that $|B(\hat{\mathbf{x}}, \mathcal{T}) \cap \mathcal{T}| = k_n + 1$ where $k_n = \lfloor k\sqrt{n} \rfloor$ and $k \in \mathbb{N}$ is a hyper-parameter. The requirement that $k_n + 1$ observations lie within the ball is due to the constatation that $\hat{\mathbf{x}}$ is in $B(\hat{\mathbf{x}}, \mathcal{T}) \cap \mathcal{T}$ by construction but it shall not be involved in the subsequent estimation of the pdf over $\hat{\mathbf{x}}$ itself, in order to prevent the very estimate from being biased. The probability $P$ that a random vector drawn from $p(\mathbf{x})$ lies within $B(\hat{\mathbf{x}}, \mathcal{T})$ is $P = \int_{B(\hat{\mathbf{x}}, \mathcal{T})} p(\mathbf{x}) d\mathbf{x}$. Since $P$ can be estimated via the usual frequentist approach as $P \simeq k_n/n$ and the integral can be approximated geometrically as $\int_{B(\hat{\mathbf{x}}, \mathcal{T})} p(\mathbf{x}) d\mathbf{x} \simeq p(\hat{\mathbf{x}}) V(B(\hat{\mathbf{x}}, \mathcal{T}))$ where $V(.)$ represents the hyper-volume of its argument, we can write [3] $p(\hat{\mathbf{x}}) \simeq \frac{k_n/n}{V(B(\hat{\mathbf{x}}, \mathcal{T}))}$. Therefore, the proposed training algorithm stems from the minimization of the (on-line) criterion function

$$C(\mathcal{T}, W, \hat{\mathbf{x}}) = \frac{1}{2} \left( \frac{k_n/n}{V(B(\hat{\mathbf{x}}, \mathcal{T}))} - \tilde{p}(\hat{\mathbf{x}}, W) \right)^2 + \frac{\rho}{2} \left( 1 - \int_S \varphi(\mathbf{x}, W) d\mathbf{x} \right)^2 \quad (1)$$

with respect to the MLP parameters $W$, in an incremental fashion upon presentation of the next training vector, say $\hat{\mathbf{x}}$. The first term in the criterion aims at normalized MLP outputs that approximate the empirical estimate of the unknown pdf, while the second term is a "soft" constraint that enforces a unit integral of $\tilde{p}(\mathbf{x}, W)$ over $S$. The hyper-parameter $\rho \in \mathbb{R}^+$ controls the importance of the constraint, and it can be used to tackle numerical issues in practical applications. A gradient descent learning rule for the generic MLP parameter $w$ can be devised as usual, prescribing a weight update $\Delta w$ defined as $\Delta w = -\eta \frac{\partial C(.)}{\partial w}$, where $\eta \in \mathbb{R}^+$ is the learning rate. Taking the partial derivative of $C(\mathcal{T}, W, \hat{\mathbf{x}})$ with respect to $w$ requires calculating the derivatives of the first and the second term in the right-hand side of Eq. (1). For the first term we have:

$$\frac{\partial}{\partial w} \left\{ \frac{1}{2} \left( \frac{k_n/n}{V(B(\hat{\mathbf{x}}, \mathcal{T}))} - \frac{\varphi(\hat{\mathbf{x}}, W)}{\int_S \varphi(\mathbf{x}, W) d\mathbf{x}} \right)^2 \right\} = \quad (2)$$

$$= - \left( \frac{k_n/n}{V(B(\hat{\mathbf{x}}, \mathcal{T}))} - \frac{\varphi(\hat{\mathbf{x}}, W)}{\int_S \varphi(\mathbf{x}, W) d\mathbf{x}} \right) \frac{\partial}{\partial w} \left\{ \frac{\varphi(\hat{\mathbf{x}}, W)}{\int_S \varphi(\mathbf{x}, W) d\mathbf{x}} \right\}$$

$$= - \left( \frac{k_n/n}{V(B(\hat{\mathbf{x}}, \mathcal{T}))} - \frac{\varphi(\hat{\mathbf{x}}, W)}{\int_S \varphi(\mathbf{x}, W) d\mathbf{x}} \right) \cdot$$

$$\cdot \left\{ \frac{1}{\int_S \varphi(\mathbf{x}, W) d\mathbf{x}} \frac{\partial \varphi(\hat{\mathbf{x}}, W)}{\partial w} - \frac{\tilde{p}(\hat{\mathbf{x}}, W)}{\int_S \varphi(\mathbf{x}, W) d\mathbf{x}} \frac{\partial}{\partial w} \int_S \varphi(\mathbf{x}, W) d\mathbf{x} \right\}$$

$$= - \frac{1}{\int_S \varphi(\mathbf{x}, W) d\mathbf{x}} \left( \frac{k_n/n}{V(B(\hat{\mathbf{x}}, \mathcal{T}))} - \frac{\varphi(\hat{\mathbf{x}}, W)}{\int_S \varphi(\mathbf{x}, W) d\mathbf{x}} \right) \cdot$$

$$\cdot \left\{ \frac{\partial \varphi(\hat{\mathbf{x}}, W)}{\partial w} - \frac{\varphi(\hat{\mathbf{x}}, W)}{\int_S \varphi(\mathbf{x}, W) d\mathbf{x}} \int_S \frac{\partial \varphi(\mathbf{x}, W)}{\partial w} d\mathbf{x} \right\}$$

where Leibniz rule was applied in the last step of the calculations, and is applied again in the calculation of the derivative of the second term as follows:

$$\frac{\partial}{\partial w}\left\{\frac{\rho}{2}\left(1 - \int_S \varphi(\mathbf{x}, W)d\mathbf{x}\right)^2\right\} = \tag{3}$$

$$= -\rho\left(1 - \int_S \varphi(\mathbf{x}, W)d\mathbf{x}\right)\frac{\partial}{\partial w}\int_S \varphi(\mathbf{x}, W)d\mathbf{x}$$

$$= -\rho\left(1 - \int_S \varphi(\mathbf{x}, W)d\mathbf{x}\right)\int_S \frac{\partial\varphi(\mathbf{x}, W)}{\partial w}d\mathbf{x}.$$

Therefore, in order to compute the right-hand side of Eqs. (2) and (3), algorithms for the computation of $\frac{\partial\varphi(\mathbf{x}, W)}{\partial w}$, $\int_S \varphi(\mathbf{x}, W)d\mathbf{x}$, and $\int_S \frac{\partial}{\partial w}\varphi(\mathbf{x}, W)d\mathbf{x}$ are needed. The quantity $\frac{\partial\varphi(\mathbf{x}, W)}{\partial w}$ is the usual partial derivative of the output of a MLP with respect to a generic parameter of the network, and is readily computed via plain backpropagation. If $w = w_{ij}$, that is the connection weight between the $j$-th unit in any hidden layer (or, in the input layer) and the $i$-th unit in the upper layer, then $\frac{\partial\varphi(\mathbf{x}, W)}{\partial w} = \delta_i f_j(a_j)$ where $\delta_i = f_i'(a_i)$ if $i$ represents the output unit, or $\delta_i = (\sum_u w_{ui}\delta_u)f_i'(a_i)$ otherwise, where the sum is extended to all units $u$ in the upper layer once the corresponding *deltas* have been properly computed according to the usual top-down order. As for the integrals, in principle any plain, deterministic numerical quadrature integration technique may do, e.g., Simpson's method, trapezoidal rule, etc. While such methods are viable if $d = 1$, they do not scale up computationally to higher dimensions ($d \geq 2$) of the feature space (especially in the light of the fact that $\int_S \frac{\partial}{\partial w}\varphi(\mathbf{x}, W)d\mathbf{x}$ has to be computed individually for each parameter of the MLP). Besides, they do not exploit the very nature of the specific function to be integrated, that is expected to be closely related to the pdf $p(\mathbf{x})$ that explains the distribution of the specific data at hand. Accounting for the pdf of the data is expected to drive the integration algorithm towards integration points that cover "interesting" (i.e., having high likelihood) regions of the domain of the integrand. For these reasons, we propose a non-deterministic, multi-dimensional integration technique which accounts for the pdf underlying the data.

The technique is an instance of Markov chain Monte Carlo [1]. Let $\phi(\mathbf{x})$ denote the integrand of interest (either $\varphi(\mathbf{x}, W)$ or $\frac{\partial\varphi(\mathbf{x}, W)}{\partial w}$). The integral of $\phi(\mathbf{x})$ over $S$ is approximated via Monte Carlo with importance sampling [10] as

$$\int_S \phi(\mathbf{x})d\mathbf{x} \simeq \frac{V(S)}{m}\sum_{\ell=1}^m \phi(\dot{\mathbf{x}}_\ell) \tag{4}$$

relying on $m$ properly sampled integration points $\dot{\mathbf{x}}_1, \ldots, \dot{\mathbf{x}}_m$. Sampling occurs by drawing $\dot{\mathbf{x}}_\ell$ at random (for $\ell = 1, \ldots, m$) from the mixture density $p_u(\mathbf{x})$ defined as

$$p_u(\mathbf{x}) = \alpha(t)u(\mathbf{x}) + (1 - \alpha(t))\tilde{p}(\mathbf{x}) \tag{5}$$

where $u(\mathbf{x})$ is the uniform distribution over $S$, and $\alpha : \mathbb{N} \to (0, 1)$ is a decaying function of the number of the MLP training epochs $t = 1, \ldots, T$ (a training epoch is a completed re-iteration of Eqs. (2) and (3) over all the observations in $\mathcal{T}$), such that $\alpha(1) \simeq 1.0$ and $\alpha(T) \simeq 0.0$. In this paper we let

$$\alpha(t) = \frac{1}{1 + e^{\frac{t/T - 1/2}{\theta}}} \tag{6}$$

and a value $\theta = 0.07$ is used in the experiments. The rationale behind Eq. (5) is twofold. On the one hand, the importance sampling mechanism entailed by Eq. (5) accounts for the (estimated) pdf $\tilde{p}(\mathbf{x})$ of the data, respecting their natural distribution and focusing integration on the relevant integration points. On the other hand, the estimates of the pdf are too noisy and unreliable during the early stage of the MLP training process, such that sampling from a uniform distribution (like in the plain Monte Carlo algorithm) is sounder at the beginning. Trust in the MLP estimates keeps growing as training proceeds, such that sampling from $\tilde{p}(\mathbf{x})$ progressively replaces sampling from $u(\mathbf{x})$, ending up in pure non-uniform importance sampling. The form of the credit-assignment function $\alpha(t)$ described by Eq. (6), shown in Fig. 1 (basically, the reflection of a sigmoid), is such that the algorithm sticks with the uniform sampling for quite some time before (rather abruptly) beginning crediting the estimated pdf, but it ends up giving most of the credit to the MLP throughout the last period of training. Noteworthily, for $t \to T$ the sampling substantially occurs from $|\varphi(\mathbf{x}, W)| / \int_S |\varphi(\mathbf{x}, W)| \, d\mathbf{x}$ (this is an implicit consequence of $\varphi(\mathbf{x}, W)$ being non-negative by construction) which forms a sufficient condition for granting that, like in the VEGAS algorithm, the variance of the estimated integral vanishes and the corresponding error goes to zero [9].

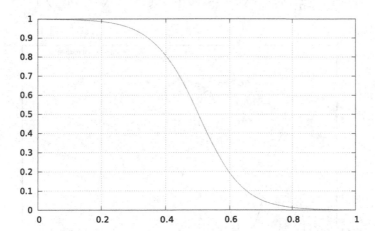

**Fig. 1.** An instance of the function $\alpha(.)$ defined by Eq. (6) with $\theta = 0.07$, plotted over the domain $[0, 1]$.

It is seen that sampling from $p_u(\mathbf{x})$ requires the application of a method for sampling from the MLP output. To this end, we resort to a specific instance of Markov chain Monte Carlo, namely the Metropolis-Hastings (M-H) algorithm [8]. The latter results in a sampling that is robust to the fact that $\varphi(\mathbf{x}, W)/\int_S \varphi(\mathbf{x}, W)d\mathbf{x}$ may not be properly normalized during the training (in fact, the estimate $\tilde{p}(\mathbf{x})$ does not necessarily respect the axioms of probability), since $\varphi(\mathbf{x}, W)$ is proportional (by construction) to the corresponding properly normalized pdf estimate [8]. Due to its ease and efficiency of sampling, the proposal pdf $q(\mathbf{x}'|\mathbf{x})$ required by M-H to generate the next candidate sample $\mathbf{x}' = (x'_1, \ldots, x'_d)$ from current sample $\mathbf{x} = (x_1, \ldots, x_d)$ is defined as a multivariate logistic pdf with radial basis, having location $\mathbf{x}$ and scale $\sigma$, that is

$$q(\mathbf{x}'|\mathbf{x}, \sigma) = \prod_{i=1}^{d} \frac{e^{(x'_i - x_i)/\sigma}}{\sigma(1 + e^{(x'_i - x_i)/\sigma})^2} \tag{7}$$

which is readily sampled via the inverse transform sampling method. Both $\sigma$ and the burn-in period for M-H are hyper-parameters to be fixed empirically via any model selection technique.

## 3    Demonstration

A qualitative demonstration of the viability of the technique is given as follows. A dataset consisting of 200 samples was randomly drawn from a mixture of 3 Fisher-Tippett distributions, namely

$$p(x) = \sum_{i=1}^{3} \frac{\Pi_i}{\beta_i} \exp\left(-\frac{x - \mu_i}{\beta_i}\right) \exp\left\{-\exp\left(-\frac{x - \mu_i}{\beta_i}\right)\right\} \tag{8}$$

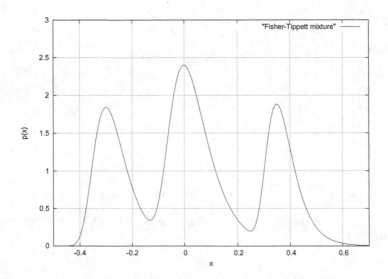

**Fig. 2.** The mixture of Fisher-Tippett pdfs used in the simulation.

where $\Pi_1 = 0.3$, $\Pi_2 = 0.45$, and $\Pi_3 = 0.25$ are the mixing parameters, and the component densities are identified by the locations $\mu_1 = -0.3$, $\mu_2 = 0.0$, $\mu_3 = 0.35$, and by the scales $\beta_1 = 0.06$, $\beta_2 = 0.07$, $\beta_3 = 0.05$, respectively. The mixture density is shown in Fig. 2, illustrating the asymmetry typical of the Fisher-Tippett pdf (a consequence of its positive skewness), which may raise some fitting issues when applying estimators based on symmetric kernels (e.g., Gaussian).

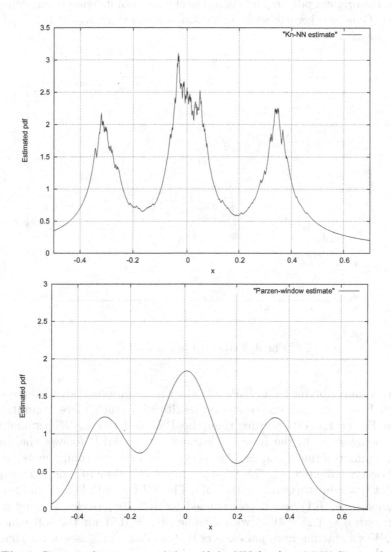

**Fig. 3.** Statistical estimates of the pdf: $k_n$-NN (top) and PW (bottom).

Figure 3 shows the estimates of the pdf obtained form the data sample using two major statistical nonparametric density estimation techniques, namely the

$k_n$-NN (top) and the PW (bottom), respectively. The $k_n$-NN density estimator
[3] relied on the usual value of $k_n$ defined as $k_n = k_1\sqrt{n}$ (to ensure unbiased
estimation [3]) where $n = 200$ is the sample size, and where we let $k_1 = 2$ (the
same value that we used in this simulation for computing Eq. (2) during the MLP
training). As expected, the $k_n$-NN results in a peaky estimate which is formally
not a pdf (not only the integral is larger than 1, but it is also unbounded). To
the contrary, the PW yields a proper pdf but it eventually misses the detailed
shape of the original pdf (in particular, the skewness of its component densities).
Standard Gaussian kernels with an unbiased bandwidth $h_n = 1/\sqrt{n}$ [3] were
adopted in the PW.

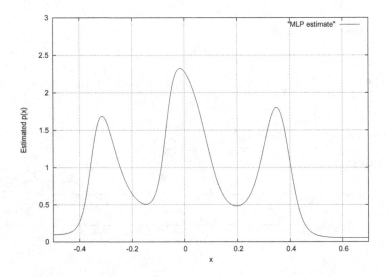

**Fig. 4.** Estimated pdf via MLP.

The estimate obtained via MLP trained with the proposed algorithm is shown
in Fig. 4. It is seen that the estimator results in a much more accurate fit to
the true Fisher-Tippett mixture than the PW and the $k_n$-NN estimates do.
Besides being smooth, the function learned by the MLP follows the form of
the underlying pdf (including the skewness of the first two components), and its
integral over $S$ (where $S = (-0.5, 0.7)$) is numerically close to satisfy the required
constraint (its value turns out to be 1.07). The MLP has 12 logistic hidden units
(with smoothness 0.1) and it was trained for $6 \times 10^4$ epochs, keeping a fixed
learning rate $\eta = 1.25 \times 10^{-4}$ with a value of $\rho = 0.01$ for the soft constraint
in $C(\mathcal{T}, W, \hat{\mathbf{x}})$, starting from parameters $W$ (weights and biases of the sigmoids)
initialized at random uniformly over the $(-0.5, 0.5)$ interval. The sampling of
$m = 200$ integration points occurred at the beginning of each training epoch,
using a scale $\sigma = 9$ for the logistic proposal pdf in M-H. The burn-in period of
the Markov chain in M-H was stretched over the first 500 states.

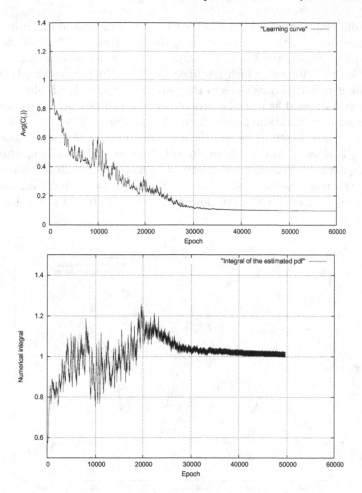

**Fig. 5.** Learning curve, i.e. avg. value of $C(\mathcal{T}, W, x)$ (top), and numerical integral of $\varphi(.)$ (bottom) as functions of the number of training epochs.

It is interesting to observe the learning curve of the MLP during the training process, shown in Fig. 5 (top). At each epoch, its value is computed by averaging the value of Eq. (1), i.e. the criterion function $C(\mathcal{T}, W, x)$, over the data in $\mathcal{T}$. The curve is substantially different from the usual learning curves we are acquainted with in regular MLP training via plain least-squares minimization, although the overall trend is still the progressive reduction of the criterion function. While it is not surprising that, due to the on-line nature of the training algorithm, there may be occasional increases in the average value of $C(\mathcal{T}, W, x)$, the irregular behavior of the curve in the first stage of the process is mostly due to two problem-specific causes. The former is in the numerical integration error occurring at each new sampling (i.e., at each iteration), since integrals computed from different random samples of integration points generally differ from each

other (significantly, at times). The latter cause is in the initial difficulty of finding a balance between the minimization of the first term and that of the second
term in Eq. (1), namely the loss (with respect to the empirical estimate) and the
constraint on the integral, which are likely to drive the parameters $W$ towards
possibly different initial directions in the parameter space. Both phenomena are
progressively mitigated as learning proceeds towards better solutions and as the
sampling of the integration points from $\varphi(.)$ gradually kicks in. This is seen
clearly from Fig. 5 (bottom), where the evolution in time of the numerical integral of $\varphi(.)$ is shown. Starting from an initial value of 0.53, the approximated
integral is quickly increased, lying mostly within the $1.0 \pm 0.25$ range for he first
half of the training process. It stabilizes markedly in the second half, reaching a
substantial equilibrium (with very slow decrease) close to 1.0 in the last period
of training.

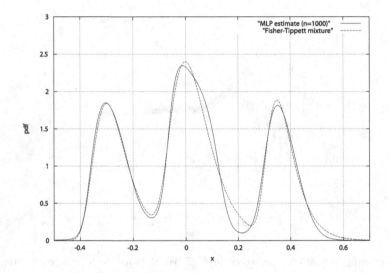

**Fig. 6.** Estimated pdf via MLP with $n = 1000$.

Finally, Fig. 6 plots the Fisher-Tippett mixture against the estimated pdf
yielded by the MLP trained via the proposed algorithm on a larger data sample
($n = 1000$). The MLP architecture and the hyper-parameters for its training were
unchanged with respect to the previous simulation. It is seen that the algorithm
results in accurate estimates, as long as the training set is large enough (i.e.,
statistically representative of the data distribution under investigation).

## 4     Preliminary Conclusions and On-Going Research

The demonstration confirms that the proposed algorithm may be a viable tool for
practitioners interested in the application of pdf estimation methods. In spite of

the relative simplicity of the data used, it is seen that statistical approaches yield unnatural estimates, while the MLP may result in a much better and smoother fit to the pdf underlying the unlabeled dataset at hand. At the time being, we are in the early stages of investigating the behavior of the technique over multi-dimensional data. Several variants of the algorithm are under investigation, as well, mostly revolving around different choices for the function $\alpha(t)$ and for the proposal function to be used within the M-H sampling.

# References

1. Andrieu, C., de Freitas, N., Doucet, A., Jordan, M.I.: An introduction to MCMC for machine learning. Mach. Learn. **50**(1–2), 5–43 (2003)
2. Beirami, A., Sardari, M., Fekri, F.: Wireless network compression via memory-enabled overhearing helpers. IEEE Trans. Wirel. Commun. **15**(1), 176–190 (2016)
3. Duda, R.O., Hart, P.E.: Pattern Classification and Scene Analysis. Wiley, New York (1973)
4. Koslicki, D., Thompson, D.: Coding sequence density estimation via topological pressure. J. Math. Biol. **70**(1/2), 45–69 (2015)
5. Liang, F., Barron, A.: Exact minimax strategies for predictive density estimation, data compression, and model selection. IEEE Trans. Inf. Theory **50**(11), 2708–2726 (2004)
6. Magdon-Ismail, M., Atiya, A.: Density estimation and random variate generation using multilayer networks. IEEE Trans. Neural Netw. **13**(3), 497–520 (2002)
7. Modha, D.S., Fainman, Y.: A learning law for density estimation. IEEE Trans. Neural Netw. **5**(3), 519–523 (1994)
8. Newman, M.E.J., Barkema, G.T.: Monte Carlo Methods in Statistical Physics. Oxford University Press, Oxford (1999)
9. Ohl, T.: VEGAS revisited: adaptive Monte Carlo integration beyond factorization. Comput. Phys. Commun. **120**, 13–19 (1999)
10. Rubinstein, R.Y., Kroese, D.P.: Simulation and the Monte Carlo Method, 2nd edn. Wiley, Hoboken (2012)
11. Trentin, E.: Networks with trainable amplitude of activation functions. Neural Netw. **14**(45), 471–493 (2001)
12. Trentin, E.: Simple and effective connectionist nonparametric estimation of probability density functions. In: Schwenker, F., Marinai, S. (eds.) ANNPR 2006. LNCS (LNAI), vol. 4087, pp. 1–10. Springer, Heidelberg (2006)
13. Trentin, E., Gori, M.: Robust combination of neural networks and hidden Markov models for speech recognition. IEEE Trans. Neural Netw. **14**(6), 1519–1531 (2003)
14. Vapnik, V.: The Nature of Statistical Learning Theory. Springer, New-York (1995)
15. Yang, Z.: Machine Learning Approaches to Bioinformatics. World Scientific Publishing Company, Singapore (2010)

# Density Based Clustering via Dominant Sets

Jian Hou[1,2]([⊠]), Weixue Liu[3], and Xu E[3]

[1] College of Engineering, Bohai University, Jinzhou 121013, China
dr.houjian@gmail.com
[2] ECLT, Università Ca' Foscari Venezia, 30124 Venezia, Italy
[3] College of Information Science, Bohai University, Jinzhou 121013, China

**Abstract.** While density based clustering algorithms are able to detect clusters of arbitrary shapes, their clustering results usually rely heavily on some user-specified parameters. In order to solve this problem, in this paper we propose to combine dominant sets and density based clustering to obtain reliable clustering results. We firstly use the dominant sets algorithm and histogram equalization to generate the initial clusters, which are usually subsets of the real clusters. In the second step the initial clusters are extended with density based clustering algorithms, where the required parameters are determined based on the initial clusters. By merging the merits of both algorithms, our approach is able to generate clusters of arbitrary shapes without user-specified parameters as input. In experiments our algorithm is shown to perform better than or comparably to some state-of-the-art algorithms with careful parameter tuning.

**Keywords:** Dominant set · Histogram equalization · Density based clustering · Cluster extension

## 1 Introduction

As an important unsupervised learning approach, clustering is widely applied in various domains, including data mining, pattern recognition and image processing. A vast amount of clustering algorithms have been proposed in the literature and many of them have found successful applications. In addition to the commonly used k-means approach, density based and spectral clustering algorithms have received much attention in recent decades. DBSCAN (Density-Based Spatial Clustering of Applications with Noise) [4] is a typical density based clustering algorithm. Given a neighborhood radius and the minimum number of data in the neighborhood, DBSCAN determines a density threshold and admits into clusters the data satisfying the density constraint. By extracting the clusters in a sequential manner, DBSCAN is able to determine the number of clusters automatically. In [10] the authors presented a density peak (DP) based clustering algorithm, which applies local density and the distance to the nearest neighbor with higher density to isolate and identify the cluster centers. Then the non-center data are assigned the same labels as their nearest neighbors with higher density. Given that the cluster centers are identified correctly, this algorithm is reported to generate excellent clustering results in [10]. Spectral clustering, e.g., the normalized

© Springer International Publishing AG 2016
F. Schwenker et al. (Eds.): ANNPR 2016, LNAI 9896, pp. 80–91, 2016.
DOI: 10.1007/978-3-319-46182-3_7

cuts (NCuts) algorithm [11], performs dimensionality reduction by means of the eigenvalues of the pairwise data similarity matrix in the first step, and then start the clustering process in the smaller dimension. Like k-means, spectral clustering algorithms require the number of clusters to be specified by user. Also with the pairwise data similarity matrix as input, the affinity propagation (AP) [1] and dominant sets (DSets) [9] algorithms accomplish the clustering process with different strategies from spectral clustering. Given the preference values of each data as a cluster center, the affinity propagation algorithm iteratively pass the affinity (similarity) messages among data and identifies the cluster centers and members gradually. With the pairwise similarity matrix as the sole input, the dominant sets algorithm defines a non-parametric concept of a cluster and extract the clusters sequentially.

One problem with many of the above-mentioned clustering algorithms is that they can only detect spherical clusters. The algorithms afflicted by this problem include k-means, spectral cluster, AP and DSets. While the density based algorithms, e.g., DBSCAN and DP, are able to generate clusters of arbitrary shapes, they involve user-specified parameters and the clustering results rely heavily on these parameters. Unfortunately, the appropriate parameters of these algorithms are usually data dependent and not easy to determine. DBSCAN uses a neighborhood radius *Eps* and the minimum number *MinPts* of data in the neighborhood to denote a density threshold, which is then used to determine if one data can be included into a cluster. Evidently inappropriate selection of the two parameters may result in over-large or over-small clusters and influence the clustering results. The DP algorithm does not involve parameter input in theory. However, the calculation of density involves a cut-off distance, which is usually determined empirically. Furthermore, the automatic identification of cluster centers seem difficult and additional parameter or even human assistance is often necessary.

In order to obtain clusters of arbitrary shapes reliably, in this paper we present a two-step approach which combines the merits of the DSets algorithm and density based clustering algorithms. Our work is based on the observation that the two kinds of algorithms have some complementary properties. The DSets algorithm uses only the pairwise data similarity matrix as input and does not involve any other parameters explicitly. However, it can only generate spherical clusters. In contrast, density based algorithms are able to generate clusters of arbitrary shapes, on condition that appropriate density related parameters are determined beforehand. Motivated by this observation, we propose to generate initial clusters with the DSets algorithm in the first step, and then perform density based clustering where the required parameters are determined based on the initial clusters. Following the DSets algorithm, our algorithm extracts the clusters in a sequential manner and determines the number of clusters automatically. Experiments on data clustering and image segmentation validate the effectiveness of our approach.

The following sections are organized as follows. In Sect. 2 we introduce the DSets algorithm and two typical density based clustering algorithms, based on which we present our two-step algorithm in details in Sect. 3. Extensive experiments are conducted to validate the proposed approach in Sect. 4, and finally the conclusions are given in Sect. 5.

## 2    Related Works

In this part we firstly introduce the DSets algorithm and discuss its properties. Then two density based clustering algorithms, i.e., DBSCAN and DP, are introduced briefly. These content serves as the basis of deriving our two-step clustering algorithm in Sect. 3.

### 2.1    Dominant Set

Dominant set is a graph-theoretic concept of a cluster [9]. In other words, a dominant set is regarded as a cluster in the DSets algorithm. In DSets clustering, the dominant sets (clusters) are extracted in a sequential manner. Given the pairwise data similarity matrix, we extract the first cluster in the whole set of data. After removing the data in the first cluster, we continue to extract the second cluster in the remaining data. In this way we are able to accomplish the clustering process and obtain the number of clusters automatically. As a promising clustering approach, the DSets algorithm has been applied successfully in various tasks [2,8,12,15].

Many clustering algorithms rely on the user-specified number of clusters to partition the data and obtain the clusters as a by-product of the partitioning process. While DBSCAN extract clusters in a region growing fashion sequentially, it also needs density related parameters to determine the cluster border. Different from these parameter-dependent clustering algorithms, the DSets algorithm defines dominant set as a non-parametric concept of a cluster. The basic idea is to maximize the internal similarity in a cluster by admitting into the cluster only the data helpful to increase the internal similarity. For this purpose, we need a measure to evaluate the relationship between one data and a cluster. Let's say that $S$ is the set of data to be clustered, $A = (a_{ij})$ is the pairwise data similarity matrix, $D$ is a non-empty subset in $S$, and $i \in D$ and $j \notin D$ are two data in $S$. We firstly define the relationship between $j$ and $i$ by

$$\phi_D(i,j) = a_{ij} - \frac{1}{|D|} \sum_{k \in D} a_{ik}, \tag{1}$$

where $|D|$ denotes the number of data in $D$. This variable compares the similarity between $j$ and $i$, with respect to the average similarity between $i$ and the data in $D$, thereby providing a connection between the data $j$ outside $D$ and those in $D$. Then we define

$$w_D(i) = \begin{cases} 1, & \text{if } |D| = 1, \\ \sum_{l \in D \setminus \{i\}} \phi_{D \setminus \{i\}}(l, i) w_{D \setminus \{i\}}(l), & \text{otherwise.} \end{cases} \tag{2}$$

This key variable is defined in a recursive form and it is not evident to see what it means. However, Eq. (2) shows that $w_D(i)$ can be approximately regarded as a weighted sum of $\phi_{D\setminus\{i\}}(l, i)$ for all $k \in D \setminus \{i\}$. Based on Eq. (1) we further see that $w_D(i)$ equals approximately to the average similarity between $i$ and those in $D \setminus \{i\}$, minus the average of pairwise similarity in $D \setminus \{i\}$, i.e.,

$$w_D(i) \approx \frac{1}{|D| - 1} \sum_{l \in D\setminus\{i\}} a_{li} - \frac{1}{(|D| - 1)^2} \sum_{l \in D\setminus\{i\}} \sum_{k \in D\setminus\{i\}} a_{lk} \qquad (3)$$

As the average of pairwise similarity is a suitable measure of the internal similarity in $D \setminus \{i\}$, here we see that $w_D(i) > 0$ means that $D$ has a higher internal similarity than $D \setminus \{i\}$ and $i$ is helpful to increase the internal similarity in $D \setminus \{i\}$. In contrast, $w_D(i) < 0$ implies that admitting $i$ into $D \setminus \{i\}$ will reduce the internal similarity in $D \setminus \{i\}$.

Now we are ready to present the formal definition of dominant set. With $W(D) = \sum_{i \in D} w_D(i)$, we call $D$ as a dominant set if

1. $W(T) > 0$, for all non-empty $T \subseteq D$.
2. $w_D(i) > 0$, for all $i \in D$.
3. $w_{D \cup \{i\}}(i) < 0$, for all $i \notin D$.

From this definition we see that only the data with positive $w_D(i)$ can be admitted into the dominant set $D$. In other words, a dominant set accepts only the data helpful to increase its internal similarity, and the dominant set extraction can be viewed as an internal similarity maximization process.

The dominant set definition requires that each data in a dominant set is able to increase the internal similarity. This condition is very strict as it means that each data in a dominant set must be very similar to all the others in the dominant set, including the closest ones and the farthest ones. As a result, the DSets algorithm tends to generate clusters of spherical shapes only.

## 2.2 Density Based Clustering Algorithms

DBSCAN is one of the most popular density based clustering approaches. With the user-specified neighborhood radius *Eps* and the minimum number *MinPts* of data in the neighborhood, DBSCAN defines the minimum density acceptable in a cluster. The data with density larger than this threshold are called core points. Staring from an arbitrary core point, DBSCAN admits all the data in its *Eps* neighborhood into the cluster. Repeating this process for all the core points in the cluster we obtain the first cluster. In the same way we can obtain the remaining clusters, and the data not included in any cluster are regarded as noise. One distinct advantage of DBSCAN is that it does not require the number of clusters to be specified, and is able to generate clusters of arbitrary shapes.

The DP algorithm is proposed to identify cluster centers in the first step and then determine the labels of other data based on the cluster centers. This algorithm uses local density $\rho$ and the distance $\delta$ to the nearest neighbor with higher density to characterize the data. Since cluster centers usually have both high $\rho$ and high $\delta$, they are isolated from other data in the $\rho$-$\delta$ decision graph and can be identified relatively easily. After the cluster centers are identified, the other data are assigned labels based on the labels of the nearest neighbors with higher density. On condition that the cluster centers are identified correctly, this algorithm can be used to generate clusters of arbitrary shapes. However, the correct identification of cluster centers usually involve the appropriate selection of density parameters, and human assistance may be necessary.

## 3    Our Approach

In the last section we see that the DSets algorithm and density based clustering algorithms have some complementary properties. The DSets algorithm uses only the pairwise similarity matrix as input and does not involve user-specified parameters explicitly. However, it generates clusters of only spherical shapes. In contrast, density base clustering algorithms are able to detect clusters of arbitrary shapes, on condition that appropriate density parameters are specified by user. This observation motivates us to combine both algorithms to make use of their respective merits. Specifically, we firstly use the DSets algorithm to generate an initial cluster, and then use a density based algorithm to obtain the final cluster, where the parameters needed by density based clustering are determined from the initial cluster. In this way our algorithm is able to generate clusters of arbitrary shapes and the involved parameters can be determined appropriately. Similar to the DSets algorithm, out algorithm extracts the clusters in a sequential manner.

### 3.1    DSets-histeq

We have known that the DSets algorithm uses only the pairwise data similarity matrix as input, and involves no user-specified parameters. However, we notice that in many cases the data are represented as points in vector space. This means that we need to construct the pairwise similarity matrix from the data represented in the form of vectors. One common measure of the similarity between two data (vectors) $x$ and $y$ is in the form of $s(x,y) = exp(-d(x,y)/\sigma)$, where $d(x,y)$ is the distance and $\sigma$ is a regulating parameter. Given a set of data to be clustered, different $\sigma$'s result in different similarity matrices, which then lead to different DSets clustering results, as illustrated in Fig. 1.

In this paper we intend to use the DSets algorithm to generate initial clusters, based on which the parameters required by density based clustering can be determined. In order to serve this purpose, we expect one initial cluster to be a not too small subset of a real cluster. The reason is as follows. If the initial cluster is too small, it doesn't contain enough data which are used to estimate density parameters reliably. In contrast, if the initial cluster is too large, it may be larger than the real cluster and contains also data from other clusters. In this case, the estimated density parameters are not accurate. Only when the initial cluster is a not too small subset of the real cluster, the density information captured in the

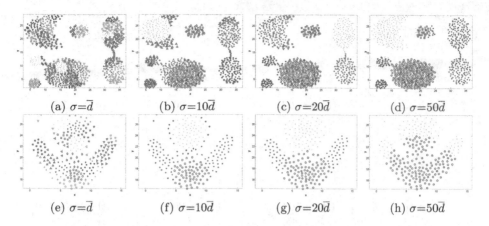

(a) $\sigma=\overline{d}$       (b) $\sigma=10\overline{d}$       (c) $\sigma=20\overline{d}$       (d) $\sigma=50\overline{d}$

(e) $\sigma=\overline{d}$       (f) $\sigma=10\overline{d}$       (g) $\sigma=20\overline{d}$       (h) $\sigma=50\overline{d}$

**Fig. 1.** The DSets clustering results on the Aggregation [6] and Flame [5] datasets with different $\sigma$, where $\overline{d}$ is the average of all pairwise distances. The top and bottom rows belong to Aggregation and Flame, respectively.

initial cluster can be viewed as an approximation of that in the real cluster, and therefore can be used to estimate the density parameters. However, Fig. 1 shows that the sizes of initial clusters vary with $\sigma$'s, and large $\sigma$'s tend to generate large clusters. In this case, if we use a very small $\sigma$, the generate clusters may be too small to support reliable density parameters estimation. On the contrary, too large $\sigma$'s may make the generated initial clusters larger than the real clusters. With a medium $\sigma$, the initial clusters may be smaller than the real ones for some datasets and greater than the real ones for others. In general, it is difficult to find a suitable $\sigma$ to generate the appropriate initial clusters.

In order to solve this problem, we propose to apply histogram equalization transformation to the similarity matrices before clustering. This transformation is adopted based on two reasons. First, by applying histogram equalization transformation to the similarity matrix, the influence of $\sigma$ on DSets clustering results can be removed. From $s(x,y) = exp(-d(x,y)/\sigma)$ we see that different $\sigma$'s only change the absolute similarity values. The relative magnitude among similarity values are determined solely by the data distances and invariant to $\sigma$'s. In other words, the ordering of similarity values in the similarity matrices are invariant to $\sigma$. Since in histogram equalization transformation the new similarity values are determined only based on the ordering of the original similarity values, we are ready to see that the new similarity matrices are invariant to $\sigma$'s. Consequently, the resulted clustering results are no longer influenced by $\sigma$'s. In practice, there will be very slight difference caused by the quantization process in histogram equalization. This is illustrated in Fig. 2, where the clustering results with different $\sigma$'s are very similar to each other. Second, histogram equalization transformation is a process to increase the overall similarity contrast in the similarity matrix. As dominant set by definition exerts a high requirement on the internal similarity, this transformation tends to reduce the size of clusters. On the other hand, as the similarity values are distributed in the range [0,1] relatively evenly

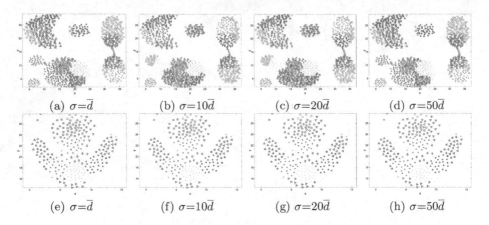

(a) $\sigma=\overline{d}$    (b) $\sigma=10\overline{d}$    (c) $\sigma=20\overline{d}$    (d) $\sigma=50\overline{d}$

(e) $\sigma=\overline{d}$    (f) $\sigma=10\overline{d}$    (g) $\sigma=20\overline{d}$    (h) $\sigma=50\overline{d}$

**Fig. 2.** The DSets clustering results on the Aggregation [6] and Flame [5] datasets with different $\sigma$, where $\overline{d}$ is the average of all pairwise distances, and the similarity matrices are transformed by histogram equalization. The top and bottom rows belong to Aggregation and Flame, respectively.

after histogram equalization transformation, the obtained clusters will not be too small, as illustrated in Fig. 2. This means that the initial clusters obtained in this way can be used to estimate density parameters reliably.

Based on the above observations, we propose to use the DSets algorithm to extract initial clusters, where the similarity matrices are transformed by histogram equalization before clustering. For ease of expression, in the following we use DSets-histeq to denote this algorithm. With DSets-histeq, we can use any $\sigma$ to build the similarity matrix, then the histogram equalization transformation enables the DSets algorithm to generate not too small initial clusters. The data in these initial clusters are then used to estimate the density in the real cluster, and further estimate the density parameters required by density based clustering algorithms.

Theoretically, we can use non-parametric similarity measures, e.g., cosine and histogram intersection, to evaluate the similarity among data and build the similarity matrix. However, we have found that with these two measures, the DSets algorithm usually generate large clusters which contain data of more than one real cluster, as illustrated in Fig. 3. For this reason, we stick to use DSets-histeq to generate the initial clusters.

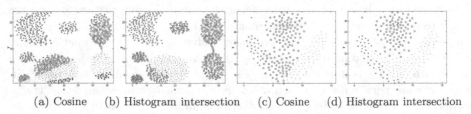

(a) Cosine    (b) Histogram intersection    (c) Cosine    (d) Histogram intersection

**Fig. 3.** The DSets clustering results on the Aggregation and Flame datasets, where the similarity matrices are built with cosine and histogram intersection similarity measures, respectively.

## 3.2    Initial Cluster Extension

As our algorithm extracts clusters in a sequential manner and all the clusters are extracted with the same procedures, in this part we describe how one cluster is obtained. With DSets-histeq we are able to extract a not too small initial cluster. This initial cluster is usually a subset of the real cluster, and contains important density information which can be used to estimate the density parameters required by density based algorithms. On the other hand, since the initial cluster is a subset of the real cluster, the density based clustering can be regarded as a density based cluster extension process. In the following we show that by making full use of the initial cluster, the cluster extension can be accomplished with very simple methods.

As stated in Sect. 2, in extracting a dominant set, we intend to maximize the internal similarity in the dominant set, and only the data helpful to increase the internal similarity can be included. As a consequence of this strict requirement of the internal similarity, the obtained dominant set (initial cluster) is usually the densest part in the real cluster. This observation motivates us to extend the initial cluster with the following two steps.

First, Eq. (3) indicates that in order to be admitted into a dominant set, one data must be very similar to all the data in the dominant set, including the nearest ones and the farthest ones. In other words, the density of these data is evaluated by taking all the data in the dominant set into account. This condition is a little too strict and enables the DSets algorithm to generate clusters of spherical shapes only. Therefore we choose to relax the condition and evaluate the data density with only its nearest neighbors. In this case, we firstly use the smallest data density in the initial cluster as the density threshold, which denotes the minimum density acceptable in the initial cluster. Then the neighboring data of the initial cluster are included if their density is above the threshold.

In the first step we have included the outside data with density above the density threshold. As mentioned above, the initial cluster is usually the densest part of the real cluster, and the density threshold in determined in the initial cluster. This means that some data with density smaller than the density threshold may also belong to the real cluster. Simply reducing the density threshold is not a good option as we don't know to which extent the density threshold should be reduced. Therefore we need a different approach to solve this problem. In the DP algorithm, after the cluster centers are identified, the non-center data are assigned labels equalling to those of their nearest neighbors with higher density. Motivated by this algorithm, we propose the following procedure to extend the initial cluster further. For each data $i$ in the initial cluster, we find its nearest neighbor $i_{nn}$. If $i_{nn}$ is outside the initial cluster and its density is smaller than that of $i$, we include $i_{nn}$ into the initial cluster. Repeating this process until no data can be included, we accomplish the cluster extension process and obtain the final cluster.

## 4    Experiments

### 4.1    Histogram Equalization

In our algorithm we use histogram equalization transformation of similarity matrices to remove the dependence on $\sigma$'s and generate not too small initial clusters for later extension. In this part we use experiments to validate the effectiveness in these two aspects respectively.

The data clustering experiments are conducted on eight datasets, including Aggregation, Compound [14], R15 [13], Jain [7], Flame and three UCI datasets, namely Thyroid, Iris and Wdbc. We use different $\sigma$'s in our algorithm and report the clustering results (F-measure and Jaccard index) in Fig. 4, where it is evident that the influence of $\sigma$'s on clustering results have been removed effectively.

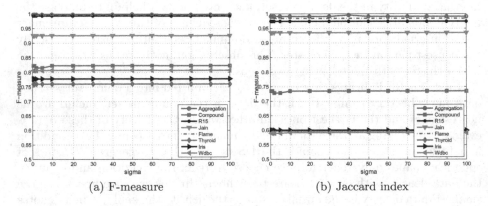

(a) F-measure                                      (b) Jaccard index

**Fig. 4.** The clustering results of our algorithm with different $\sigma$'s on eight datasets. The real $\sigma$'s are equal to the product of $\overline{d}$ and the horizontal axes.

In our algorithm we use histogram equalization transformation to generate not too small initial clusters, which are used in later cluster extension. As small $\sigma$'s can also be used to generate small initial clusters, it is interesting to compare these two methods. Specifically, we conduct two sets of experiments with our algorithm, where the only difference is that in one set the similarity matrices are transformed by histogram equalization, and in the other set the similarity matrices are not transformed. The comparison of average clustering results on the eight datasets are reported in Fig. 5. From the comparison we observe that histogram equalization transformation performs better than any fixed $\sigma$'s. This confirms that it is difficult to find a fixed $\sigma$ which is able to generate not too small initial clusters for different datasets. In this case, histogram equalization transformation of similarity matrices becomes a better option.

### 4.2    Data Clustering

In this part we use data clustering experiments to compare our algorithm with some state-of-the-art algorithms, including the original DSets algorithm,

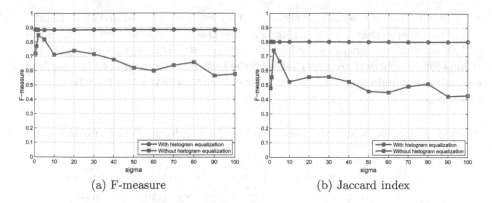

(a) F-measure                                    (b) Jaccard index

**Fig. 5.** The average clustering results of our algorithm on eight datasets with and without histogram equalization of similarity matrices. The real $\sigma$'s are equal to the product of $\bar{d}$ and the horizontal axes.

k-means, DBSCAN, NCuts, AP, SPRG [16], DP-cutoff (the DP algorithm with the cutoff kernel) and DP-Gaussian (the DP algorithm with the Gaussian kernel). The experimental setups of these algorithms are as follows.

1. With DSets, we manually select the best-performing $\sigma$ from $0.5\bar{d}$, $\bar{d}$, $2\bar{d}$, $5\bar{d}$, $10\bar{d}$, $20\bar{d}$, $\cdots$, $100\bar{d}$, for each dataset separately. This means that the reported results are approximately the best possible results obtainable from the DSets algorithm.
2. As k-means, NCuts and SPRG requires the number of clusters as input, we feed the ground truth numbers of clusters to them. Considering that their clustering results are influenced by random initial cluster centers, we report the average results of five runs.
3. The AP algorithm requires the preference value of each data as input, and the authors of [10] have published the code to calculate the range $[p_{min}, p_{max}]$ of this parameter. In experiments we manually select the best-performing $p$ from $p_{min} + step$, $p_{min} + 2step$, $\cdots$, $p_{min} + 9step$, $p_{min} + 9.1step$, $p_{min} + 9.2step$, $\cdots$, $p_{min} + 9.9step$, with $step = (p_{max} - p_{min})/10$, for each dataset separately. Like in the case of DSets, the reported results represent approximately the best possible ones from the AP algorithm.
4. With DBSCAN, we manually select the best performing $MinPts$ from 2, 3, $\cdots$, 10, for each dataset separately, and $Eps$ is determined based on $MinPts$ with the method presented in [3]. The reported results represent approximately the best possible ones from DBSCAN.
5. The original DP algorithm involves manual selection of cluster centers. In order to avoid the influence of human factors, with DP-cutoff and DP-Gaussian we feed the ground truth numbers of clusters and select the cluster centers based on $\gamma = \rho\delta$, where $\rho$ denotes and local density and $\delta$ represents the distance to the nearest neighbor with higher density.

The experimental results on the eight datasets are reported in Tables 1 and 2, where we observe that our algorithm performs the best or near-best in four out of eight datasets, and generates the best average result. Noticing that the eight algorithms for comparison are reported with the best possible results or benefit from ground truth numbers of clusters, we believe these results validate the effectiveness of our algorithm.

**Table 1.** Clustering quality (F-measure) comparison of different clustering algorithms on eight datasets.

|  | 1 | 2 | 3 | 4 | 5 | 6 | 7 | 8 | Average |
|---|---|---|---|---|---|---|---|---|---|
| DSets | 0.95 | 0.81 | 0.98 | 0.86 | 0.82 | 0.82 | 0.79 | 0.87 | 0.86 |
| k-means | 0.80 | 0.78 | 0.81 | 0.79 | 0.84 | 0.81 | 0.80 | 0.84 | 0.81 |
| DBSCAN | 0.94 | **0.97** | 0.77 | **0.95** | **0.99** | 0.72 | 0.78 | 0.69 | 0.85 |
| NCuts | **0.99** | 0.70 | **0.99** | 0.63 | **0.99** | 0.63 | 0.92 | 0.84 | 0.84 |
| AP | 0.86 | 0.77 | **0.99** | 0.76 | 0.85 | 0.65 | **0.93** | 0.83 | 0.83 |
| SPRG | 0.67 | 0.63 | 0.95 | 0.84 | 0.71 | **0.96** | 0.88 | **0.91** | 0.82 |
| DP-cutoff | **0.99** | 0.70 | **0.99** | 0.91 | 0.84 | 0.52 | 0.70 | 0.65 | 0.79 |
| DP-Gaussian | 0.85 | 0.69 | **0.99** | 0.91 | 0.79 | 0.51 | 0.90 | 0.77 | 0.80 |
| Ours | **0.99** | 0.82 | **0.99** | 0.93 | **0.99** | 0.76 | 0.78 | 0.81 | **0.88** |

**Table 2.** Clustering quality (Jaccard index) comparison of different clustering algorithms on eight datasets.

|  | 1 | 2 | 3 | 4 | 5 | 6 | 7 | 8 | Average |
|---|---|---|---|---|---|---|---|---|---|
| DSets | 0.92 | 0.73 | 0.93 | 0.61 | 0.54 | 0.71 | 0.60 | 0.68 | 0.71 |
| k-means | 0.59 | 0.65 | 0.63 | 0.53 | 0.58 | 0.60 | 0.61 | 0.65 | 0.61 |
| DBSCAN | 0.87 | **0.94** | 0.42 | **0.95** | 0.96 | 0.58 | 0.60 | 0.53 | 0.73 |
| NCuts | 0.98 | 0.46 | **0.99** | 0.42 | 0.97 | 0.40 | **0.77** | 0.65 | 0.70 |
| AP [1] | 0.73 | 0.69 | **0.99** | 0.62 | 0.61 | 0.53 | **0.77** | 0.56 | 0.69 |
| SPRG [16] | 0.42 | 0.38 | 0.87 | 0.61 | 0.47 | **0.88** | 0.68 | **0.73** | 0.63 |
| DP-cutoff [10] | **0.99** | 0.48 | 0.97 | 0.74 | 0.58 | 0.30 | 0.51 | 0.45 | 0.63 |
| DP-Gaussian [10] | 0.66 | 0.46 | **0.99** | 0.74 | 0.52 | 0.29 | 0.73 | 0.58 | 0.62 |
| Ours | **0.99** | 0.74 | 0.97 | 0.94 | **0.98** | 0.60 | 0.60 | 0.59 | **0.80** |

# 5 Conclusions

In this paper we present a new clustering algorithm based on dominant set and density based clustering algorithms. By means of histogram equalization transformation of similarity matrices, the dominant sets algorithm is used to generate not too small initial clusters. In the next step we propose two methods to make use of the density information in initial clusters in extending initial clusters to final ones. By merging the merits of dominant sets algorithm and density based clustering algorithms, our algorithm requires no parameter input, and is able to generate clusters of arbitrary shapes. In data clustering results our algorithm performs comparably to or better than other algorithms with parameter tuning.

**Acknowledgement.** This work is supported in part by the National Natural Science Foundation of China under Grant No. 61473045 and by China Scholarship Council.

# References

1. Brendan, J.F., Delbert, D.: Clustering by passing messages between data points. Science **315**, 972–976 (2007)
2. Bulo, S.R., Torsello, A., Pelillo, M.: A game-theoretic approach to partial clique enumeration. Image Vis. Comput. **27**(7), 911–922 (2009)
3. Daszykowski, M., Walczak, B., Massart, D.L.: Looking for natural patterns in data: Part 1. Density-based approach. Chemom. Intell. Lab. Syst. **56**(2), 83–92 (2001)
4. Ester, M., Kriegel, H.P., Sander, J., Xu, X.W.: A density-based algorithm for discovering clusters in large spatial databases with noise. In: International Conference on Knowledge Discovery and Data Mining, pp. 226–231 (1996)
5. Fu, L., Medico, E.: Flame, a novel fuzzy clustering method for the analysis of DNA microarray data. BMC Bioinform. **8**(1), 1–17 (2007)
6. Gionis, A., Mannila, H., Tsaparas, P.: Clustering aggregation. ACM Trans. Knowl. Discov. Data **1**(1), 1–30 (2007)
7. Jain, A.K., Law, M.H.C.: Data clustering: a user's dilemma. In: Pal, S.K., Bandyopadhyay, S., Biswas, S. (eds.) PReMI 2005. LNCS, vol. 3776, pp. 1–10. Springer, Heidelberg (2005)
8. Pavan, M., Pelillo, M.: Efficient out-of-sample extension of dominant-set clusters. In: Advances in Neural Information Processing Systems, pp. 1057–1064 (2005)
9. Pavan, M., Pelillo, M.: Dominant sets and pairwise clustering. IEEE Trans. Pattern Anal. Mach. Intell. **29**(1), 167–172 (2007)
10. Rodriguez, A., Laio, A.: Clustering by fast search and find of density peaks. Science **344**, 1492–1496 (2014)
11. Shi, J., Malik, J.: Normalized cuts and image segmentation. IEEE Trans. Pattern Anal. Mach. Intell. **22**(8), 167–172 (2000)
12. Tripodi, R., Pelillo, M.: Document clustering games. In: The 5th International Conference on Pattern Recognition Applications and Methods, pp. 109–118 (2016)
13. Veenman, C.J., Reinders, M., Backer, E.: A maximum variance cluster algorithm. IEEE Trans. Pattern Anal. Mach. Intell. **24**(9), 1273–1280 (2002)
14. Zahn, C.T.: Graph-theoretical methods for detecting and describing gestalt clusters. IEEE Trans. Comput. **20**(1), 68–86 (1971)
15. Zemene, E., Pelillo, M.: Path-based dominant-set clustering. In: Murino, V., Puppo, E. (eds.) ICIAP 2015. LNCS, vol. 9279, pp. 150–160. Springer, Heidelberg (2015)
16. Zhu, X., Loy, C.C., Gong, S.: Constructing robust affinity graphs for spectral clustering. In: IEEE International Conference on Computer Vision and Pattern Recognition, pp. 1450–1457 (2014)

# Co-training with Credal Models

Yann Soullard$^{(\boxtimes)}$, Sébastien Destercke, and Indira Thouvenin

Sorbonne University, Université de Technologie de Compiègne,
CNRS UMR 7253 Heudiasyc, CS 60 319, 60 203 Compiègne Cedex, France
{yann.soullard,sebastien.destercke,indira.thouvenin}@hds.utc.fr

**Abstract.** So-called credal classifiers offer an interesting approach when the reliability or robustness of predictions have to be guaranteed. Through the use of convex probability sets, they can select multiple classes as prediction when information is insufficient and predict a unique class only when the available information is rich enough. The goal of this paper is to explore whether this particular feature can be used advantageously in the setting of co-training, in which a classifier strengthen another one by feeding it with new labeled data. We propose several co-training strategies to exploit the potential indeterminacy of credal classifiers and test them on several UCI datasets. We then compare the best strategy to the standard co-training process to check its efficiency.

**Keywords:** Co-training · Imprecise probabilities · Semi-supervised learning · Ensemble models

## 1 Introduction

There are many application fields (gesture, human activity, finance, ...) where extracting numerous unlabeled data is easy, but where labeling them reliably require costly human efforts or an expertise that may be rare and expensive. In this case, getting a large labeled dataset is not possible, making the task of training an efficient classifier from labeled data alone difficult. The general goal of semi-supervised learning techniques [1,7,28] is to solve this issue by exploiting the information contained in unlabeled data. It includes different approaches such as the adaptation of training criteria [13,14,16], active learning methods [18] and co-training-like approaches [6,19,22].

In this paper, we focus on the co-training framework. This approach aims at training two classifiers in parallel, and each model then attempts to strengthen the other by labeling a selection of unlabeled data. We will call *trainer* the classifier providing new labeled instances and *learner* the classifier using it as new training data. In the standard co-training approach [6,22], the trainer provides to the learner the data about which it gets the most confident labels. However, those labels are predicted with high confidence by the trainer but it is not guaranteed that the new labeled instances will be informative for the learner, in the sense that it may not help him to improve its accuracy.

© Springer International Publishing AG 2016
F. Schwenker et al. (Eds.): ANNPR 2016, LNAI 9896, pp. 92–104, 2016.
DOI: 10.1007/978-3-319-46182-3_8

To solve this issue, we propose a new co-training approach using credal classifiers. Such classifiers, through the use of convex sets of probabilities, can predict
a set of labels when training data are insufficiently conclusive. It means they will
produce a single label as prediction only when the information is enough (i.e.,
when the probability set is small enough). The basic idea of our approach is to
select as potential new training data for the learner those instances for which
the (credal) trainer has predicted a single label and the learner multiple ones.

## 2    Co-training Framework

We assume that samples are elements of a space $\mathcal{X} \times \mathcal{Y}$, where $\mathcal{X}$ is the input
space and $\mathcal{Y}$ the output space of classes. In a co-training setting, it is assumed
that the input space $\mathcal{X}$ can be split into two different views $\mathcal{X}_1 \times \mathcal{X}_2$ and that
classifiers can be learned from each of those views. That is, an instance $\mathbf{x} \in \mathcal{X}$
can be split into a couple $(\mathbf{x}_1, \mathbf{x}_2)$ with $\mathbf{x}_j \in \mathcal{X}_j$. In the first works introducing
co-training [6,22], it is assumed that each view is sufficient for a correct classifi-
cation and is conditionally independent to the other given the class label. However, it has been shown that co-training can also be an efficient semi-supervised
techniques when the views are insufficient or when labels are noisy [24]. In addition, many studies provide theoretical results on co-training: [4] shows that the
assumption of conditional independence can be relaxed to some extent; [11] gives
some theoretical justifications of the co-training algorithm, providing a bound on
the generalization error that is related on the empirical agreement between the
two classifiers; [23] analyzes the sufficient and necessary condition for co-training
to succeed through a graph view of co-training. Besides, in [26], the authors propose to estimate the labeling confidence using data editing techniques, allowing
especially to identify an appropriate number of predicted examples to pass on
to the other classifier.

We will denote by $\mathcal{L}_j$ the set of labeled examples from which a model $h_j$
is learned i.e. $\mathcal{L}_j = \{(\mathbf{x}_j^{(i)}, y^{(i)}), i \in \{1, ..., m_j\}\}$ where $\mathbf{x}_j^{(i)}$ is the $j^{\text{th}}$ view of
the instance $\mathbf{x}^{(i)}$ and $m_j$ denotes the number of labeled examples in $\mathcal{L}_j$. Co-
training starts with a (usually small) common set of labeled examples, i.e. $\mathcal{L}_1 = \{(\mathbf{x}_1^{(i)}, y^{(i)}), i \in \{1, ..., m\}\}$ and $\mathcal{L}_2 = \{(\mathbf{x}_2^{(i)}, y^{(i)}), i \in \{1, ..., m\}\}$, and a pool
of unlabeled examples $\mathcal{U} = \{(\mathbf{x}_1^{(i)}, \mathbf{x}_2^{(i)}), i \in \{m+1, ..., n\}\}$. Based on previous
works of [22], the standard co-training method that we will adapt to imprecise
probability setting goes as follow: at each step of the co-training process, two
classifiers $h_j : \mathcal{X}_j \to \mathcal{Y}$ ($j \in \{1,2\}$) are learned from the learning set $\mathcal{L}_j$. We also
assume that classifier $h_j$ uses an estimated conditional probability distribution
$p_j(\cdot|\mathbf{x}) : \mathcal{Y} \to [0,1]$ to take its decision, i.e.

$$h_j(\mathbf{x}_j) = \arg\max_{y \in \mathcal{Y}} p_j(y|\mathbf{x}_j). \tag{1}$$

The $n_u$ examples labeled by $h_j$ with the most confidence (the highest probability,
in our case) are then added to the set $\mathcal{L}_k$ of training examples of $h_k, k \neq j$ and

**Input:** $h_1, h_2$ learned from $\mathcal{L}_1, \mathcal{L}_2$, set $\mathcal{U}$, number $n_u$ of added learning samples;
**Output:** Updated sets $\mathcal{L}_1, \mathcal{L}_2$
n=1;
**repeat**

> set $\mathbf{x}^* = \arg\max_{\mathbf{x}^{(i)} \in \mathcal{U}} p_1(h_1(\mathbf{x}_1^{(i)})|\mathbf{x}_1^{(i)})$ ;
> $\mathcal{L}_2 \leftarrow \mathcal{L}_2 \cup (\mathbf{x}^*, h_1(\mathbf{x}^*))$ ;
> set $\tilde{\mathbf{x}}^* = \arg\max_{\mathbf{x}^{(i)} \in \mathcal{U}} p_2(h_2(\mathbf{x}_2^{(i)})|\mathbf{x}_2^{(i)})$ ;
> $\mathcal{L}_1 \leftarrow \mathcal{L}_1 \cup (\tilde{\mathbf{x}}^*, h_2(\tilde{\mathbf{x}}^*))$ ;
> $\mathcal{U} \leftarrow \mathcal{U} \setminus \{\mathbf{x}^*, \tilde{\mathbf{x}}^*\}$ ;
> $n \leftarrow n+1$ ;

**until** $n = n_u$;

**Algorithm 1.** Standard co-training procedure

removed from $\mathcal{U}$. One iteration of this process is summarized by Algorithm 1. The procedure is then iterated a number of pre-defined times.

Note that co-training can also be used with two different classifiers using the same view $\mathcal{X}$. [22] provide a theoretical analysis demonstrating that two learners can be improved in such a procedure provided they have a large difference. This view is taken further by [27] that studies ensemble learning (with more than two classifiers) in a semi-supervised framework.

## 3   Basics of Credal Models

We introduce the basic elements we need about imprecise probabilities. Interested readers are referred to [3] for further details.

### 3.1   Imprecise Probabilities and Decision

Let $\mathcal{Y}$ be a finite space (e.g., of classes) and $\Sigma_{\mathcal{Y}}$ be the set of all probability mass functions over $\mathcal{Y}$. In imprecise probability theory, the uncertainty about a variable $Y$ is described by a convex set $\mathcal{P} \subseteq \Sigma_{\mathcal{Y}}$ of probabilities which is usually called *credal set*. When this convex set depends on some input data $\mathbf{x}$, as in classification problems, we will denote it by $\mathcal{P}_{\mathbf{x}}$. Given $\mathbf{x}$, lower and upper probabilities of elements of $y \in \mathcal{Y}$ can then be defined as:

$$\underline{p}(y|\mathbf{x}) = \inf_{p(\cdot|\mathbf{x}) \in \mathcal{P}_{\mathbf{x}}} p(y|\mathbf{x}) \text{ and } \overline{p}(y|\mathbf{x}) = \sup_{p(\cdot|\mathbf{x}) \in \mathcal{P}_{\mathbf{x}}} p(y|\mathbf{x}) \qquad (2)$$

and when $\mathcal{P}_{\mathbf{x}}$ is reduced to a singleton, we retrieve the probabilistic case where $\underline{p}(y|\mathbf{x}) = \overline{p}(y|\mathbf{x})$ for any $y \in \mathcal{Y}$. Probabilities of events and expected values can be extended in the same way, by considering boundary values.

Many different ways have been proposed for extending the classical decision criterion $h(\mathbf{x}) = \arg\max_{y \in \mathcal{Y}} p(y|\mathbf{x})$ within imprecise probability theory [20]. Some of them, such as the maximin that replaces $p(y|\mathbf{x})$ by $\underline{p}(y|\mathbf{x})$ in Eq. (1), also produce unique predictions. Others, that we will use here, can produce sets of possible predictions, the size of the set reflecting the lack of information. Such a decision rule is then a mapping[1] $H : \mathcal{X} \to 2^{\mathcal{Y}}$ from the input set to the power set of classes.

In this paper, we will focus on two of the most popular decision rules that are the so-called *interval dominance* and the *maximality* criteria. The *interval dominance* decision rule is defined as:

$$H(\mathbf{x}) = \{y \in \mathcal{Y} | \; \not\exists \, y' s.t. \quad \underline{p}(y'|\mathbf{x}) > \overline{p}(y|\mathbf{x})\}. \tag{3}$$

The idea of this rule is that a class is rejected as a possible prediction when its upper bound of probability is lower than the lower bound of at least one other class. However, this means that two classes $y, y'$ may be compared according to different precise probabilities within $\mathcal{P}_\mathbf{x}$ (as their boundary probabilities can be obtained at different points). This is not the case of the *maximality* decision rule, that rejects a class which is certainly less probable (according to any $p \in \mathcal{P}_\mathbf{x}$) than the others:

$$H(\mathbf{x}) = \{y \in \mathcal{Y} | \; \not\exists \, y' s.t. \quad p(y'|\mathbf{x}) > p(y|\mathbf{x}) \forall \, p(\cdot|\mathbf{x}) \in \mathcal{P}_\mathbf{x}\}. \tag{4}$$

Those decision rules are reduced to (1) in a precise framework, as it consists in choosing the top element of the order induced by the probability weights. The set (4) can be computed in the following way: starting from the whole set $\mathcal{Y}$, if the value

$$\inf_{p(\cdot|\mathbf{x}) \in \mathcal{P}_\mathbf{x}} p(y|\mathbf{x}) - p(y'|\mathbf{x}) \tag{5}$$

is strictly positive for a given pair of classes $y, y'$, then $y'$ can be removed from $H(\mathbf{x})$, since if (5) is positive, $p(y|\mathbf{x})$ is strictly greater than $p(y'|\mathbf{x})$ for all $p(\cdot|\mathbf{x})$. The set (4) can then be obtained by iterating this procedure over all pairs of classes, removing those that are not optimal. Note that this approach will produce set predictions both in case of ambiguity ($\mathcal{P}_\mathbf{x}$ may be small, but may contain probabilities whose higher values are similar) and of lack of information ($\mathcal{P}_\mathbf{x}$ is large because few training data were available), and will therefore produce precise predictions (the cardinality $|H(\mathbf{x})| = 1$) only when being very confident. It should be noted that the set produced by Eq. (3) will always include (hence will be more precise) the one produced by Eq. (4).

### 3.2   Learning Credal Models

A common way to learn credal models is to extend Bayesian models by considering sets of priors. After learning, they provide a set of posterior probabilities which converges towards a single probability when the training dataset size increases. The most well-known example is the naive credal classifier [25], that extends the naive Bayes classifier by learning set of conditional probabilities, using the so-called imprecise Dirichlet model [5].

---

[1] We use capital letter to denote the fact that the returned prediction may be a set of classes.

Some works combine sets of credal classifiers to extend popular techniques such as Bayesian model averaging [8], binary decomposition [12] or boosting [21], as well as there are some preliminary works that exploit credal approaches in semi-supervised settings [2,17]. However, we are not aware of any work trying to exploit credal models to mutually enrich sets of classifiers.

## 4 Co-training with Credal Models

We propose a new co-training approach based on credal models that extends the standard co-training framework recalled in Sect. 2. We will refer to it as *credal co-training*.

### 4.1 Motivation and Idea

Working with a small labeled training set may cause several issues in semi-supervised learning, one major issue being the possible bias resulting from few samples [15,27]. Second, in the standard co-training framework recalled in Sect. 2, there is no guarantee that the data selected by the trainer will actually be useful to the learner, in the sense that the learner may already be quite accurate on those data.

We think our proposal tackles, at least partially, both problems. Working with sets of priors and with sets of probabilities whose sizes depend on the number of training data is a way to be less affected by possible bias. Second, the fact that predictions are set-valued can help to identify on which data the trainer is really confident (those for whose it makes a unique prediction) and the learner is not (those for which it makes set-valued prediction). Based on this distinction, our approach consists in modifying Algorithm 1 in two different aspects:

- Select data from a subset of $\mathcal{U}$, denoted $\mathcal{S}_{H_i \to H_j} \subseteq \mathcal{U}$, corresponding to data for which $H_i$ is confident and $H_j$ is not (Sect. 4.2).
- Adapt the notion of confidence to imprecise probabilities, to choose specific data from $\mathcal{S}_{H_i \to H_j}$ (Sect. 4.3).

The resulting co-training process, that we will detail in the next sections, is illustrated in Fig. 1.

### 4.2 Dataset Selection Among the Unlabeled Instances

In this section, we define the set $\mathcal{S}_{H_i \to H_j}$ containing the pool of unlabeled data that may be labeled by the trainer $H_i$ for the learner $H_j$. The idea is that it should contain data for which $H_i$ (the trainer) is confident and $H_j$ (the learner) is not. We denote $\mathcal{S}^c_{H_i}$ the dataset on which $H_i$ provides precise classifications and $\mathcal{S}^u_{H_j}$ the one on which $H_j$ provides indeterminate predictions. The set $\mathcal{S}_{H_i \to H_j}$ is defined as follow:

$$S_{H_i \to H_j} = S^c_{H_i} \cap S^u_{H_j} := \{\mathbf{x} \in \mathcal{U} \mid |H_i(\mathbf{x})| = 1 \wedge |H_j(\mathbf{x})| > 1\} \qquad (6)$$

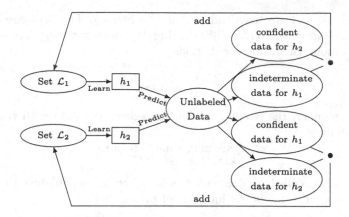

**Fig. 1.** Co-training with credal models.

It contains the unlabeled data for which $H_i$ predicts a unique value and $H_j$ multiple ones. As this set may be empty, we reduce the imprecision of the trainer so that it converges towards the precise framework. This guarantees that the trainer have confident instances that may be selected for labeling, but this may reduce the trainer confidence about its predictions. We will discuss in details this strategy in our experimentations (see Sect. 5.2). Note that when the learner $H_j$ is confident on $\mathcal{U}$, the pool of unlabeled data is reduced to the instances for which the trainer is confident, i.e. $S_{H_i \to H_j} = S^c_{H_i}$.

### 4.3 Data Selection for Labeling

Given a set of unlabeled data $\mathcal{U}$ and two models $H_1$ and $H_2$ learned from $\mathcal{L}_1$ and $\mathcal{L}_2$ respectively, we define several strategies for selecting a data in the unlabeled dataset $S_{H_i \to H_j}$. A first one, that we call *uncertain strategy*, consists in choosing the data for which the learner is the less confident:

$$\mathbf{x}^\star = \arg\max_{\mathbf{x} \in S_{H_i \to H_j}} \sum_{y \in \mathcal{Y}} \left( \overline{p}_j(y|\mathbf{x}) - \underline{p}_j(y|\mathbf{x}) \right) \tag{7}$$

where $\underline{p}_j$ and $\overline{p}_j$ are the lower and upper probabilities induced from $\mathcal{L}_j$. We therefore replace Line 3 of Algorithm 1 by Eq. (7) with $i = 1, j = 2$, and Line 5 likewise with $i = 2, j = 1$ (the same will apply to all strategies). The idea of the uncertain strategy is that $H_j$ gains information where it is the least informed. We refine this strategy by focusing on the classes predicted by the learner $H_j$ and on data for which $H_j$ is the most uncertain, what we call the *indeterminate strategy*:

$$\mathbf{x}^\star = \arg\max_{\mathbf{x} \in S_{H_i \to H_j}} \left( \mathbf{1}_{|H_j(\mathbf{x})| = \max_{\mathbf{x}' \in \mathcal{U}} |H_j(\mathbf{x}')|} \times \sum_{y \in H_j(\mathbf{x})} \left( \overline{p}_j(y|\mathbf{x}) - \underline{p}_j(y|\mathbf{x}) \right) \right) \tag{8}$$

Those two strategies are not extensions of the standard framework, in the sense that we do not retrieve Algorithm 1 when $\underline{p} = \overline{p}$. The next strategies are such extensions, and are based on probability bounds. The first one, that we call *optimistic strategy*, consists in selecting the data with the highest upper probability among the trainer predictions:

$$\mathbf{x}^{\star} = \underset{\mathbf{x} \in S_{H_i \to H_j}}{\arg\max} \ \underset{y \in H_i(\mathbf{x})}{\max} \ \overline{p}_i(y|\mathbf{x}) \qquad (9)$$

In a similar manner, the *pessimistic strategy* selects the data with the highest lower probability:

$$\mathbf{x}^{\star} = \underset{\mathbf{x} \in S_{H_i \to H_j}}{\arg\max} \ \underset{y \in H_i(\mathbf{x})}{\max} \ \underline{p}_i(y|\mathbf{x}) \qquad (10)$$

The last strategy, called *median*, consists in selecting the unlabeled data with the highest median value in the intervals of probability:

$$\mathbf{x}^{\star} = \underset{\mathbf{x} \in S_{H_i \to H_j}}{\arg\max} \ \underset{y \in H_i(\mathbf{x})}{\max} \ \frac{1}{2}\big(\underline{p}_i(y|\mathbf{x}) + \overline{p}_i(y|\mathbf{x})\big). \qquad (11)$$

Those last strategies correspond to common decision criteria used in imprecise probability theory (and robust methods in general), that are respectively the maximax, maximin and Hurwicz criteria [20]. As in the standard framework, these strategies aim at giving informative samples to the learner, but intend to exploit the robustness of credal models.

## 5    Experimentations

We experiment our proposed approach on various UCI datasets described in Table 1. We use a Naive Credal Classifier [9,10,25] to test our approach. For each dataset, we split the feature set in two distinct parts where the first half serve as the first view and the second half as the second view. Although we have no guarantee that the two views are sufficient, recent results [24] suggest that co-training can also work in such a setting. Thus, we expect that our approach will be able to improve the supervised case and that it will overcome the standard co-training framework.

### 5.1    Naive Credal Classifier (NCC)

The Naive Credal Classifier (NCC) [10,25] is an extension of Naive Bayes Classifier (NBC) to imprecise probabilities. Let $\mathbf{x} = (x_1, .., x_K)$ be an instance and let us denote $\mathcal{P}_y$ a credal set of prior distributions $p(y)$ on the classes, and $\mathcal{P}_{x_k}^y$ a credal set of conditional distributions $p(x_k|y)$. The model is characterized by a set of joint probability distributions $p(y, \mathbf{x})$ which satisfies the assumption that, for a given class $y$, the value of a feature $x_k$ for $k \in \{1, ..., K\}$ is independent of the value of any other feature. According to this assumption, the model is defined as follow:

$$p(y, \mathbf{x}) = p(y, x_1, .., x_K) = p(y) \prod_{k=1}^{K} p(x_k|y) \qquad (12)$$

Table 1. UCI datasets, with the number of data, of features and of classes.

| Name | #Samples | #Features | #Classes |
|------|----------|-----------|----------|
| Diabetes | 768 | 8 | 2 |
| Haberman | 306 | 3 | 2 |
| Ionosphere | 351 | 34 | 2 |
| Iris | 150 | 4 | 3 |
| KDD synth. | 600 | 60 | 6 |
| Kr vs. kp | 3196 | 36 | 2 |
| Mfeat morph. | 2000 | 6 | 10 |
| Opdigits | 5620 | 64 | 10 |
| Page-blocks | 5473 | 10 | 5 |
| Segment | 2310 | 19 | 7 |
| Spambase | 4601 | 57 | 2 |
| Wine | 178 | 13 | 3 |

with $p(y) \in \mathcal{P}_{\mathcal{Y}}$ and $p(x_k|y) \in \mathcal{P}^y_{x_k}$ for $k \in \{1, ..., K\}$ and $y \in \mathcal{Y}$. This results in a set of posterior probabilities. Conditional credal sets $\mathcal{P}^y_{x_k}$ are typically learned using the Imprecise Dirichlet model [5,25] provides an efficient procedure to compute the sets (3) and (4) respectively based on the interval dominance and maximality decision rules, we refer to this work for details.

## 5.2   Comparison of Data Selection Strategies

We first compare the various data selection strategies defined in Sect. 4.3. The five strategies are compared to the supervised framework. The supervised case provides the initial performances on each view obtained without using $\mathcal{U}$. The co-training is performed during 50 training iterations and, at each iteration, one instance is labeled per model. Table 2 shows some results on a 10-fold cross validation where, for each fold, 10 % of the data are used for the test, 40 % of them are labeled instances used for training the models[2] and the rest is considered to be unlabeled and may be selected by the models using one of the strategies. As defined by [25], we use a NCC hyper-parameter for controlling at which speed the credal set converge towards a unique probability. In our experiments, this hyper-parameter, called $s$ value (see [25] for details) is equal to 5 for the trainer and 2 for the learner. At each co-training iteration, we decrease the $s$ value of the trainer if the pool of unlabeled data (6) defined in Sect. 4.2 is empty until $s = 0$ (the precise setting). If the learner has no indeterminate predictions (i.e. $S^u_{h_L} = \emptyset$), a data is selected in the pool of data for which the trainer is confident.

---

[2] We use 40 % of them as labeled instances to get a compromise between having a large part of data as unlabeled and having a sufficiently large labeled dataset to reduce the sampling bias.

**Table 2.** Performances of the uncertainty (*UNC*), indeterminate (*IND*), optimistic (*OPT*), pessimistic (*PES*) and median (*MED*) strategies of the credal co-training (with maximality) compared to the supervised setting (*SUP*).

| DATASET | FIRST MODEL | | | | | | SECOND MODEL | | | | | |
|---|---|---|---|---|---|---|---|---|---|---|---|---|
| | SUP. | UNC. | IND. | OPT. | PES. | MED. | SUP. | UNC. | IND. | OPT. | PES. | MED. |
| Diabetes | 73.9% | 74.0% | 74.0% | **74.2%** | 73.7% | 73.7% | 70.5% | **72.5%** | **72.5%** | 71.9% | 72.1% | 72.4% |
| Ionosphere | 89.8% | **90.1%** | **90.1%** | **90.1%** | **90.1%** | 89.8% | 79.9% | 81.5% | 81.5% | **82.9%** | 81.7% | 81.7% |
| Iris | 71.3% | 76.7% | 75.3% | **77.3%** | 76.0% | 75.3% | 92.7% | 93.3% | 92.7% | **93.3%** | **93.3%** | **93.3%** |
| KDD synthetic control | 75.2% | 77.3% | **77.7%** | 77.3% | 77.5% | 77.5% | 84.3% | 85.3% | 85.2% | **85.7%** | 85.2% | 85.2% |
| Page-blocks | **92.9%** | 92.2% | 91.4% | **92.9%** | 92.3% | 92.3% | **86.8%** | 85.5% | 85.5% | 86.1% | 86.1% | 86.2% |
| Segment | 57.7% | 58.5% | **58.7%** | 58.2% | 58.5% | 58.5% | 81.8% | **83.3%** | 82.5% | 82.5% | 81.6% | 81.5% |
| Spambase | 86.4% | **86.6%** | **86.6%** | **86.6%** | **86.6%** | **86.6%** | 84.5% | **84.6%** | **84.6%** | 84.4% | 84.5% | 84.5% |
| Wine | 82.0% | **88.5%** | **88.5%** | **88.5%** | 87.9% | **88.5%** | 93.2% | 95.0% | 95.4% | **96.3%** | **96.3%** | **96.3%** |

Results of Table 2 show that our approach generally improves the supervised framework. This confirms the interest of exploiting unlabeled data by a co-training process as the one we propose. Both the *uncertain* and *indeterminate* strategies provide similar results: they often get robust performances but when the credal co-training is not relevant, i.e. when the credal co-training is weaker than the simple supervised setting whatever the selection method, they are less efficient than the other strategies.

In contrast, other strategies (*optimistic*, *pessimistic* and *median*) seems to be more robust. They have close performances but the *optimistic* strategy presents overall the best performances. As using an optimistic strategy is common in semi-supervised setting, we will use only the optimistic strategy in the next experiments comparing our approach to the standard one (Sect. 5.3).

## 5.3    Comparison with Standard Co-training

Having defined our data selection strategy for credal co-training, we now compare it to the standard co-training strategy recalled in Sect. 2 to confirm that it performs at least as well as this latter one (Table 3). To do so, we start from the same initial data sets and run the two co-training settings in parallel. Once those co-training processes are done, the final learning sets are used to learn standard Naive Bayes Classifiers (producing determinate predictions), whose their usual accuracies are then compared. Thus, credal models are used in the credal co-training phase, but not to obtain a final predictive model.

The experimental setting is the same as in Sect. 5.2. Here, STD stands for the standard method, while PACC stands for the precise accuracy of the Naive Bayes Classifier learned after the credal co-training process. We experiment the credal co-training with the *interval dominance* (INT) and *maximality* (MAX) decision rules.

The co-training with credal models almost always improve the supervised case and it is generally better than the standard co-training framework. A standard Wilcoxon test comparing the credal co-training performances with those of the standard co-training give a p-value of 0.13 for PACC.-INT. (interval

**Table 3.** Comparison of Naive Bayes Classifier performances after a credal co-training process (PREC-ACC), a standard co-training (STD) and a supervised training (SUP).

| DATASET | FIRST MODEL | | | | SECOND MODEL | | | |
|---|---|---|---|---|---|---|---|---|
| | SUP. | STD | PACC.-INT. | PACC.-MAX. | SUP. | STD | PACC.-INT. | PACC.-MAX. |
| Diabetes | 73.9% ± 3.9 | 73.7% ± 4.3 | **74.2% ± 4.7** | **74.2% ± 4.7** | 70.5% ± 3.4 | **72.1% ± 4.7** | 71.9% ± 3.3 | 71.9% ± 3.3 |
| Haberman | **73.5% ± 0.5** | **73.5% ± 0.5** | **73.5% ± 0.5** | **73.5% ± 0.5** | 74.0% ± 6.7 | **74.5% ± 3.0** | 74.2% ± 1.4 | 74.2% ± 1.4 |
| Ionosphere | 89.8% ± 5.9 | **90.2% ± 6.1** | 90.1% ± 5.3 | 90.1% ± 5.3 | 79.9% ± 7.8 | 82.3% ± 6.6 | **82.9% ± 6.1** | **82.9% ± 6.1** |
| Iris | 71.3% ± 11.2 | 75.3% ± 10.8 | 74.7% ± 11.1 | **77.3% ± 10.4** | 92.7% ± 5.5 | 93.3% ± 6.0 | **94.0% ± 5.5** | 93.3% ± 5.2 |
| KDD synthetic control | 75.2% ± 4.9 | 74.8% ± 5.4 | **77.3% ± 5.3** | **77.3% ± 5.3** | 84.3% ± 6.6 | 82.0% ± 6.0 | **85.7% ± 5.4** | **85.7% ± 5.4** |
| Kr vs kp | **70.6% ± 2.1** | 70.0% ± 2.3 | 70.2% ± 2.4 | 70.2% ± 2.4 | 80.9% ± 1.3 | **81.1% ± 1.2** | 80.8% ± 1.9 | 80.8% ± 1.9 |
| Mfeat morphological | **46.3% ± 1.2** | 46.2% ± 1.5 | 45.8% ± 0.9 | 46.0% ± 1.0 | 44.2% ± 4.2 | 43.2% ± 4.0 | 44.3% ± 3.6 | **44.4% ± 3.9** |
| Optdigits | 78.8% ± 1.9 | 78.8% ± 1.8 | **79.2% ± 1.9** | 79.1% ± 1.9 | 78.0% ± 2.5 | 78.1% ± 2.4 | **78.3% ± 2.3** | 78.2% ± 2.3 |
| Page-blocks | **92.9% ± 1.2** | **92.9% ± 1.2** | 92.3% ± 1.1 | 92.9% ± 1.0 | **86.8% ± 1.5** | 86.7% ± 1.4 | 86.3% ± 1.4 | 86.1% ± 1.5 |
| Segment | 57.7% ± 2.8 | 58.1% ± 2.8 | **58.2% ± 2.7** | **58.2% ± 2.8** | 81.8% ± 2.0 | 82.0% ± 1.8 | **82.6% ± 2.0** | 82.5% ± 1.9 |
| Spambase | 86.4% ± 1.0 | **86.6% ± 1.0** | 86.5% ± 1.2 | **86.6% ± 1.1** | **84.5% ± 1.4** | 84.1% ± 1.3 | 84.4% ± 1.3 | 84.4% ± 1.3 |
| Wine | 82.0% ± 9.0 | 86.0% ± 6.1 | **88.5% ± 6.0** | **88.5% ± 6.0** | 93.2% ± 5.7 | 94.5% ± 4.9 | 95.4% ± 4.5 | **96.3% ± 3.0** |

dominance) and of $0, 07$ for PACC.-MAX., indicating that credal co-training with maximality would be statistically higher if the significance were set to 0.1 threshold. Moreover, there are few cases in which co-training is harmful to the performances: this is probably due to too insufficient view, in which case co-training performances may suffer of label noise or sampling bias as mentioned by [24].

### 5.4 Behaviours of Credal Models During the Co-training Iterations

Having confirmed the interest of a credal co-training approach, we now examine more closely the behaviour of this approach when new data are labeled and added to the training sets. In addition to tracking the evolution of the standard and precise accuracy (computed as in Sect. 5.3), we also train, after each iteration of the co-training procedures, a Naive Credal Classifier (with $s = 2$) that can produce set-valued predictions. We then compute values commonly investigated in the assessment of credal approaches [9]: the *single accuracy* which is the percentage of good classification when the decision is determinate; the *set accuracy* which is the percentage of indeterminate predictions that contain the true class; the *determinacy* which is the percentage of confident data (i.e. for which the model decision is determinate).

Figures 2 shows the average curves of the various terms described above according to the number of training iterations, for all the UCI datasets we experiment in this paper. It should be recalled that, in our experiments, we add one instance per iteration in the training set of each model and that this instance is labeled by the other model. To have smoother curves, we compute the terms every 5 iterations. We illustrate the determinacy, the single accuracy and the set accuracy only for the maximality decision rule since we get similar curves with the interval dominance.

A first thing we can notice is that the precise accuracy (whatever the decision rule), compared to the standard accuracy, increases in a steadier and steeper way. This, again, indicates that credal co-training can be more robust than its

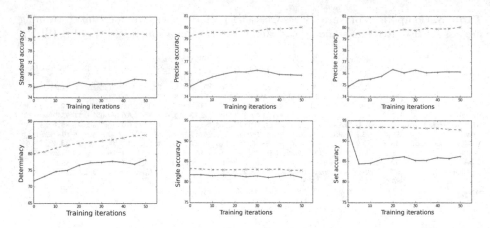

**Fig. 2.** Average performances on all data sets. Full blue and red dotted lines correspond to the first and second model, respectively. First line (from left to right): NBC
accuracy with standard co-training, with credal co-training using interval dominance
(middle) and maximality (on the right). Second line: determinacy, single accuracy and
set accuracy of the NCC resulting from credal co-training with maximality decision
rule. (Color figure online)

standard counterpart. The fact that the single-accuracy is high confirms that
the set $S_{H_i \to H_j}$ will contain informative data and that the data for which the
credal models give unique predictions during the co-training process are very reliable. Similarly, the high set accuracy suggest that the indeterminate predictions
generally contain the true labels.

In addition, the increase of determinacy shows that each classifier becomes
more confident as labelled data accumulates, yet it remains cautious on some
instances. The fact that the determinacy tends to stabilize after a while suggests
that the proposed approach is mainly interesting when starting the process, that
is when the information is minimal. This confirms our intuition that one of the
main interest of the credal approach is to avoid possible prior bias.

## 6    Conclusion

In this paper, we propose an extension of the standard co-training process to
credal models. Combining the co-training process with the imprecise probability
framework enables to define new strategies for selecting informative instances in
a pool of unlabeled data. The idea of these strategies is to use the ability of credal
models to produce unique predictions only when having enough information, and
set-valued predictions when being too uncertain.

We experiment on several UCI datasets the various selection strategies we
propose and we compare the credal co-training with the standard co-training
process and the supervised framework. Experiments confirm that the co-training

with credal models is generally more efficient and more reliable than the standard co-training framework and the supervised case.

**Acknowledgments.** This work is founded by the European Union and the French region Picardie. Europe acts in Picardie with the *European Regional Development Fund (ERDF)*.

# References

1. Amini, M., Usunier, N.: Learning with Partially Labeled and Interdependent Data. Springer, Switzerland (2015)
2. Antonucci, A., Corani, G., Gabaglio, S.: Active learning by the naive credal classifier. In: Sixth European Workshop on Probabilistic Graphical Models (PGM 2012), pp. 3–10 (2012)
3. Augustin, T., Coolen, F.P., de Cooman, G., Troffaes, M.C.: Introduction to Imprecise Probabilities. Wiley, Chichester (2014)
4. Balcan, M.F., Blum, A., Yang, K.: Co-training and expansion: towards bridging theory and practice (2004)
5. Bernard, J.M.: An introduction to the imprecise Dirichlet model for multinomial data. Int. J. Approx. Reason. **39**(2), 123–150 (2005)
6. Blum, A., Mitchell, T.: Combining labeled and unlabeled data with co-training. In: Proceedings of the Eleventh Annual Conference on Computational Learning Theory, COLT 1998, pp. 92–100. ACM (1998)
7. Chapelle, O., Schlkopf, B., Zien, A.: Semi-supervised Learning. MIT Press, Cambridge (2006)
8. Corani, G., Zaffalon, M.: Credal model averaging: an extension of Bayesian model averaging to imprecise probabilities. In: Daelemans, W., Goethals, B., Morik, K. (eds.) ECML PKDD 2008, Part I. LNCS (LNAI), vol. 5211, pp. 257–271. Springer, Heidelberg (2008)
9. Corani, G., Zaffalon, M.: Learning reliable classifiers from small or incomplete data sets: the naive credal classifier 2. J. Mach. Learn. Res. **9**, 581–621 (2008)
10. Corani, G., Zaffalon, M.: Naive credal classifier 2: an extension of naive bayes for delivering robust classifications. DMIN **8**, 84–90 (2008). CSREA Press
11. Dasgupta, S., Littman, M.L., McAllester, D.A.: PAC generalization bounds for co-training. In: Dietterich, T., Becker, S., Ghahramani, Z. (eds.) Advances in Neural Information Processing Systems, vol. 14, pp. 375–382. MIT Press, Cambridge (2002)
12. Destercke, S., Quost, B.: Combining binary classifiers with imprecise probabilities. In: Tang, Y., Huynh, V.-N., Lawry, J. (eds.) IUKM 2011. LNCS, vol. 7027, pp. 219–230. Springer, Heidelberg (2011)
13. Grandvalet, Y., Bengio, Y.: Semi-supervised learning by entropy minimization. Network **17**(5), 529–536 (2005)
14. Kingma, D.P., Rezende, D.J., Mohamed, S., Welling, M.: Semi-supervised learning with deep generative models. CoRR abs/1406.5298 (2014)
15. Liu, A., Reyzin, L., Ziebart, B.D.: Shift-pessimistic active learning using robust bias-aware prediction, pp. 1–7 (2015)
16. Nigam, K., McCallum, A., Thrun, S., Mitchell, T.M.: Text classification from labeled and unlabeled documents using EM. Mach. Learn. **39**(2/3), 103–134 (2000)

17. Qi, R.H., Yang, D.L., Li, H.F.: A two-stage semi-supervised weighted naive credal classification model. Innov. Comput. Inf. Control J. **5**(2), 503–508 (2011)
18. Settles, B.: Active Learning. Synthesis Lectures on Artificial Intelligence and Machine Learning. Morgan and Claypool Publishers, San Rafael (2012)
19. Soullard, Y., Saveski, M., Artieres, T.: Joint semi-supervised learning of hidden conditional random fields and hidden Markov models. Pattern Recogn. Lett. (PRL) **37**, 161–171 (2013)
20. Troffaes, M.C.: Decision making under uncertainty using imprecise probabilities. Int. J. Approx. Reason. **45**(1), 17–29 (2007)
21. Utkin, L.V.: The imprecise Dirichlet model as a basis for a new boosting classification algorithm. Neurocomputing **151**, 1374–1383 (2015)
22. Wang, W., Zhou, Z.-H.: Analyzing co-training style algorithms. In: Kok, J.N., Koronacki, J., Lopez de Mantaras, R., Matwin, S., Mladenič, D., Skowron, A. (eds.) ECML 2007. LNCS (LNAI), vol. 4701, pp. 454–465. Springer, Heidelberg (2007)
23. Wang, W., Zhou, Z.H.: A new analysis of co-training. In: Proceedings of the 27th International Conference on Machine Learning (ICML 2010), pp. 1135–1142 (2010)
24. Wang, W., Zhou, Z.H.: Co-training with insufficient views. In: Asian Conference on Machine Learning, pp. 467–482 (2013)
25. Zaffalon, M.: The naive credal classifier. J. Stat. Plan. Inference **105**(1), 5–21 (2002). Imprecise Probability Models and their Applications
26. Zhang, M.L., Zhou, Z.H.: Cotrade: confident co-training with data editing. IEEE Trans. Syst. Man Cybern. Part B (Cybern.) **41**(6), 1612–1626 (2011)
27. Zhou, Z.H.: When semi-supervised learning meets ensemble learning. Front. Electr. Electron. Eng. China **6**(1), 6–16 (2011)
28. Zhu, X., Goldberg, A.B., Brachman, R., Dietterich, T.: Introduction to Semi-Supervised Learning. Morgan and Claypool Publishers, San Francisco (2009)

# Interpretable Classifiers in Precision Medicine: Feature Selection and Multi-class Categorization

Lyn-Rouven Schirra[1,2], Florian Schmid[1], Hans A. Kestler[1(✉)],
and Ludwig Lausser[1]

[1] Institute of Medical Systems Biology, Ulm University, 89069 Ulm, Germany
hans.kestler@uni-ulm.de
[2] Institute of Number Theory and Probability Theory, Ulm University,
89069 Ulm, Germany

**Abstract.** Growing insight into the molecular nature of diseases leads
to the definition of finer grained diagnostic classes. Allowing for better adapted drugs and treatments this change also alters the diagnostic
task from binary to multi-categorial decisions. Keeping the corresponding multi-class architectures accurate and interpretable is currently one
of the key tasks in molecular diagnostics.

In this work, we specifically address the question to which extent biomarkers that characterize pairwise differences among classes, correspond
to biomarkers that discriminate one class from all remaining. We compare one-against-one and one-against-all architectures of feature selecting
base classifiers. They are validated for their classification performance
and their stability of feature selection.

## 1 Introduction

The analysis of molecular profiles adds a new instrument to the toolbox of medical diagnoses. It allows for a deeper insight in the molecular processes of a
cell or a tissue. Due to their high dimensionality, the interpretation of these
profiles is often quite challenging. Comprising tens of thousands of molecular
measurements, the size of a profile typically exceeds the possibility of a direct
visual inspection. Computer-aided classification algorithms are needed for diagnostic purposes [9,18,22]. Training these models often incorporates an internal
feature selection process [14,21], which basically yields at a limitation of the
measurements in the final prediction [8,19]. The resulting feature signature typically optimizes heuristic criteria [12]. It is often constructed in a purely data-
driven or model-driven procedure [2]. Alternatively, feature selection can also be
conducted from (prior) domain knowledge about the subject or the measuring
process of an experiment [15].

One of the most important findings from the analysis of molecular profiles,
is the insight that an observable phenotype or disease that was thought to be

---

L.-R. Schirra and F. Schmid—Contributed equally.

H.A. Kestler and L. Lausser—Joint senior authors.

© Springer International Publishing AG 2016
F. Schwenker et al. (Eds.): ANNPR 2016, LNAI 9896, pp. 105–116, 2016.
DOI: 10.1007/978-3-319-46182-3_9

a uniform entity can be evoked by varying molecular causes [11]. These refinements of the traditional phenotypes bring up the possibility of more specific treatments and can be seen as a starting point for the field of precision medicine or personalized medicine [4]. From a diagnostic point of view, the challenge of identifying a correct phenotype has changed due to the increased number of diagnostic classes [7]. Primarily designed for binary categorization problems many classification models cannot be directly applied to such a multi-class scenario [6,20]. Fusion architectures for combining an ensemble of binary classifiers are needed [16]. These combining techniques can either follow a predefined scheme [5] or are adapted during the overall training phase of the classifier system [10]. Fusion architectures can also incorporate known relationships among the diagnostic categories [13].

In this work, we focus on the interactions between data-driven feature selection strategies and predefined multi-class architectures. We analyze the influence of the one-against-one and the one-against-all training on the selection strategy of feature selecting base classifiers [16]. We compare the multi-class architectures according to their classification performance and evaluate the selection stability of the individual feature sets.

## 2   Methods

We denote classification as the task of predicting a category $y \in \mathcal{Y}$ of an object according to a vector of measurements $\mathbf{x} \in \mathcal{X}$. Here, $\mathcal{Y}$ denotes a finite label space with $|\mathcal{Y}| \geq 2$ and $\mathcal{X} \in \mathbb{R}^n$ denotes the feature space. The single categories will be represented by natural numbers $\mathcal{Y} = \{1, \ldots, |\mathcal{Y}|\}$. The ordering of these numbers is not assumed to reflect a known ordering of categories. The elements of a feature vector will be denoted by $\mathbf{x} = (x^{(1)}, \ldots, x^{(n)})^T$. A classification function, a classifier, will be seen as a function mapping

$$c : \mathcal{X} \to \mathcal{Y}. \tag{1}$$

A classifier is typically chosen from a predefined concept class $\mathcal{C}$ and adapted in a data-driven procedure $l$. Here, the classifier is trained on a set of labeled training sampled $\mathcal{T} = \{(\mathbf{x}_i, y_i)\}_{i=1}^m$,

$$l : \mathcal{C} \times \mathcal{T} \to c_{\mathcal{T}}. \tag{2}$$

After this initial training phase, the classifier can be applied for predicting the class label of new unseen samples. Its generalization ability is typically quantified on a separate set of validation samples $\mathcal{V} = \{(\mathbf{x}_i', y_i')\}_{i=1}^{m'}$, $\mathcal{T} \cap \mathcal{V} = \emptyset$. We will mainly focus on the overall accuracy of a classifier $c(\mathbf{x})$

$$\mathrm{acc}_{\mathcal{V}} = \frac{1}{|\mathcal{V}|} \sum_{(\mathbf{x},y) \in \mathcal{V}} \mathbb{I}_{[c(\mathbf{x})=y]} \tag{3}$$

and the multi-class extensions of the sensitivity $se_{\mathcal{V}}(y)$ and specificity $sp_{\mathcal{V}}(y)$

$$se_{\mathcal{V}}(y) = \frac{1}{|\mathcal{V}_y|} \sum_{(\mathbf{x},y') \in \mathcal{V}_y} \mathbb{I}_{[c(\mathbf{x})=y]} \quad \text{and} \quad sp_{\mathcal{V}}(y) = \frac{1}{|\mathcal{V} \setminus \mathcal{V}_y|} \sum_{(\mathbf{x},y') \in \mathcal{V} \setminus \mathcal{V}_y} \mathbb{I}_{[c(\mathbf{x}) \neq y]}.$$

(4)

Here, $\mathcal{V}_y = \{(\mathbf{x},y') \in \mathcal{V} \,|\, y' = y\}$ denotes those samples in $\mathcal{V}$ with class label $y$ and $\mathbb{I}_{[]}$ denotes the indicator function.

## 2.1 Multi-class Classification

In this work, we are mainly interested in training strategies that construct multi-class classifiers ($|\mathcal{Y}| > 2$) by the combination of binary ones. These strategies are allowed to train an ensemble of base classifiers $\mathcal{E} \subset \mathcal{C}$

$$\mathcal{E} = \{c_i : \mathcal{X} \to \mathcal{Y}_i\}_{i=1}^{|\mathcal{E}|}$$

(5)

that are finally combined via a subsequently applied fusion strategy

$$h_{\mathcal{E}} : \mathcal{Y}_1 \times ... \times \mathcal{Y}_{|\mathcal{E}|} \to \mathcal{Y} \text{ with } h_{\mathcal{E}}(c_1(\boldsymbol{x}), ..., c_{|\mathcal{E}|}(\mathbf{x})) = y.$$

(6)

We will restrict ourselves to untrainable fusion strategies in the following.

The elements of $\mathcal{E}$ can in principle be trained on different subsets of the overall training set and they can also map to distinct label spaces with distinct interpretation. A base classifier will be denoted by

$$c_{(y,y')} : \mathcal{X} \to \{y, y'\},$$

(7)

if its label space is of interest. In the following, we will restrict ourselves to the well known one-against-one scheme and the one-against-all scheme.

**One-against-one Scheme (OaO).** The one-against-one (OaO) architecture comprises ensemble members for each pairwise classification between the elements in $\mathcal{Y}$, therefore it consists of $|\mathcal{E}| = \frac{|\mathcal{Y}|(|\mathcal{Y}|-1)}{2}$ base classifiers. Each ensemble member utilizes solely the training samples $\mathcal{T}_y \cup \mathcal{T}_{y'}$ of the selected classes

$$c_{(y,y')} : \mathcal{X} \to \{y, y'\} \quad \text{for all} \quad y, y' \in \mathcal{Y}, \, y < y'.$$

(8)

For the prediction of a multi-class label $y \in \mathcal{Y}$, all base classifiers of the OaO ensemble are applied simultaneously. Their individual predictions are finally combined via an unweighted majority vote

$$h_{OaO}(\mathbf{x}) = \arg\max_{y \in \mathcal{Y}} \sum_{c \in \mathcal{E}} \mathbb{I}_{[c(\mathbf{x})=y]}.$$

(9)

**One-against-all Scheme (OaA).** The one-against-all (OaA) architecture can be seen as an ensemble of one-class detectors. It consists of $|\mathcal{E}| = |\mathcal{Y}|$ members. Each base classifier $c \in \mathcal{E}$ is trained to separate a single class $y \in \mathcal{Y}$ from the remaining ones

$$c_{(y,\bar{y})} : \mathcal{X} \to \{y, \bar{y}\} \quad \text{for all} \quad y \in \mathcal{Y}, \, \bar{y} = \mathcal{Y} \setminus y.$$

(10)

In this context, the artificial complement class $\bar{y}$ of $y$ is often referred to as the rest class of $y$. As a class label $y \in \mathcal{Y}$ can only be predicted by one member of an OaA ensemble, an unweighted majority-vote would be prone to ties. It is typically replaced by some kind of weighting scheme of the single decisions. In the basic version of the OaA, it is assumed that the chosen type of base classifier $\mathcal{C}$ is able to provide an additional certainty measure $p_{(y,\bar{y})}(\mathbf{x})$ of a prediction $c_{(y,\bar{y})}(\mathbf{x}) = y$. The final OaA multi-class decision is then given by

$$h_{OaA}(\mathbf{x}) = \arg\max_{y \in \mathcal{Y}} p_{(y,\bar{y})}(\mathbf{x}). \qquad (11)$$

## 2.2  Feature Selection for Multi-class Classification

The training of classification algorithms can incorporate a feature selection phase in which the number of available measurements is reduced to a small number of important markers. Especially in high-dimensional settings, such as the analysis of gene expression profiles, a reduction to a small, interpretable and measurable set of markers is of interest. The selected features can lead to new hypotheses on the causes for a certain class difference.

Formally a feature selection process can be seen as a function

$$f : \mathcal{C} \times \mathcal{T} \to \mathcal{I} = \{\mathbf{i} \in \mathbb{N}^{\hat{n} \leq n} | i_k < i_{k+1}, 1 \leq i_k \leq n\}, \qquad (12)$$

which maps to the space of index vectors $\mathbf{i} \in \mathcal{I}$ of maximal length $n$. The element of $\mathbf{i} = (i^{(1)}, \ldots, i^{(\hat{n})})^T$ indicate the features $\mathbf{x}^{(\mathbf{i})} = (x^{(i^{(1)})}, \ldots, x^{(i^{(\hat{n})})})$ that will be passed to a subsequent processing step.

In the context of multi-class ensembles, feature selection can be applied in different ways. It can either be seen as an initial step of the overall ensemble training (TYPE I) or as a component of the individual training procedures (TYPE II) of the corresponding base classifiers. If feature selection is incorporated in the overall ensemble training, one common feature signature is extracted for all base classifiers. A suitable (supervised) selection criterion for this type of feature selection must be able to handle multi-class labels.

If feature selection is incorporated in the training of the base classifiers, an individual feature signature $\mathcal{I}_{\mathcal{E}} = \{\mathbf{i}_1, \ldots, \mathbf{i}_{|\mathcal{E}|}\}$ is extracted for each of the ensemble members $\mathcal{E} = \{c_1, \ldots, c_{|\mathcal{E}|}\}$. Here, the feature selection process does not only provide a reduction of the initial feature set. The process also distributes the selected features among the base classifiers. As the ensemble members are designed for binary classification tasks, feature selection criteria for two-class scenarios can be applied in this framework.

As basic two-class feature selection criteria we have chosen the Threshold Number of Misclassification (TNoM) score proposed by Ben-Dor et al. [1], the Pearson's Correlation Coefficient (PCC) and the T-Test (TT).

*Threshold Number of Misclassificaitons (TNoM):* The TNoM is a univariate filter criterion which applies the (re-)classification error of a single threshold classifier $c_t(\mathbf{x})$ as selection score $s_i$

$$s_i = \frac{1}{n} \sum_{(\mathbf{x},y)\in\mathcal{T}} \mathbb{I}_{[c_t(\mathbf{x})=y]} \quad \text{with} \quad c_t(\mathbf{x}) = \begin{cases} y' & \text{if} \quad d(x^{(i)} - t) \geq 0 \\ y'' & \text{otherwise.} \end{cases} \quad (13)$$

The parameters $t \in \mathbb{R}$ and $d \in \{-1, +1\}$ are initially optimized for minimizing the classification error (TNoM) on $\mathcal{T}$ or the mean classwise error (TNoM$_{cw}$).

*Pearson's Correlation Coefficient (PCC):* The PCC is designed for the detection of linear correlations of two measurements. If applied to a single feature and the corresponding class label, it can be utilized as a score for univariate feature selection. We will utilize the absolute PCC in the following

$$s_i = \left| \frac{\sum_{(\mathbf{x},y)\in\mathcal{T}}(x^{(i)} - \bar{x}^{(i)})(y - \bar{y})}{\sqrt{\sum_{(\mathbf{x},y)\in\mathcal{T}}(x^{(i)} - \bar{x}^{(i)})^2 \sum_{(\mathbf{x},y)\in\mathcal{T}}(y - \bar{y})^2}} \right|. \quad (14)$$

Here $\bar{x}^{(i)}$ and $\bar{y}$ denote the average feature value of the $i$th feature and the average class label.

*T-Test (TT):* The T-statistic is designed for detecting differences in the mean value of two normally distributed classes $y'$ and $y''$. In its empirical version it can be used as a univariate feature selection criterion

$$s_i = \frac{\bar{x}_{y'}^{(i)} - \bar{x}_{y''}^{(i)}}{\sqrt{(\sigma_{y'}^{(i)})^2/|\mathcal{T}_{y'}| + (\sigma_{y''}^{(i)})^2/|\mathcal{T}_{y''}|}}. \quad (15)$$

Here $\bar{x}_y^{(i)}$ and $\sigma_y^{(i)}$ denote the classwise and featurewise mean value and standard deviation.

## 3   Experiments

In the following we are mainly interested in multi-class feature selection algorithms of TYPE II. We apply this kind of feature selection in experiments with both OaO-ensembles and OaA-ensembles and give a direct comparison of their results. As a base classifier the linear support vector machine was chosen [20]. It constructs linear decision rules of type

$$c_{(y,y')}(\mathbf{x}) = \begin{cases} y & \text{if} \quad p_{(y,y')}(\mathbf{x}) \geq 0 \\ y' & \text{else} \end{cases} \quad \text{and} \quad p_{(y,y')}(\mathbf{x}) = \mathbf{w}^t\mathbf{x} - t, \quad (16)$$

where $\mathbf{w} \in \mathbb{R}^n$ and $t \in \mathbb{R}$ are optimized to maximize the margin between the samples of $y$ and $y'$.

For each base classifier, the top-$k$ features with the best scores will be selected. We have chosen $k = 100$ for our experiments. All experiments have been designed as $10 \times 10$ cross-validation experiments. They have been conducted in the TunePareto framework [17].

As an example dataset we utilize the collection of pediatric acute leukemia entities provided by Yeoh et al. [23]. The dataset consists of gene expression profiles of 360 patients which are categorized into 6 risk classes (Table 1). The corresponding gene expression profiles comprise $n = 12558$ measurements.

**Table 1.** Classwise composition of the leukemia dataset [23].

|  | $y = 1$ | $y = 2$ | $y = 3$ | $y = 4$ | $y = 5$ | $y = 6$ |
|---|---|---|---|---|---|---|
| Risk class | BCR-ABL | E2A-PBX1 | Hyperdip. | MLL | T-ALL | TEL-AML |
| Samples | 15 | 27 | 64 | 20 | 43 | 79 |

## 4  Results

A summary of the classification performance of the tested multi-class classifier systems can be found in Tables 2 (overall) and 3 (classwise). Without feature selection (noFS) the OaA ensemble outperforms the OaO ensemble in terms of the overall accuracy (OaA: 97.0 %, OaO: 91.2 %). It also achieves the highest mean sensitivity in our experiments (OaA: 97.8 %). Coupled to a TNoM, $TNoM_{cw}$, PCC feature selection both ensemble types achieve mean overall accuracies over 97.0 %. The best overall accuracy of 97.7 % was achieved for the TNoM-OaO ensemble. The application of the TT slightly decreased the overall accuracy (OaO: 95.6 %, OaA: 96.5 %). The best mean specificity was observed for the TNoM-OaO (99.6 %).

On a classwise level, highest differences in sensitivity can be observed for class $y = 1$. Here, the sensitivity of the feature selecting OaO ensembles is up to 15.4 % higher than the sensitivity of their OaA counterparts. A similar tendency can be seen for class $y = 4$. A maximal sensitivity difference of 11.5 % was observed. For classes $y = 3$ and $y = 6$ higher sensitivities can be seen for the feature selecting OaA ensemble. The highest observed differences in these classes is 5.6 %.

**Table 2.** Overall accuracy, mean sensitivity, mean specificity (mean ± standard deviation in %) achieved in the $10 \times 10$ cross-validation experiments on the leukemia dataset. For each performance measure, the highest values are highlighted.

|  | Overall accuracy | | Mean sensitivty | | Mean specificity | |
|---|---|---|---|---|---|---|
|  | OaO | OaA | OaO | OaA | OaO | OaA |
| noFS | $91.2 \pm 0.8$ | $97.0 \pm 0.4$ | $93.1 \pm 1.2$ | $\mathbf{97.8 \pm 0.8}$ | $98.4 \pm 0.1$ | $99.5 \pm 0.1$ |
| TNoM | $\mathbf{97.7 \pm 0.5}$ | $97.4 \pm 0.8$ | $97.6 \pm 1.1$ | $95.1 \pm 1.4$ | $\mathbf{99.6 \pm 0.1}$ | $99.5 \pm 0.2$ |
| $TNoM_{cw}$ | $97.2 \pm 0.6$ | $97.1 \pm 0.4$ | $97.1 \pm 0.9$ | $94.8 \pm 1.0$ | $99.5 \pm 0.1$ | $99.4 \pm 0.1$ |
| PCC | $97.5 \pm 0.7$ | $97.5 \pm 0.7$ | $97.6 \pm 1.1$ | $95.6 \pm 1.1$ | $99.5 \pm 0.1$ | $99.5 \pm 0.1$ |
| TT | $95.6 \pm 0.8$ | $96.5 \pm 0.8$ | $95.5 \pm 0.9$ | $93.6 \pm 1.2$ | $99.2 \pm 0.1$ | $99.3 \pm 0.2$ |

**Table 3.** Classwise sensitivities and specificities (mean in %) achieved in the $10 \times 10$ cross-validation experiments on the leukemia dataset. For each performance measure, the highest values are highlighted.

| | $y = 1$ | | $y = 2$ | | $y = 3$ | | $y = 4$ | | $y = 5$ | | $y = 6$ | |
| | OaO | OaA | OaO | OaA | OaO | OaA | OaO | OaA | OaO | OaA | OaO | OaA |
|---|---|---|---|---|---|---|---|---|---|---|---|---|
| Sensitivity | | | | | | | | | | | | |
| noFS | 98.0 | **98.7** | 96.7 | **100.0** | 81.2 | 92.2 | 95.0 | 99.5 | 94.4 | 97.7 | 93.5 | 98.5 |
| TNoM | 92.7 | 78.0 | **100.0** | **100.0** | 93.6 | **96.2** | **100.0** | 96.5 | **100.0** | **100.0** | 99.4 | **100.0** |
| TNoM$_{cw}$ | 90.7 | 75.3 | **100.0** | **100.0** | 93.3 | 94.8 | **100.0** | 98.5 | **100.0** | **100.0** | 98.5 | **100.0** |
| PCC | 93.3 | 81.3 | **100.0** | **100.0** | 92.7 | 96.1 | **100.0** | 96.0 | **100.0** | **100.0** | 99.4 | **100.0** |
| TT | 86.0 | 78.0 | **100.0** | **100.0** | 89.7 | 95.3 | **100.0** | 88.5 | **100.0** | **100.0** | 97.1 | 99.9 |
| Specificity | | | | | | | | | | | | |
| noFS | 91.5 | 98.1 | 99.9 | 99.5 | 99.8 | **99.9** | 99.6 | 99.2 | **100.0** | **100.0** | 99.8 | **100.0** |
| TNoM | 98.2 | 99.4 | **100.0** | 99.6 | 99.8 | 98.3 | 99.5 | 99.7 | **100.0** | 99.8 | **100.0** | **100.0** |
| TNoM$_{cw}$ | 97.9 | 99.0 | 100.0 | 100.0 | 99.6 | 97.9 | 99.5 | **99.9** | **100.0** | 99.8 | **100.0** | 99.8 |
| PCC | 98.2 | 99.2 | **100.0** | 99.6 | 99.8 | 98.6 | 99.3 | 99.6 | **100.0** | **100.0** | **100.0** | **100.0** |
| TT | 96.9 | **99.8** | **100.0** | 99.9 | 99.0 | 97.6 | 99.2 | 99.4 | 100.0 | 99.0 | **100.0** | **100.0** |

The selection stability of the feature selecting base classifier is summarized in Fig. 1. As a basic measure we apply our stability score presented in Lausser et al. [14]. It is used to characterize the stability of a feature selection strategy in cross-validation experiments. A score near 1 indicates a perfectly stable (fixed) feature selection; a score near zero indicate perfectly instable feature selection (distinct features for each experiment). The scores achieved by the single base classifiers are organized classwise.

In general, all achieved stability scores lie in the range of 0.70 to 0.95. Similar to the classification performance, classwise differences in the selection stability can be observed. For all feature selection strategies, the lowest selection stability of an OaA base classifier is reported for class $y = 1$, which is the class with the lowest number of samples. For this class ($y = 1$) all selection strategies also show the lowest classwise mean stability over all OaO base classifiers. Among the remaining classes, similar observation can be found for class $y = 4$. For this second smallest class, the stability of the OaA base classifier is smaller than for the remaining four classes for TNoM, TNoM$_{cw}$ and PCC. The mean classwise stability of the OaO base classifiers is smaller for all four feature selection strategies. Higher selection stabilities are observed for the larger classes. The highest stabilities of OaA classifiers are observed for the largest class $y = 6$. The highest classwise mean stabilities of OaO base classifiers are observed for class $y = 5$ in three of four cases.

Figure 2 shows the feature selections of the TNoM score. For each class $y \in \mathcal{Y}$, a panel reports on the features of all base classifiers that where trained for recognizing $y$. The corresponding features are sorted according to the number of

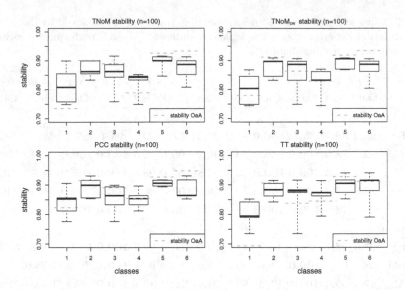

**Fig. 1.** Selection stabilities of the $10 \times 10$ cross validation experiments: The figure gives a comparison of the stability scores of the OaA ensemble and the OaO ensemble. For each of the class $y$, the dashed lines give the stability score of the corresponding OaA base classifiers. The boxplots summarize the stability scores of the OaO base classifiers that are trained to predict $y$.

base classifiers $c_{(y,y')}(\mathbf{x})$, $y' \in \mathcal{Y} \setminus \{y\}$ for which they were selected. For each base classifier $c_{(y,y')}(\mathbf{x})$, those features are reported that are selected in at least 70 % of all training runs. The result of the TNoM score can be seen as exemplarily for the other feature selections.

In this classwise view, the classes $y \in \mathcal{Y}$ are represented by $n = 232$ ($y = 5$) up to $n = 368$ ($y = 3$) features. Most of these features are selected in a single two class comparison in the OaO ensembles ($n = 112$ ($y = 5$) up to $n = 249$ ($y = 3$)). The number of selected features decreases with the number of selections per class. At most $n = 21$ features were selected in all five base classifiers $c_{(y,y')}(\mathbf{x})$ of one class $y$ ($y = 5$) in the OaO ensemble. For two classes ($y = 3$ and $y = 4$), no feature was selected in all corresponding OaO base classifiers. The color of a feature indicates its selection frequency in the base classifiers $c_{(y',y'')}(\mathbf{x})$, $y'' \in \mathcal{Y} \setminus \{y'\}$ of the reference class $y'$ of $c_{y,y'}(\mathbf{x})$. Most of these features are selected only once in the OaO base classifiers of $y$, which indicates a preference for $y'$.

The selections of the OaA ensemble members $c_{(y,\bar{y})}(\mathbf{x})$ are shown in the rightmost column of each panel. It can be seen that they have a large overlap to the selections of the OaO base classifiers. Only 20.2 % ($y = 5$) up to 32.6 % ($y = 1$) of the OaA selections are not included in the frequently selected features of the OaO base classifiers. For classes $y = 1$, $y = 3$, $y = 5$ no OaA feature was selected four or five times by the corresponding OaO base classifiers. For class $y = 6$, only one OaA feature was selected more than twice by the OaO base classifiers. These observations can also be seen for the features that were

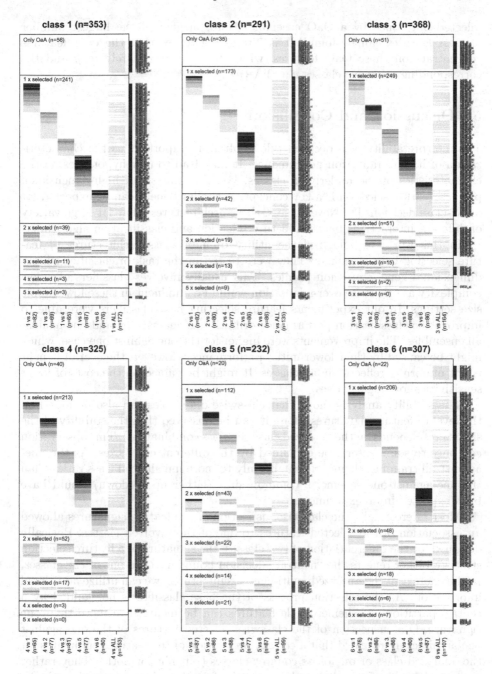

**Fig. 2.** Overlap of feature signatures (TNoM score): The figure shows the frequently selected features ($\geq 70\%$) of the $10 \times 10$ cross-validation experiments with the YEOH dataset. The signatures are grouped according to the six classes $y \in \mathcal{Y}$ and sorted according to their selection frequency in the base classifiers $c_{y,y'}(\mathbf{x})$, $y' \in \mathcal{Y} \setminus \{y\}$. The color of the features indicates its selection frequency in the OaO base classifiers $c_{y',y''}(\mathbf{x})$, $y' \in \mathcal{Y} \setminus \{y\}$. A light gray color indicates a selection in one base classifier of $y'$ a black color indicates a selection in all base classifiers of $y'$.

selected only once by a OaO ensemble. Here, in general the largest number of OaA features can be found. It is interesting to see that the OaA features concentrate on those OaO features, which were selected rarely in $y$ and the corresponding reference classes $y' \in \mathcal{Y} \setminus \{y\}$.

## 5   Discussion and Conclusion

The interpretability of a decision rule is often as important as the characterization of its generalization performance. It may lead to new hypotheses on the characteristics of the underlying classes. For the analysis of high-dimensional profiles, feature selection is a key concept. A similar benefit can be expected for multi-class decisions [3]. Nevertheless, these architectures allow for large variety of different interactions between feature selection and classification algorithms.

In this work, we analyzed the utility of multi-class architectures of feature selecting base classifiers. They were tested in the task of classifying high-dimensional gene expression profiles. This usually involves classifiers of a low complexity as otherwise over-adaptation would be imminent in these low sample size settings. In our experiments feature selecting base classifiers were able to improve the classification performance of both one against one and one against all ensembles. The improvements were higher for the one against one case, which might be caused by their lower initial performance. However, the overall benefit is not uniformly reflected in all classes. It might be gained on the costs of lower sensitivities for single classes.

The stability analysis shows that classwise differences can also be found on the level of feature signatures. Here, it can be observed that the stability of the signatures depends on the analyzed class (or class combination). This observation can only to some extend be explained by the different class sizes. As the one-against-all training scheme leads uniformly to more imbalanced class ratios than the one-against-one scheme, a common shift (either up or down) would have been expected in all experiments.

Most interestingly, the classwise inspection of the feature signatures allowed an association of the selected features to one of the involved classes. Basically designed for the differentiation of two classes, the signature of a feature selection strategy rather reflects the "neutral" decision boundary than one of the classes. In the context of the analyzed multi-class architectures, we can utilize additional information. As the selection process of the base classifier is not guided by a central partitioning scheme, single features can be included in more than one signature. Our inspection of the classwise selected features of the one against one architecture revealed that a certain proportion of the features cumulates for the analyzed class or one of its counter classes (but not for both). They rather characterize one of the classes than a (pairwise) class difference and might be considered as class attributes. In this line of argumentation, we would have expected an high overlap between these attributes and the features selected by an one against all ensemble. Counterintuitively, this seems not to be the case. The features that were selected in both strategies rather correspond to those features which are involved in one particular two class decision. They might be seen as a holistic view on the pairwise differences. Nevertheless the base classifiers of the one against all strategy select a certain amount of features, which is not considered by the one-against-one scheme. These features might also contain class attributes but can not be evaluated as mentioned above.

**Acknowledgements.** The research leading to these results has received funding from the European Community's Seventh Framework Programme (FP7/20072013) under grant agreement n°602783, the German Research Foundation (DFG, SFB 1074 project Z1), and the Federal Ministry of Education and Research (BMBF, Gerontosys II, Forschungskern SyStaR, ID 0315894A and e:Med, SYMBOL-HF, ID 01ZX1407A) all to HAK.

# References

1. Ben-Dor, A., Bruhn, L., Friedman, N., Nachman, I., Schummer, M., Yakhini, Z.: Tissue classification with gene expression profiles. J. Comput. Biol. **7**(3–4), 559–583 (2000)
2. Blum, A., Langley, P.: Selection of relevant features and examples in machine learning. Artif. Intell. **97**(1–2), 245–271 (1997)
3. Bolón-Canedo, V., Sánchez-Maroño, N., Alonso-Betanzos, A., Benítez, J., Herrera, F.: A review of microarray datasets and applied feature selection methods. Inf. Sci. **282**, 111–135 (2014)
4. Chow, S.C., Song, F.: Some thoughts on precision medicine. J. Biom. Biostat. **6**(5), 1–2 (2015)
5. Dietterich, T.G., Bariki, G.: Solving multiclass problems via error-correcting output codes. J. Artif. Intell. Res. **2**, 263–286 (1995)
6. Freund, Y., Schapire, R.E.: A decision-theoretic generalization of on-line learning and an application to boosting. In: Vitányi, P.M.B. (ed.) EuroCOLT 1995. LNCS, vol. 904, pp. 23–37. Springer, Heidelberg (1995)
7. Gress, T.M., Kestler, H.A., Lausser, L., Fiedler, L., Sipos, B., Michalski, C.W., Werner, J., Giese, N., Scarpa, A., Buchholz, M.: Differentiation of multiple types of pancreatico-biliary tumors by molecular analysis of clinical specimens. J. Mol. Med. **90**(4), 457–464 (2011)
8. Guyon, I., Elisseeff, A.: An introduction to variable and feature selection. J. Mach. Learn. Res. **3**, 1157–1182 (2003)
9. Hastie, T., Tibshirani, R., Friedman, J.H.: The Elements of Statistical Learning. Springer, New York (2001)
10. Huang, Y., Suen, C.: A method of combining multiple experts for the recognition of unconstrained handwritten numerals. IEEE Trans. Pattern Anal. Mach. Intell. **17**, 90–94 (1995)
11. Khan, J., Wei, J., Ringner, M., Saal, L., Westermann, F., Berthold, F., Schwab, M., Antonesco, C., Peterson, C., Meltzer, P.: Classification and diagnostic prediction of cancer using gene expression profiling and artificial neural networks. Nat. Med. **6**, 673–679 (2001)
12. Kraus, J., Lausser, L., Kestler, H.A.: Exhaustive k-nearest-neighbour subspace clustering. J. Stat. Comput. Simul. **85**(1), 30–46 (2015)

13. Lattke, R., Lausser, L., Müssel, C., Kestler, H.A.: Detecting ordinal class structures. In: Schwenker, F., Roli, F., Kittler, J. (eds.) MCS 2015. LNCS, vol. 9132, pp. 100–111. Springer, Heidelberg (2015)
14. Lausser, L., Müssel, C., Maucher, M., Kestler, H.A.: Measuring and visualizing the stability of biomarker selection techniques. Comput. Stat. **28**(1), 51–65 (2013)
15. Lausser, L., Schmid, F., Platzer, M., Sillanpää, M.J., Kestler, H.A.: Semantic multi-classifier systems for the analysis of gene expression profiles. Arch. Data Sci. Ser. A (Online First) **1**(1), 1–19 (2016)
16. Lorena, A., de Carvalho, A., Gama, J.: A review on the combination of binary classifiers in multiclass problems. Artif. Intell. Rev. **30**, 19–37 (2008)
17. Müssel, C., Lausser, L., Maucher, M., Kestler, H.A.: Multi-objective parameter selection for classifiers. J. Stat. Softw. **46**(5), 1–27 (2012)
18. Ripley, B.D.: Pattern Recognition and Neural Networks. Cambridge University Press, Cambridge (1996)
19. Saeys, Y., Iñza, I., Larrañaga, P.: A review of feature selection techniques in bioinformatics. Bioinformatics **23**(19), 2507–2517 (2007)
20. Vapnik, V.: Statistical Learning Theory. Wiley, New York (1998)
21. Völkel, G., Lausser, L., Schmid, F., Kraus, J.M., Kestler, H.A.: Sputnik: ad hoc distributed computation. Bioinformatics **31**(8), 1298–1301 (2015)
22. Webb, A.R.: Statistical Pattern Recognition, 2nd edn. Wiley, New York (2002)
23. Yeoh, E.J., Ross, M.E., Shurtleff, S.A., Williams, W.K., Patel, D., Mahfouz, R., Behm, F.G., Raimondi, S.C., Relling, M.V., Patel, A., Cheng, C., Campana, D., Wilkins, D., Zhou, X., Li, J., Liu, H., Pui, C.H., Evans, W.E., Naeve, C., Wong, L., Downing, J.R.: Classification, subtype discovery, and prediction of outcome in pediatric acute lymphoblastic leukemia by gene expression profiling. Cancer Cell **1**(2), 133–143 (2002)

# On the Evaluation of Tensor-Based Representations for Optimum-Path Forest Classification

Ricardo Lopes[1], Kelton Costa[2], and João Papa[2(✉)]

[1] Instituto de Pesquisas Eldorado, Campinas, Brazil
ricardoriccilopes@gmail.com
[2] Department of Computing, São Paulo State University, São Paulo, Brazil
kelton.costa@gmail.com, papa@fc.unesp.br

**Abstract.** Tensor-based representations have been widely pursued in the last years due to the increasing number of high-dimensional datasets, which might be better described by the multilinear algebra. In this paper, we introduced a recent pattern recognition technique called Optimum-Path Forest (OPF) in the context of tensor-oriented applications, as well as we evaluated its robustness to space transformations using Multilinear Principal Component Analysis in both face and human action recognition tasks considering image and video datasets. We have shown OPF can obtain more accurate recognition rates in some situations when working on tensor-oriented feature spaces.

**Keywords:** Optimum-Path Forest · Tensors · Gait and face recognition

## 1 Introduction

Methodologies for data representation have been widely pursued in the last decades, being most part of them based on *Vector Space Models* (VSM). Roughly speaking, given a dataset $\mathcal{X}$ e a label set $\mathcal{Y}$, each sample $\mathbf{x}_i \in \mathcal{X}$ is represented as an $n$-dimensional point on that space with a label associated to it, i.e., each sample can be modelled as a pair $(\mathbf{x}_i, y_i)$, $y_i \in \mathbb{N}$ and $i = 1, 2, \ldots, |\mathcal{X}|$. Therefore, a machine learning algorithm aims at finding a decision function $f : \mathcal{X} \to \mathcal{Y}$ that leads to the best feature space partition [1].

Advances in storage technologies have fostered an increasing number of large data repositories, composed mainly of high-resolution images and videos. Such new environments now require more efficient and effective data representation and classification approaches, which shall take into account the high-dimensionality nature of the data [5]. Images acquired through cell phones, for instance, may contain thousands of hundreds of pixels, being 2-dimensional data by nature. As such, a specific segment of researchers have devoted a considerable effort to study more natural data representation systems. One of the most actively data description approaches is related to the well-known *Tensor Space Model* (TSM), in which a dataset sample is represented as a tensor instead of a

© Springer International Publishing AG 2016
F. Schwenker et al. (Eds.): ANNPR 2016, LNAI 9896, pp. 117–125, 2016.
DOI: 10.1007/978-3-319-46182-3_10

regular point, being the properties of such space ruled by the multilinear algebra. Roughly speaking, we can consider an image as a 2-order tensor (matrix), a video as a 3-order tensor (cube), and a scalar number is considered an 1-order tensor. Therefore, tensor-based representations can be understood as a generalization of vector-space models.

Although one can find a number of tensor-based machine learning works out there, they are responsible for only a few percentage of the published literature. Vasilescu and Terzopoulos [13], for instance, used tensorial models to dimensionality reduction in the context of face-oriented person identification. Other works focused on the extension of some well-known techniques in computer vision, i.e., Principal Component Analysis and Linear Discriminant Analysis, to tensor models [4]. In addition, Sutskever et al. [12] employed tensorial factorization together with Bayesian clustering to learn relational structures in text recognition, and Ranzato et al. [10] used a deep learning-based approach parameterized by tensors in the context of image processing.

Later on, Cai et al. [2] introduced the Support Tensor Machines (STMs) for text categorization, which is a tensor-based variant of the so-called Support Vector Machines (SVMs) classifier. The original samples were mapped as 2-order tensors, and the problem of identifying the hyperplane with maximum margin in the vector space was then changed to a tensor space. A tensor-oriented neural network was also considered for text recognition by Socher et al. [11]. As aforementioned, one can notice a lack of research regarding tensor-based machine learning, since only a few techniques have been considered in such context.

Some years ago, Papa et al. [6,7] introduced a new pattern recognition technique called Optimum-Path Forest, which models the problem of pattern classification as a graph partition task, in which each dataset sample is encoded as a graph node and connected to others through an adjacency relation. The main idea is to rule a competition process among some key samples (*prototypes*) that try to conquer the remaining nodes in order to partition the graph into optimum-path trees, each one rooted at one prototype. In this paper, we introduce OPF in the context of tensor-space learning, since it has never been evaluated in such representation model so far. We present some results in the context of face recognition using an image-oriented dataset, as well as human action recognition in video data. The remainder of this paper is organized as follows. Sections 2 and 3 present a theoretical background regarding OPF and the methodology and experiments, respectively. Finally, Sect. 4 states conclusions and future works.

## 2   Optimum-Path Forest Classification

Let $\mathcal{D} = \mathcal{D}_1 \cup \mathcal{D}_2$ be a labeled dataset, such that $\mathcal{D}_1$ and $\mathcal{D}_2$ stands for the training and test sets, respectively. Let $\mathcal{S} \subset \mathcal{D}_1$ be a set of prototypes of all classes (i.e., key samples that best represent the classes). Let $(\mathcal{D}_1, A)$ be a complete graph whose nodes are the samples in $\mathcal{D}_1$, and any pair of samples defines an arc in $A = \mathcal{D}_1 \times \mathcal{D}_1$. Additionally, let $\pi_s$ be a path in $(\mathcal{D}_1, A)$ with terminus at sample $s \in D_1$.

The OPF algorithm proposed by Papa et al. [6,7] employs the path-cost function $f_{max}$ due to its theoretical properties for estimating prototypes (Sect. 2.1 gives further details about this procedure):

$$f_{max}(\langle s \rangle) = \begin{cases} 0 & \text{if } s \in S \\ +\infty & \text{otherwise,} \end{cases}$$
$$f_{max}(\pi_s \cdot \langle s, t \rangle) = \max\{f_{max}(\pi_s), d(s,t)\}, \tag{1}$$

where $d(s,t)$ stands for a distance between nodes $s$ and $t$, such that $s,t \in \mathcal{D}_1$. Therefore, $f_{max}(\pi_s)$ computes the maximum distance between adjacent samples in $\pi_s$, when $\pi_s$ is not a trivial path. In short, the OPF algorithm tries to minimize $f_{max}(\pi_t)$, $\forall t \in \mathcal{D}_1$.

## 2.1 Training

We say that $S^*$ is an optimum set of prototypes when the OPF algorithm minimizes the classification errors for every $s \in \mathcal{D}_1$. We have that $S^*$ can be found by exploiting the theoretical relation between the minimum-spanning tree and the optimum-path tree for $f_{max}$. The training essentially consists of finding $S^*$ and an OPF classifier rooted at $S^*$. By computing a Minimum Spanning Tree (MST) in the complete graph $(\mathcal{D}_1, A)$, one obtain a connected acyclic graph whose nodes are all samples of $\mathcal{D}_1$ and the arcs are undirected and weighted by the distances $d$ between adjacent samples. In the MST, every pair of samples is connected by a single path, which is optimum according to $f_{max}$. Hence, the minimum-spanning tree contains one optimum-path tree for any selected root node.

The optimum prototypes are the closest elements of the MST with different labels in $\mathcal{D}_1$ (i.e., elements that fall in the frontier of the classes). By removing the arcs between different classes, their adjacent samples become prototypes in $S^*$, and the OPF algorithm can define an optimum-path forest with minimum classification errors in $\mathcal{D}_1$.

## 2.2 Classification

For any sample $t \in \mathcal{D}_2$, we consider all arcs connecting $t$ with samples $s \in \mathcal{D}_1$, as though $t$ were part of the training graph. Considering all possible paths from $S^*$ to $t$, we find the optimum path $P^*(t)$ from $S^*$ and label $t$ with the class $\lambda(R(t))$ of its most strongly connected prototype $R(t) \in S^*$. This path can be identified incrementally, by evaluating the optimum cost $C(t)$ as follows:

$$C(t) = \min\{\max\{C(s), d(s,t)\}\}, \ \forall s \in D_1. \tag{2}$$

Let the node $s^* \in \mathcal{D}_1$ be the one that satisfies Eq. 2 (i.e., the predecessor $P(t)$ in the optimum path $P^*(t)$). Given that $L(s^*) = \lambda(R(t))$, the classification simply assigns $L(s^*)$ as the class of $t$. An error occurs when $L(s^*) \neq \lambda(t)$.

# 3    Experimental Evaluation

In this section, we present the methodology and experiments used to validate OPF in the context of tensor-based feature representation.

## 3.1    Datasets

We considered two public datasets, as follows:

- Gait-based Human ID[1]: this dataset comprises 1,870 sequences from 122 individuals aiming at the automatic identification of humans from gait. Since this dataset is composed of videos, it has an interesting scenario for the application of tensor representations; and
- AT&T Face Dataset[2]: formerly "ORL Dataset", it comprises $92 \times 112$ images of human faces from 40 subjects.

The "Gait-based Human ID" dataset has seven fixed scenarios, called PrbA, PrbB, PrbC, PrbD, PrbE, PrbF and PrbG. In such case, the algorithms are trained in a "Gallery" set, and then tested on each scenario. In addition to that, we employed a cross-validation procedure over the entire dataset (i.e., PrbA $\cup$ PrbB $\cup \ldots \cup$ PrbG) for comparison results. Notice the cross-validation has been applied to "AT&T Face dataset" as well. Figure 1 displays some examples of dataset samples.

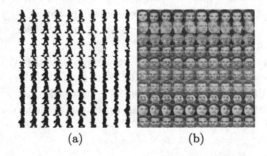

(a)                                    (b)

**Fig. 1.** Some dataset samples from (a) Gait-based Human ID and (b) AT&T Face datasets.

## 3.2    Experiments

In this work, we compared OPF [8] in two distinct scenarios, VSM and TSM, i.e., vector- and tensor-space models, respectively. Additionally, we evaluated SVM to the same context using RBF (Radial Basis Function), Linear, Polynomial (Poly) and sigmoid (Sig) kernels, as well as SVM without kernel mapping (SVM-noKernel). In regard to SVM with kernel functions, we employed the well-known

---

[1] http://figment.csee.usf.edu/GaitBaseline/.

[2] http://www.cl.cam.ac.uk/research/dtg/attarchive/facedatabase.html.

LIBSVM [3], and with respect to SVM without kernel mapping, we employed LIBLINEAR library [9]. Finally, we used an accuracy rate proposed by Papa et al. [7] that considers unbalanced datasets.

**Gait-Based Human ID.** In this section, we present the results considering the Gait-based Human ID dataset in two distinct rounds of experiments: (i) in the first one, called "fixed data", we used the very same aforementioned configuration composed of seven fixed scenarios; and (ii) in the latter experiment, we performed a cross-validation with random generated folds (we called here "random data"). Table 1 presents the results considering the "fixed data" experiment. The techniques labeled with "PCA" stand for vector-space models, and the ones labeled with "MPCA" and "MPCA-LDA" denote the tensor-space modeling. In the latter approach, a Linear Discriminant Analysis was performed after MPCA. The most accurate techniques are highlighted in bold.

**Table 1.** Mean recognition rates considering the "fixed data" experiment for the Gait-based Human ID dataset.

| | PrbA | PrbB | PrbC | PrbD | PrbE | PrbF | PrbG |
|---|---|---|---|---|---|---|---|
| OPF-PCA | 50.59 % | 49.59 % | 49.91 % | 50.59 % | 49.65 % | 50.36 % | 50.56 % |
| OPF-MPCA | 74.55 % | 73.73 % | 63.70 % | 59.50 % | 50.00 % | 54.66 % | 54.62 % |
| OPF-MPCA-LDA | 86.39 % | 50.84 % | 70.74 % | 60.66 % | 49.82 % | **57.18 %** | **56.67 %** |
| SVM-noKernel-PCA | 50,64 % | 49.66 % | 49.86 % | 50.55 % | 50.03 % | 50.38 % | 50.09 % |
| SVM-Linear-PCA | 51.32 % | 49.37 % | 49.93 % | 50.68 % | 49.66 % | 50.67 % | 51.26 % |
| SVM-Poly-PCA | 50.63 % | 49.52 % | 49.65 % | 50.16 % | 49.79 % | 50.85 % | 50.30 % |
| SVM-RBF-PCA | 51.32 % | 49.37 % | 49.93 % | 50.68 % | 49.66 % | 50.67 % | 51.27 % |
| SVM-Sig-PCA | 51.32 % | 49.37 % | 49.93 % | 50.68 % | 49.66 % | 50.67 % | 51.26 % |
| SVM-noKernel-MPCA | 86.22 % | **79.52 %** | 70.81 % | 58.81 % | 50.27 % | 56.82 % | 54.51 % |
| SVM-Linear-MPCA | 87.20 % | 78.93 % | 70.81 % | 60.68 % | 50.10 % | 56.69 % | 55.86 % |
| SVM-Poly-MPCA | 70.28 % | 67.08 % | 59.92 % | 56.31 % | 49.61 % | 52.86 % | 52.61 % |
| SVM-RBF-MPCA | 87.06 % | 79.06 % | 70.81 % | 60.68 % | 50.10 % | 56.63 % | 55.87 % |
| SVM-Sig-MPCA | 87.20 % | 78.93 % | 70.81 % | 60.68 % | 50.10 % | 56.69 % | 55.86 % |
| SVM-noKernel-MPCA-LDA | 84.86 % | 51.29 % | 69.34 % | 59.63 % | 50.10 % | 56.17 % | 54.82 % |
| SVM-Linear-MPCA-LDA | 88.52 % | 50.92 % | 71.09 % | 61.58 % | 50.30 % | 56.64 % | 55.26 % |
| SVM-Poly-MPCA-LDA | 68.78 % | 49.91 % | 55.07 % | 55.32 % | 50.16 % | 53.30 % | 52.00 % |
| SVM-RBF-MPCA-LDA | **88.68 %** | 51.03 % | 70.99 % | **61.73 %** | 50.30 % | 56.69 % | 55.26 % |
| SVM-Sig-MPCA-LDA | 88.20 % | 50.90 % | **71.25 %** | 61.51 % | **50.60 %** | 56.64 % | 55.73 % |

From those results, some interesting conclusions can be drawn: (i) tensor-space models have obtained the best results for both classifiers, i.e., OPF and SVM, (ii) SVM obtained the best results in 5 out 7 folds, and (iii) OPF can benefit from tensor-space models, which is the main contribution of this paper.

In addition, OPF results were very close to SVM ones, but being much faster for training, since it is parameterless and thus not require an optimization procedure.

In the second round of experiments, we evaluated OPF for tensor-space models using randomly generated folds in two distinct configurations: the first one employs 10 % of the whole dataset for training, and the another one that uses 50 % of the samples to train the classifiers. Table 2 presents the results considering the aforementioned configurations. Since we have considered a cross-validation procedure over 10 runnings, we performed a statistical validation through Wilcoxon signed-rank test [14].

**Table 2.** Mean recognition rates considering the "random data" experiment for the Gait-based Human ID dataset. The values in bold stand for the most accurate techniques according to that test.

|                        | 10 %              | 50 %              |
| ---------------------- | ----------------- | ----------------- |
| OPF-PCA                | 60.86 % ± 0.41    | 76.92 % ± 0.50    |
| OPF-MPCA               | 67.77 % ± 0.47    | 80.55 % ± 0.45    |
| OPF-MPCA-LDA           | 63.72 % ± 0.47    | 73.15 % ± 0.42    |
| SVM-noKernel-PCA       | 55.68 % ± 0.21    | 56.07 % ± 0.26    |
| SVM-Linear-PCA         | 57.79 % ± 0.30    | 65.03 % ± 0.50    |
| SVM-Poly-PCA           | 55.27 % ± 0.55    | 70.25 % ± 0.44    |
| SVM-RBF-PCA            | 59.06 % ± 0.38    | 73.81 % ± 0.48    |
| SVM-Sig-PCA            | 57.62 % ± 0.40    | 65.04 % ± 0.58    |
| SVM-noKernel-MPCA      | 72.16 % ± 0.57    | 86.89 % ± 0.42    |
| SVM-Linear-MPCA        | **74.06 % ± 0.66** | **91.42 % ± 0.48** |
| SVM-Poly-MPCA          | 61.07 % ± 1.04    | 83.29 % ± 0.39    |
| SVM-RBF-MPCA           | 73.95 % ± 0.61    | **91.41 % ± 0.44** |
| SVM-Sig-MPCA           | 74.01 % ± 0.69    | **91.38 % ± 0.43** |
| SVM-noKernel-MPCA-LDA  | 66.55 % ± 0.58    | 76.24 % ± 0.41    |
| SVM-Linear-MPCA-LDA    | 67.89 % ± 0.54    | 79.98 % ± 0.56    |
| SVM-Poly-MPCA-LDA      | 53.77 % ± 1.61    | 69.93 % ± 0.83    |
| SVM-RBF-MPCA-LDA       | 68.57 % ± 0.57    | 81.68 % ± 0.52    |
| SVM-Sig-MPCA-LDA       | 67.95 % ± 0.60    | 79.59 % ± 0.68    |

Considering this experiment, we can clearly observe SVM-Linear-MPCA with tensor-space modeling has obtained better results using 10 % of the entire dataset for training purposes. However, if we take into account 50 % of the data for training, only OPF with MPCA outperformed the standard vector-space modeling (i.e., OPF-PCA), since OPF-MPCA-LDA did not achieve better results than OPF-PCA. This is might be due to the poor mapping performed by LDA when considering a bigger training set. Finally, SVM also benefit from tensor-based features, achieving better results than OPF as well.

**AT&T Face Dataset.** In this section, we evaluated vector- and tensor-space models considering the task of face recognition. Once again, we employed two distinct configurations, with the first one using 10 % of the dataset samples for training, and the another one using 50 % to train the classifiers. Table 3 presents the mean recognition rates through a cross-validation procedure with 10 runnings. Similar techniques according to Wilcoxon statistical test are highlighted in bold.

**Table 3.** Mean recognition rates considering AT&T Face Dataset.

|  | 10 % | 50 % |
|---|---|---|
| OPF-PCA | **83.97 % ± 1.34** | 96.54 % ± 0.74 |
| OPF-MPCA | **84.52 % ± 0.85** | 96.49 % ± 0.94 |
| OPF-MPCA-LDA | 61.11 % ± 1.52 | 77.54 % ± 1.34 |
| SVM-noKernel-PCA | **83.93 % ± 0.93** | 96.28 % ± 0.75 |
| SVM-Linear-PCA | **83.97 % ± 1.34** | **97.90 % ± 0.55** |
| SVM-Poly-PCA | 59.95 % ± 2.48 | 88.49 % ± 1.27 |
| SVM-RBF-PCA | **83.97 % ± 1.34** | **97.46 % ± 1.16** |
| SVM-Sig-PCA | **83.62 % ± 1.98** | **97.79 % ± 0.76** |
| SVM-noKernel-MPCA | 83.46 % ± 1.23 | 96.10 % ± 0.86 |
| SVM-Linear-MPCA | **84.17 % ± 0.97** | **97.85 % ± 0.63** |
| SVM-Poly-MPCA | 65.61 % ± 1.32 | 92.51 % ± 1.20 |
| SVM-RBF-MPCA | 50.00 % ± 0.00 | 94.49 % ± 2.13 |
| SVM-Sig-MPCA | 51.14 % ± 0.50 | 52.03 % ± 0.92 |
| SVM-noKernel-MPCA-LDA | 57.95 % ± 1.10 | 73.72 % ± 1.26 |
| SVM-Linear-MPCA-LDA | 61.11 % ± 1.52 | 79.00 % ± 1.29 |
| SVM-Poly-MPCA-LDA | 53.02 % ± 0.44 | 68.92 % ± 1.66 |
| SVM-RBF-MPCA-LDA | 54.39 % ± 1.05 | 77.10 % ± 1.76 |
| SVM-Sig-MPCA-LDA | 51.13 % ± 0.55 | 62.85 % ± 0.95 |

The results showed OPF-MPCA has obtained the best results using 10 % of the dataset for training, but with OPF-PCA e some SVM variations with similar results. Considering 50 % of the dataset, SVM achieved the best results, but it seems the tensor-based representation did not play an important role in this experiment, although it has achieved very good results. A possible idea to handle such problem would be to extract features from images, and then map such features to tensor-space models, since in this work we adopted a holistic-based face recognition, i.e., we used the pixels' intensities for pattern classification purposes.

# 4    Conclusions

Tensor-based representations considering machine learning-oriented applications have been pursued in the last years aiming to obtain a more realistic and natural representation of high-dimensional data. In this paper, we evaluated the performance of OPF classifier in the context of tensor-space models. The experiments involved two distinct scenarios: gait classification in video images, and face recognition in gray-scale images. We can conclude OPF classifier can benefit from such tensor-based feature space representations.

**Acknowledgment.** The authors would like to thank CNPq grants #306166/2014-3 and #470571/2013-6, and FAPESP grants #2014/16250-9 and #2015/00801-9.

# References

1. Cai, D., He, X., Han, J.: Learning with tensor representation. Technical report, Department of Computer Science, University of Illinois at Urbana-Champaign (2006)
2. Cai, D., He, X., Wen, J.-R., Han, J., Ma, W.-Y.: Suport tensor machines for text categorization. Technical report, Department of Computer Science, University of Illinois at Urbana-Champaign (2006)
3. Chang, C.-C., Lin, C.-J.: LIBSVM: a library for support vector machines. ACM Trans. Intell. Syst. Technol. **2**, 27:1–27:27 (2011). http://www.csie.ntu.edu.tw/~cjlin/libsvm
4. He, X., Cai, D., Niyogi, P.: Tensor subspace analysis. In: Weiss, Y., Schölkopf, B., Platt, J. (eds.) Advances in Neural Information Processing Systems, vol. 18, pp. 499–506. MIT Press, Cambridge (2006)
5. Lu, H., Plataniotis, K.N., Venetsanopoulos, A.N.: A survey of multilinear subspace learning for tensor data. Pattern Recogn. **44**(7), 1540–1551 (2011)
6. Papa, J.P., Falcão, A.X., Albuquerque, V.H.C., Tavares, J.M.R.S.: Efficient supervised optimum-path forest classification for large datasets. Pattern Recogn. **45**(1), 512–520 (2012)
7. Papa, J.P., Falcão, A.X., Suzuki, C.T.N.: Supervised pattern classification based on optimum-path forest. Int. J. Imaging Syst. Technol. **19**, 120–131 (2009)
8. Papa, J.P., Suzuki, C.T.N., Falcão, A.X.: LibOPF: A library for the design of optimum-path forest classifiers (2014). http://www.ic.unicamp.br/~afalcao/LibOPF
9. Fan, R.-E., Chang, K.-W., Hsieh, C.-J., Wang, X.-R., Lin, C.-J.: LIBLINEAR: a library for large linear classification. J. Mach. Learn. Res. **9**, 1871–1874 (2008)
10. Ranzato, M., Krizhevsky, A., Hinton, G.E.: Factored 3-way restricted Boltzmann machines for modeling natural images. In: Thirteenth International Conference on Artificial Intelligence and Statistics. JMLR Proceedings, vol. 9, pp. 621–628. JMLR.org (2010)
11. Socher, R., Perelygin, A., Wu, J.Y., Chuang, J., Manning, C.D., Ng, A.Y., Potts, C.: Recursive deep models for semantic compositionality over a sentiment treebank. In: Conference on Empirical Methods in Natural Language Processing, pp. 1631–1642 (2013)

12. Sutskever, I., Tenenbaum, J.B., Salakhutdinov, R.: Modelling relational data using bayesian clustered tensor factorization. In: Bengio, Y., Schuurmans, D., Lafferty, J., Williams, C., Culotta, A. (eds.) Advances in Neural Information Processing Systems, vol. 22, pp. 1821–1828. Curran Associates, Inc., New York (2009)
13. Vasilescu, M.A.O., Terzopoulos, D.: Multilinear subspace analysis of image ensembles. In: IEEE Conference on Computer Vision and Pattern Recognition, vol. 2, pp. 93–99 (2003)
14. Wilcoxon, F.: Individual comparisons by ranking methods. Biom. Bull. 1(6), 80–83 (1945)

# On the Harmony Search Using Quaternions

João Papa[1(✉)], Danillo Pereira[1], Alexandre Baldassin[1], and Xin-She Yang[2]

[1] Department of Computing, São Paulo State University, São Paulo, Brazil
papa@fc.unesp.br, danilopereira@unoeste.br
[2] School of Science and Technology, Middlesex University, London, UK
x.yang@mdx.ac.uk

**Abstract.** Euclidean-based search spaces have been extensively studied to drive optimization techniques to the search for better solutions. However, in high dimensional spaces, non-convex functions might become too tricky to be optimized, thus requiring different representations aiming at smoother fitness landscapes. In this paper, we present a variant of the Harmony Search algorithm based on quaternions, which extend complex numbers and have been shown to be suitable to handle optimization problems in high dimensional spaces. The experimental results in a number of benchmark functions against standard Harmony Search, Improved Harmony Search and Particle Swarm Optimization showed the robustness of the proposed approach. Additionally, we demonstrated the robustness of the proposed approach in the context of fine-tuning parameters in Restricted Boltzmann Machines.

**Keywords:** Harmony Search · Quaternions · Optimization

## 1 Introduction

Function optimization plays an important role in a number of applications, ranging from simulation in aerodynamics to fine-tuning machine learning algorithms. Although one can find several problems that can be modeled by convex functions, most applications out there pose a bigger challenge, since they are usually encoded by non-convex functions, which means they may contain both local and global optima. In light of that, one may handle such sort of functions by means of two approaches: (i) the first one concerns with trying different optimization techniques, and (ii) the second one aims at finding a different representation of the search space in order to deal with smoother landscape functions.

In the last decades, the scientific community has focused even more on optimization techniques based on the nature, the so-called meta-heuristics [23]. Such techniques are based on different mechanisms that address the problem of optimization, such as evolutionary processes, mimetism, and swarm intelligence, just to name a few. These techniques have been in the spotlight mainly due to their elegance and easiness of implementation, as well as solid results in a number of well-known problems in the literature [1].

© Springer International Publishing AG 2016
F. Schwenker et al. (Eds.): ANNPR 2016, LNAI 9896, pp. 126–137, 2016.
DOI: 10.1007/978-3-319-46182-3_11

Recently, Malan and Engelbrecht [11] presented an interesting study that aimed at predicting possible situations in which the well-known Particle Swarm Optimization (PSO) algorithm would fail based on the fitness landscape. Basically, depending on the "smoothness degree" of the fitness function landscape, one can expect a probable performance of the algorithm. Once again, the fitness function plays an important role in the optimization problem, and finding suitable representations of that function in different search spaces is of great importance to obtain more accurate results. Years before their work, Merz and Freisleben [12] presented a study about fitness landscape analysis concerning memetic algorithms, and Humeau et al. [9] conducted a similar study, but in the context of local search algorithms.

Some years ago, Fister et al. [4] presented a modified version of the Firefly Algorithm based on quaternions, and later on Fister et al. [3] proposed a similar approach to the Bat Algorithm. Roughly speaking, the quaternion algebra extends the complex numbers by representing a real number using four variables [6], being widely used in areas when one needs to perform rotations with minimal computation, such as spacecraft controllers, for instance. Usually, standard representations of meta-heuristic techniques, i.e., the ones based on Euclidean spaces, tend to get trapped from local optima in higher dimensional spaces, since the fitness landscapes may become even more complicated. Therefore, that is the main motivation in using quaternion-based algebra.

One of the main drawbacks related to swarm-driven meta-heuristics concerns with their computational burden, since the fitness function needs to be evaluate whenever a possible solution changes its position. Since all decisions of a movement are taken into account collectively, a single movement of a possible solution may affect all remaining ones, thus requiring their new positioning. Harmony Search (HS) is a technique that updates one solution at each iteration only, making it one of the fastest meta-heuristic optimization techniques [5]. The idea is to model each decision variable as an instrument, being the best combination of them the solution to the chosen. By best combination we mean the one which provides "the music with optimal harmony".

One can refer to a number of different variants of HS, such as Improved Harmony Search [10], Global-best Harmony Search [13], and Self-adaptive Global-best Harmony Search [14], just to name a few. However, to the best of our knowledge, there is no Harmony Search implementation based on quaternions up to date. Therefore, the main contribution of this paper is to propose a quaternion-oriented Harmony Search, hereinafter called Quaternion Harmony Search (QHS). The proposed approach is compared against PSO, HS and Improved Harmony Search (IHS) for the task of benchmark function optimization. Also, we validate the proposed approach to fine-tune parameters of Restricted Boltzmann Machines to the task of binary image reconstruction, which turns out to be another contribution of this work.

The remainder of the paper ir organized as follows. Sections 2 and 3 present the theoretical background about quaternions and Harmony Search, respectively. Sections 4 and 5 present the methodology and experiments, respectively. Finally, Sect. 6 states conclusions and future works.

## 2  Quaternion Algebra

A quaternion $q$ is composed of real and complex numbers, i.e., $q = x_0 + x_1 i + x_2 j + x_3 k$, where $x_0, x_1, x_2, x_3 \in \Re$ and $i, j, k$ are imaginary numbers following the next set of equations:

$$ij = k, \tag{1}$$
$$jk = i, \tag{2}$$
$$ki = j, \tag{3}$$
$$ji = -k, \tag{4}$$
$$kj = -i, \tag{5}$$
$$ik = -j, \tag{6}$$

and

$$i^2 = j^2 = k^2 = -1. \tag{7}$$

Roughly speaking, a quaternion $q$ is represented in a 4-dimensional space over the real numbers, i.e., $\Re^4$. Actually, we can consider the real numbers only, since most applications do not consider the imaginary part, as the one addressed in this work.

Given two quaternions $q_1 = x_0 + x_1 i + x_2 j + x_3 k$ and $q_2 = y_0 + y_1 i + y_2 j + y_3 k$, the quaternion algebra defines a set of main operations [2]. The addition, for instance, can be defined by:

$$q_1 + q_2 = (x_0 + x_1 i + x_2 j + x_3 k) + (y_0 + y_1 i + y_2 j + y_3 k) \tag{8}$$
$$= (x_0 + y_0) + (x_1 + y_1)i + (x_2 + y_2)j + (x_3 + y_3)k,$$

while the subtraction is defined as follows:

$$q_1 - q_2 = (x_0 + x_1 i + x_2 j + x_3 k) - (y_0 + y_1 i + y_2 j + y_3 k) \tag{9}$$
$$= (x_0 - y_0) + (x_1 - y_1)i + (x_2 - y_2)j + (x_3 - y_3)k.$$

Another important operation is the norm, which maps a given quaternion to a real-valued number, as follows:

$$N(q_1) = N(x_0 + x_1 i + x_2 j + x_3 k) \tag{10}$$
$$= \sqrt{x_0^2 + x_1^2 + x_2^2 + x_3^2}.$$

Finally, Fister et al. [3,4] introduced two other operations, $qrand$ and $qzero$. The former initializes a given quaternion with values drawn from a Gaussian distribution, and it can be defined as follows:

$$qrand() = \{x_i = \mathcal{N}(0,1) | i \in \{0,1,2,3\}\}. \tag{11}$$

The latter function initialized a quaternion with zero values, as follows:

$$qzero() = \{x_i = 0 | i \in \{0,1,2,3\}\}. \tag{12}$$

Although there are other operations, we defined only the ones employed in this work.

## 3  Harmony Search

Harmony Search is a meta-heuristic algorithm inspired in the improvisation process of music players. Musicians often improvise the pitches of their instruments searching for a perfect state of harmony [5]. The main idea is to use the same process adopted by musicians to create new songs to obtain a near-optimal solution according to some fitness function. Each possible solution is modelled as a harmony, and each musician corresponds to one decision variable.

Let $\phi = (\phi_1, \phi_2, \ldots, \phi_N)$ be a set of harmonies that compose the so-called "Harmony Memory", such that $\phi_i \in \Re^M$. The HS algorithm generates after each iteration a new harmony vector $\hat{\phi}$ based on memory considerations, pitch adjustments, and randomization (music improvisation). Further, the new harmony vector $\hat{\phi}$ is evaluated in order to be accepted in the harmony memory: if $\hat{\phi}$ is better than the worst harmony, the latter is then replaced by the new harmony. Roughly speaking, HS algorithm basically rules the process of creating and evaluating new harmonies until some convergence criterion is met.

In regard to the memory consideration step, the idea is to model the process of creating songs, in which the musician can use his/her memories of good musical notes to create a new song. This process is modelled by the Harmony Memory Considering Rate ($HMCR$) parameter, which is the probability of choosing one value from the historic values stored in the harmony memory, being $(1 - HMCR)$ the probability of randomly choosing one feasible value[1], as follows:

$$\hat{\phi}^j = \begin{cases} \phi_A^j & \text{with probability HMCR} \\ \theta \in \Phi_j & \text{with probability (1-HMCR),} \end{cases} \qquad (13)$$

where $A \sim \mathcal{U}(1, 2, \ldots, N)$, and $\Phi = \{\Phi_1, \Phi_2, \ldots, \Phi_M\}$ stands for the set of feasible values for each decision variable[2].

Further, every component $j$ of the new harmony vector $\hat{\phi}$ is examined to determine whether it should be pitch-adjusted or not, which is controlled by the Pitch Adjusting Rate (PAR) variable, according to Eq. 14:

$$\hat{\phi}^j = \begin{cases} \hat{\phi}^j \pm \varphi_j \varrho & \text{with probability PAR} \\ \hat{\phi}^j & \text{with probability (1-PAR).} \end{cases} \qquad (14)$$

The pitch adjustment is often used to improve solutions and to escape from local optima. This mechanism concerns shifting the neighbouring values of some decision variable in the harmony, where $\varrho$ is an arbitrary distance bandwidth, and $\varphi_j \sim \mathcal{U}(0, 1)$. In the following, we briefly present the Improved Harmony Search technique, which has been used in the experimental section.

### 3.1  Improved Harmony Search

The Improved Harmony Search (IHS) [10] differs from traditional HS by updating the PAR and $\varrho$ values dynamically. The PAR updating formulation at time

---

[1] The term "feasible value" means the value that falls in the range of a given decision variable.

[2] Variable $A$ denotes a harmony index randomly chosen from the harmony memory.

step $t$ is given by:

$$PAR^t = PAR_{min} + \frac{PAR_{max} - PAR_{min}}{T} t, \tag{15}$$

where $T$ stands for the number of iterations, and $PAR_{min}$ and $PAR_{max}$ denote the minimum and maximum PAR values, respectively. In regard to the bandwidth value at time step $t$, it is computed as follows:

$$\varrho^t = \varrho_{max} \exp \frac{\ln(\varrho_{min}/\varrho_{max})}{T} t, \tag{16}$$

where $\varrho_{min}$ and $\varrho_{max}$ stand for the minimum and maximum values of $\varrho$, respectively.

### 3.2 Quaternion Harmony Search

The proposed approach aims at mapping the problem of optimizing variables in the Euclidean space to the quaternions space. As aforementioned, the idea is to obtain smoother representations of the fitness landscape, thus making the problem easier to handle.

In the standard Harmony Search, each harmony $\phi_i \in \Re^M$, $i = 1, 2, \ldots, N$ is modeled as an array containing $M$ variables to be optimized, such that $\phi_{ij} \in \Re$. In QHS, each decision variable $j$ is now represented by a quaternion $q_j \in \Re^4$, such that each harmony can be seen as a matrix $\phi'_i \in \Re^{4 \times N}$. Therefore, each harmony is no longer an array of decision variables, but a matrix instead.

However, we can map each quaternion to a real-valued number in order to use standard HS. Basically, one has to compute $\phi_{ij} = N(\phi'_{ij})$, $i = 1, 2 \ldots, N$ and $j = 1, 2, \ldots, M$. Further, the standard HS procedure can be executed as usual. But note the optimization process is conducted at quaternion level, which means QHS aims finding the quaternions for each decision variable such that their norm values minimizes the fitness function. An ordinary HS aims at learning values for each decision variable that minimizes the fitness function.

## 4    Methodology

In this section, we present the methodology employed in this work to validate the proposed approach. The next sections describe the techniques used for comparison purposes, benchmark functions and the statistical analysis.

### 4.1    Functions

The benchmark functions used in the experiments are the following:

- **Sphere Function:**
  $f(x, y) = x^2 + y^2$
  The minimum is $f(0, 0) = 0$, and the domain is
  $-\infty < x, y < \infty$.

- **Ackley's Function:**
  $$f(x,y) = -20e^{(-0.2\sqrt{0.5(x^2+y^2)})} - e^{(0.5(cos(2\pi x)+cos(2\pi y)))} + e + 20$$
  The minimum is $f(0,0) = 0$, and the domain is $-5 \leq x,y \leq 5$.
- **Rosenbrock's Function:**
  $$f(x,y) = 100\left(y - x^2\right)^2 + (x-1)^2$$
  The minimum is $f(1,1) = 0$, and the domain is
  $-\infty < x,y < \infty$.
- **Beale's Function:**
  $$f(x,y) = (1.5 - x + xy)^2 + \left(2.25 - x + xy^2\right)^2 + \left(2.625 - x + xy^3\right)^2$$
  The minimum is $f(3,0.5) = 0$, and the domain is $-4.5 < x,y < 4.5$.
- **Matyas Function:**
  $$f(x,y) = 0.26\left(x^2 + y^2\right) - 0.48xy$$
  The minimum is $f(0,0) = 0$, and the domain is
  $-10 \leq x,y \leq 10$.
- **Levi Function N.13:**
  $$f(x,y) = \sin^2(3\pi x) + (x-1)^2\left(1 + \sin^2(3\pi y)\right) + (y-1)^2\left(1 + \sin^2(2\pi y)\right)$$
  The minimum is $f(1,1) = 0$, and the domain is
  $-10 < x,y < 10$.
- **Three-Hump Camel Function:**
  $$f(x,y) = 2x^2 - 1.05x^4 + \frac{x^6}{6} + xy + y^2$$
  The minimum is $f(0,0) = 0$, and the domain is
  $-5 \leq x,y \leq 5$.
- **Easom Function:**
  $$f(x,y) = -\cos(x)\cos(y)\,e^{(-((x-\pi)^2 + (y-\pi)^2))}$$
  The minimum is $f(\pi,\pi) = -1$, and the domain is
  $-100 < x,y < 100$.
- **S. Tang:**
  $$f(x,y) = 0.5((x^4 - 16x^2 + 5x) + (y^4 - 16y^2 + 5y)0$$
  The minimum is $f(-2.9035, -2.9035) = -78.3254$, and the domain is
  $-5 < x,y < 5$.
- **Schaffer Function N2:**
  $$f(x,y) = 0.5 + (\sin^2(x^2 - y^2) - 0.5)/(1 + 0.001 * (x^x + y^y)^2)$$
  The minimum is $f(0,0) = 0$, and the domain is $-100 < x,y < 100$.

## 4.2   Meta-Heuristic Techniques

We compare the efficiency and effectiveness of QHS against three other methods, say that: naïve HS, IHS and PSO. For each optimization method and function, we conducted a cross-validation with 20 runs. Table 1 displays the parameters setup, being their values empirically chosen [18]. Notice we employed 20 agents and 15,000 iterations for all meta-heuristic techniques, and both $x$ and $y$ decision variables of the aforementioned functions were initialized within the range $[-10, 10]$.

**Table 1.** Parameter values used for all optimization techniques.

| Technique | Parameters |
|-----------|-----------|
| HS | HMCR = 0.7, PAR = 0.7 |
| IHS | HMCR = 0.7, $PAR_{min}$ = 0.1, $PAR_{max}$ = 0.9 |
| PSO | $c1 = c2 = 2.0$, $w = 0.9$ |
| QHS | HMCR = 0.7, PAR = 0.7 |

## 5    Experimental Evaluation

In this section, we present the experimental results considering the aforementioned optimization techniques in the context of benchmark functions (Sect. 5.1), as well as to fine-tune Restricted Boltzmann Machines (Sect. 5.2).

### 5.1    Benchmark Functions

As aforementioned in Sect. 4.1, we considered 10 benchmark functions to evaluate the robustness of QHS. Table 2 displays the mean values obtained by all techniques. Since we are dealing with minimization problems, the smaller the values, the better the techniques. The values in bold stand for the best techniques according to the Wilcoxon signed-rank test [22] with significance level of 0.05.

Clearly, we can observe QHS obtained the best results in 9 out 10 functions, and it has been the sole technique in 4 functions that achieved the best values. In fact, if one take into account the convergence of the techniques, one can realize HS took longer to converge, even on a convex function like Sphere, for instance. Figure 1a depicts the convergence process with respect to Sphere function. We constrained the convergence up to 600 iterations to have a better look at the beginning of the process. One can observe QHS converges at the very first iterations, while HS and IHS need a few hundred of them.

**Table 2.** Mean values obtained through cross-validation.

| Function | PSO | IHS | HS | QHS |
|----------|-----|-----|-----|-----|
| Sphere | **0.00 ± 0.00** | **0.00 ± 0.00** | **0.00 ± 0.00** | **0.00 ± 0.00** |
| Ackley's | 0.24 ± 0.77 | **0.00 ± 0.00** | **0.00 ± 0.00** | **0.00 ± 0.00** |
| Rosebrock's | 0.06 ± 1.96 | 0.01 ± 0.00 | 0.13 ± 0.24 | **0.00 ± 0.00** |
| Beale's | 0.07 ± 0.19 | 0.10 ± 0.06 | 0.09 ± 0.30 | **0.04 ± 0.05** |
| Matyas | 0.42 ± 1.07 | 0.19 ± 0.07 | 0.18 ± 0.00 | **0.00 ± 0.00** |
| Levi | 0.02 ± 0.07 | 0.09 ± 0.01 | 0.08 ± 0.04 | **0.01 ± 0.01** |
| Three Hump | 0.01 ± 0.03 | **0.00 ± 0.00** | **0.00 ± 0.00** | **0.00 ± 0.00** |
| Easom | −0.99 ± 0.36 | −0.95 ± 0.03 | −0.93 ± 0.03 | **−0.99 ± 0.01** |
| S. Tang | **−57.76 ± 2.78** | −51.06 ± 21.34 | −51.81 ± 13.45 | −50.00 ± 0.01 |
| Schaffer | 0.004 ± 0.001 | **0.00 ± 0.00** | **0.00 ± 0.00** | **0.00 ± 0.00** |

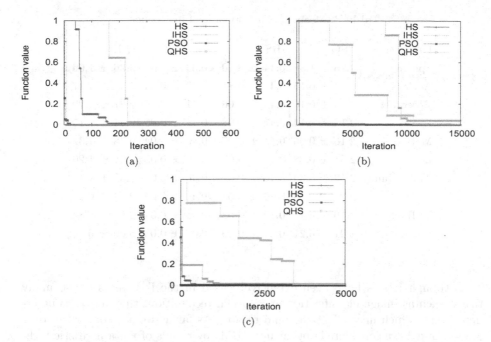

**Fig. 1.** Convergence analysis considering (a) Sphere, (b) Rosenbrock's and (c) Beale's benchmark functions.

Figure 1b displays the convergence study over Rosenbrock's function. Almost all techniques, except PSO, achieved the global minimum, being HS the fastest to converge, followed by QHS and IHS. Finally, in regard to Beale's function (Fig. 1c), QHS took longer than all techniques to converge, but it obtained the best results so far. In fact, HS-based techniques usually suffer from slow convergence when compared to swarm-based ones, since Harmony Search produces only one new solution at each iteration. We can also observe the convergence behaviour of QHS considering Rosenbrock's and Beale's functions, which follows a "stair pattern", which means QHS can get trapped from local optima (e.g., plateaus in the convergence curve), but also has the ability to escape from them.

Table 3 presents the mean computational load in seconds considering all techniques and functions. Clearly, QHS is the slowest one, since we need an extra loop to access the content of each decision variable (each variable is now represented as a 4-dimensional vector). In fact, for some datasets, QHS can be faster if we employ a different convergence criterion, since we fixed the number of iterations for all techniques in this work.

## 5.2 Fine-Tuning Restricted Boltzmann Machines

Restricted Boltzmann Machines (RBMs) are stochastic neural networks that aim at learning input representation by means of latent variables placed in a hidden layer [19]. They have become extremely used in the last years due to the "deep learning phenomenon", where we can stack RBMs on top of each other and obtain the so-called Deep Belief Networks.

**Table 3.** Mean computational load in seconds.

| Function | PSO | IHS | HS | Q-HS |
|---|---|---|---|---|
| Sphere | 2.76 ± 1.58 | 0.44 ± 0.10 | 0.39 ± 0.06 | 9.26 ± 1.06 |
| Ackley's | 2.12 ± 0.15 | 0.45 ± 0.11 | 0.39 ± 0.04 | 9.49 ± 1.17 |
| Rosebrock's | 2.14 ± 0.16 | 0.45 ± 0.06 | 0.41 ± 0.05 | 9.27 ± 1.20 |
| Beale's | 2.18 ± 0.14 | 0.47 ± 0.08 | 0.41 ± 0.05 | 9.41 ± 1.15 |
| Matyas | 2.16 ± 0.06 | 0.44 ± 0.07 | 0.40 ± 0.04 | 9.48 ± 1.32 |
| Levi | 2.51 ± 0.18 | 0.46 ± 0.076 | 0.45 ± 0.06 | 9.51 ± 1.26 |
| Three Hump | 2.18 ± 0.23 | 0.46 ± 0.09 | 0.40 ± 0.05 | 9.39 ± 1.16 |
| Easom | 2.20 ± 0.23 | 0.46 ± 0.08 | 0.40 ± 0.04 | 9.44 ± 0.91 |
| S. Tang | 2.16 ± 0.18 | 0.47 ± 0.08 | 0.41 ± 0.04 | 9.5 ± 0.78 |
| Schaffer | 2.21 ± 0.21 | 0.44 ± 0.09 | 0.40 ± 0.05 | 9.09 ± 0.70 |

Although RBMs have been extensively used in the last years for so many tasks, such as image classification and speech recognition, they are parameter-dependent, which may affect the final results. As far as we are concerned, only a few works have considered optimizing RBMs by means of meta-heuristic techniques [15–17]. In addition, we have not observed any work that attempted at using quaternion-based optimization in the context of RBMs fine-tuning.

In this section, we considered optimizing four parameters concerning the task of binary image reconstruction: $n \in [5, 100]$, $\eta \in [0.1, 0.9]$, $\lambda \in [0.1, 0.9]$ and $\varphi \in [0.00001, 0.01]$. In this case, $n$ stands for the number of hidden neurons, $\eta$ is the learning rate, $\lambda$ denotes the weight decay parameter, and $\varphi$ stands for the momentum [8]. The learning rate is used to control the convergence, and $\lambda$ and $\varphi$ are used to avoid oscillations and to keep the values of weights small enough for the learning process.

Therefore, we have a four-dimensional search space with three real-valued variables, as well as the integer-valued number of hidden units. Roughly speaking, the proposed approach aims at selecting the set of RBM parameters that minimizes the minimum squared error (MSE) of the reconstructed images from the training set. After that, the selected set of parameters is thus applied to reconstruct the images of the test set. We employed the well-known MNIST dataset[3], which is composed of images of handwritten digits. The original version contains a training set with 60,000 images from digits '0'–'9', as well as a test set with 10,000 images. Due to the computational burden for RBM model selection, we decided to employ the original test set together with a reduced version of the training set (we used 2 % of the original training set).

We conducted a cross-validation with 20 runnings, 10 iterations for the learning procedure of each RBM, and mini-batches of size 20. In addition, we also considered three learning algorithms: Contrastive Divergence (CD) [7],

---

[3] http://yann.lecun.com/exdb/mnist/.

Persistent Contrastive Divergence (PCD) [20] and Fast Persistent Contrastive Divergence (FPCD) [21]. Finally, the Wilcoxon signed-rank test [22] with significance of 0.05 was used for statistical validation purposes. Table 4 presents the results concerning RBM fine-tuning, where PSO, HS, IHS and a random search (RS) are compared against QHS.

**Table 4.** Average MSE values considering MNIST dataset.

|     | CD | PCD | FPCD |
| --- | --- | --- | --- |
| PSO | 0.1057 | 0.1058 | 0.1057 |
| HS | 0.1059 | 0.1325 | 0.1324 |
| IHS | 0.0903 | 0.0879 | **0.0882** |
| RS | 0.1105 | 0.1101 | 0.1102 |
| QHS | **0.0876** | **0.0876** | 0.0899 |

One can observe QHS obtained the best results using CD and PCD as the training algorithms, meanwhile IHS obtained the lowest errors using FPCD. As a matter of fact, FPCD is used to control possible deviations of the probability distribution of the input data, and it may require more iterations for convergence. Probably, QHS would require more iterations than IHS for some applications, as displayed in Fig. 1. However, the best results were obtained by means of CD and PCD using the proposed QHS technique.

## 6   Conclusions

In this paper, we proposed a Harmony Search approach based on quaternions, being inspired by previous works that proposed modifications of the well-known Firefly Algorithm and Bat Algorithm in the same context. The Quaternion Harmony Search was compared against naïve Harmony Search, Improved Harmony Search and Particle Swarm Optimization in 10 benchmark functions, and it has demonstrated to be able to escape from local optima as well as to convergence faster (for some datasets). However, QHS pays the price of a higher computational burden, since it maps each decision variable onto a 4-dimensional space. Therefore, given an optimization problem with $N$ variables, each possible solution can be seen as a tensor of dimensions $4 \times N$ in this space. Also, we demonstrated the validity of quaternion-based optimization in the context of RBM parameter fine-tuning.

In regard to future works, we intend to implement other variants of the Harmony Search using quaternions, as well as other swarm-based meta-heuristics, such as Particle Swarm Optimization and Cuckoo Search. Additionally, we aim at studying the behavior of QHS in high dimensional spaces, and also parallel implementations of quaternion-based optimization techniques in both CPU and GPU devices.

**Acknowledgment.** The authors would like to thank Capes PROCAD 2966/2014, CNPq grants #306166/2014-3 and #470571/2013-6, and FAPESP grants #2014/16 250-9 and #2015/00801-9.

# References

1. Boussaïd, I., Lepagnot, J., Siarry, P.: A survey on optimization metaheuristics. Inf. Sci. **237**(10), 82–117 (2013)
2. Eberly, D.: Quaternion algebra and calculus. Technical report, Magic Software (2002)
3. Fister, I., Brest Jr., J., Fister, I., Yang, X.S.: Modified bat algorithm with quaternion representation. In: IEEE Congress on Evolutionary Computation, pp. 491–498 (2015)
4. Fister, I., Yang, X.S., Breast, J., Fister Jr., I.: Modified firefly algorithm using quaternion representation. Expert Syst. Appl. **40**(18), 7220–7230 (2013)
5. Geem, Z.W.: Music-Inspired Harmony Search Algorithm: Theory and Applications, 1st edn. Springer Publishing Company, Incorporated, Berlin (2009)
6. Hamilton, W.R.: Elements of Quaternions. 2nd edn (1899)
7. Hinton, G.E.: Training products of experts by minimizing contrastive divergence. Neural Comput. **14**(8), 1771–1800 (2002)
8. Hinton, G.E., Li, D., Dong, Y., Dahl, G.E., Mohamed, A., Jaitly, N., Senior, A., Vanhoucke, V., Nguyen, P., Sainath, T.N., Kingsbury, B.: Deep neural networks for acoustic modeling in speech recognition: the shared views of four research groups. IEEE Sig. Process. Mag. **29**(6), 82–97 (2012)
9. Humeau, J., Liefooghe, A., Talbi, E.G., Verel, S.: ParadisEO-MO: from fitness landscape analysis to efficient local search algorithms. J. Heuristics **19**(6), 881–915 (2013)
10. Mahdavi, M., Fesanghary, M., Damangir, E.: An improved harmony search algorithm for solving optimization problems. Appl. Math. Comput. **188**(2), 1567–1579 (2007)
11. Malan, K., Engelbrecht, A.: Particle swarm optimisation failure prediction based on fitness landscape characteristics. In: IEEE Symposium on Swarm Intelligence, pp. 1–9 (2014)
12. Merz, P., Freisleben, B.: Fitness landscape analysis and memetic algorithms for the quadratic assignment problem. IEEE Trans. Evol. Comput. **4**(4), 337–352 (2000)
13. Omran, M.G., Mahdavi, M.: Global-best harmony search. Appl. Math. Comput. **198**(2), 643–656 (2008)
14. Pan, Q.K., Suganthan, P., Tasgetiren, M.F., Liang, J.: A self-adaptive global best harmony search algorithm for continuous optimization problems. Appl. Math. Comput. **216**(3), 830–848 (2010)
15. Papa, J.P., Rosa, G.H., Costa, K.A.P., Marana, A.N., Scheirer, W., Cox, D.D.: On the model selection of Bernoulli restricted Boltzmann machines through harmony search. In: Proceedings of the Genetic and Evolutionary Computation Conference, GECCO 2015, pp. 1449–1450. ACM, New York (2015)
16. Papa, J.P., Rosa, G.H., Marana, A.N., Scheirer, W., Cox, D.D.: Model selection for discriminative restricted Boltzmann machines through meta-heuristic techniques. J. Comput. Sci. **9**, 14–18 (2015)
17. Papa, J.P., Scheirer, W., Cox, D.D.: Fine-tuning deep belief networks using harmony search. Appl. Soft Comput. **46**, 875–885 (2015)

18. Pereira, D.R., Delpiano, J., Papa, J.P.: On the optical flow model selection through metaheuristics. EURASIP J. Image Video Process. **2015**, 11 (2015). http://dx.doi.org/10.1186/s13640-015-0066-5
19. Smolensky, P.: Information processing in dynamical systems: foundations of harmony theory. In: Parallel Distributed Processing: Explorations in the Microstructure of Cognition, vol. 1, pp. 194–281. MIT Press, Cambridge (1986)
20. Tieleman, T.: Training restricted Boltzmann machines using approximations to the likelihood gradient. In: Proceedings of the 25th International Conference on Machine Learning, pp. 1064–1071. ACM, New York (2008)
21. Tieleman, T., Hinton, G.E.: Using fast weights to improve persistent contrastive divergence. In: Proceedings of the 26th Annual International Conference on Machine Learning, pp. 1033–1040. ACM, New York (2009)
22. Wilcoxon, F.: Individual comparisons by ranking methods. Biom. Bull. **1**(6), 80–83 (1945)
23. Yang, X.S.: Nature-Inspired Metaheuristic Algorithms, 2nd edn. Luniver Press, Beckington (2010)

# Learning Parameters in Deep Belief Networks Through Firefly Algorithm

Gustavo Rosa[1], João Papa[1(✉)], Kelton Costa[1], Leandro Passos[2],
Clayton Pereira[2], and Xin-She Yang[3]

[1] Department of Computing, São Paulo State University, São Paulo, Brazil
gth.rosa@uol.com.br, papa@fc.unesp.br, kelton.costa@gmail.com
[2] Department of Computing, Federal University of São Carlos, São Carlos, Brazil
leandropassosjr@gmail.com, claytontey@gmail.com
[3] School of Science and Technology, Middlesex University, London, UK
x.yang@mdx.ac.uk

**Abstract.** Restricted Boltzmann Machines (RBMs) are among the most widely pursed techniques in the context of deep learning-based applications. Their usage enables sundry parallel implementations, which have become pivotal in nowadays large-scale-oriented applications. In this paper, we propose to address the main shortcoming of such models, i.e. how to properly fine-tune their parameters, by means of the Firefly Algorithm, as well as we also consider Deep Belief Networks, a stacked-driven version of the RBMs. Additionally, we also take into account Harmony Search, Improved Harmony Search and the well-known Particle Swarm Optimization for comparison purposes. The results obtained showed the Firefly Algorithm is suitable to the context addressed in this paper, since it obtained the best results in all datasets.

**Keywords:** Deep Belief Networks · Deep learning · Firefly algorithm

## 1 Introduction

Even today, there are still some open computer vision-related problems concerning on how to create and produce good representations of the real world, such as machine learning systems that can detect and further classify objects [4]. These techniques have been paramount during the last years, since there is an increasing number of applications that require intelligent-based decision-making processes. An attractive skill of pattern recognition techniques related to deep learning has drawn a considerable amount of interest in the last years [3], since their outstanding results have settled a hallmark for several applications, such as speech, face and emotion recognition, among others.

Roughly speaking, deep learning algorithms are shaped by means of several layers of a predefined set of operations. Restricted Boltzmann Machines (RBMs), for instance, have attracted considerable focus in the last years due to their simplicity, high level of parallelism and strong representation ability [7]. RBMs can

© Springer International Publishing AG 2016
F. Schwenker et al. (Eds.): ANNPR 2016, LNAI 9896, pp. 138–149, 2016.
DOI: 10.1007/978-3-319-46182-3_12

be interpreted as stochastic neural networks, being mainly used for image reconstruction and collaborative filtering through unsupervised learning [2]. Later on, Hinton et al. [8] realized one can obtain more complex representations by stacking a few RBMs on top of each other, thus leading to the so-called Deep Belief Networks (DBNs).

One of the main loopholes of DBNs concerns with the proper calibration of their parameters. The task of fine-tuning parameters in machine learning aims at finding suitable values for that parameters in order to maximize some fitness function, such as a classifier's recognition accuracy when dealing with supervised problems, for instance. Auto-learning tools are often used to handle this problem [16], which usually combine parameter fine-tuning with feature selection techniques. On the other hand, meta-heuristic techniques are among the most used ones for optimization problems, since they provide simple and elegant solutions in a wide range of applications. Such techniques may comprehend swarm- and population-based algorithms, as well as stochastic and nature-inspired solutions.

Some years ago, Yang [20] proposed the Firefly Algorithm (FFA), which tumbles in the pitch of meta-heuristic optimization techniques. Elementarily, the idea of the Firefly Algorithm is to solve optimization problems based on the way fireflies communicate with themselves. In other words, the idea is to model the process in which fireflies attract their mates using their flashing lights, which are produced by bioluminescence processes. The rate of flashing, amount of time and the rhythmic flash are part of their signal system, helping its association with some fitness function in order to optimize it.

However, the reader may find just a few recent works that face the problem of calibrating the parameters of both RBMs and DBNs by means of meta-heuristic techniques. Huang et al. [9], for instance, employed the well-known Particle Swarm Optimization (PSO) to optimize the number of input (visible) and hidden neurons, as well as the RBM learning rate in the context of time series forecasting prediction. Later on, Liu et al. [10] applied Genetic Algorithm to optimize the parameters of Deep Boltzmann Machines. Papa et al. [13] proposed to optimize RBMs by means of Harmony Search, and Papa et al. [14] fine-tuned the parameters of a Discriminative Restricted Boltzmann Machines. Very recently, Papa et al. [15] proposed to optimize DBNs by means of Harmony Search-based techniques.

As far as we know, FFA has never been applied for DBN parameter calibration. Therefore, the main contribution of this paper is twofold: (i) to introduce FFA to the context of DBN parameter calibration for binary image reconstruction purposes, and (ii) to fill the lack of research regarding DBN parameter optimization by means of meta-heuristic techniques. The remainder of this paper is organized as follows. Sections 2 and 3 present some theoretical background with respect to DBN and FFA, respectively. The methodology is discussed in Sect. 4, while Sect. 5 presents the experimental results. Finally, Sect. 6 states conclusions and future works.

## 2  Deep Belief Networks

In this section, we describe the main concepts related to Deep Belief Networks, but with a special attention to the theoretical background of RBMs, which are the basis for DBN understanding.

### 2.1  Restricted Boltzmann Machines

Restricted Boltzmann Machines are energy-based stochastic neural networks composed by two layers of neurons (visible and hidden), in which the learning phase is conducted by means of an unsupervised fashion. The RBM is similar to the classical Boltzmann Machine [1], except that no connections between neurons of the same layer are allowed. The architecture of a Restricted Boltzmann Machine is comprised by a visible layer $\mathbf{v}$ with $m$ units and a hidden layer $\mathbf{h}$ with $n$ units. The real-valued $m \times n$ matrix $\mathbf{W}$ models the weights between visible and hidden neurons, where $w_{ij}$ stands for the weight between the visible unit $v_i$ and the hidden unit $h_j$.

Let us assume $\mathbf{v}$ and $\mathbf{h}$ as the binary visible and hidden units, respectively. In other words, $\mathbf{v} \in \{0,1\}^m$ and $\mathbf{h} \in \{0,1\}^n$. The energy function of a Bernoulli Restricted Boltzmann Machine is given by:

$$E(\mathbf{v}, \mathbf{h}) = -\sum_{i=1}^{m} a_i v_i - \sum_{j=1}^{n} b_j h_j - \sum_{i=1}^{m} \sum_{j=1}^{n} v_i h_j w_{ij}, \tag{1}$$

where $\mathbf{a}$ and $\mathbf{b}$ stand for the biases of visible and hidden units, respectively. The probability of a configuration $(\mathbf{v}, \mathbf{h})$ is computed as follows:

$$P(\mathbf{v}, \mathbf{h}) = \frac{e^{-E(\mathbf{v},\mathbf{h})}}{\sum_{\mathbf{v},\mathbf{h}} e^{-E(\mathbf{v},\mathbf{h})}}, \tag{2}$$

where the denominator of above equation is a normalization factor that stands for all possible configurations involving the visible and hidden units. In short, the RBM learning algorithm aims at estimating $\mathbf{W}$, $\mathbf{a}$ and $\mathbf{b}$. The next section describes in more details this procedure.

### 2.2  Learning Algorithm

The parameters of an RBM can be optimized by performing stochastic gradient ascent on the log-likelihood of training patterns. Given a training sample (visible unit), its probability is computed over all possible hidden units, as follows:

$$P(\mathbf{v}) = \frac{\sum_{\mathbf{h}} e^{-E(\mathbf{v},\mathbf{h})}}{\sum_{\mathbf{v},\mathbf{h}} e^{-E(\mathbf{v},\mathbf{h})}}. \tag{3}$$

In order to update the weights and biases, it is necessary to compute the following derivatives:

$$\frac{\partial \log P(\mathbf{v})}{\partial w_{ij}} = E[h_j v_i]^{data} - E[h_j v_i]^{model}, \tag{4}$$

$$\frac{\partial \log P(\mathbf{v})}{\partial a_i} = v_i - E[v_i]^{model}, \tag{5}$$

$$\frac{\partial \log P(\mathbf{v})}{\partial b_j} = E[h_j]^{data} - E[h_j]^{model}, \tag{6}$$

where $E[\cdot]$ stands for the expectation operation, and $E[\cdot]^{data}$ and $E[\cdot]^{model}$ correspond to the data-driven and the reconstructed-data-driven probabilities, respectively.

In practical terms, we can compute $E[h_j v_i]^{data}$ considering $\mathbf{h}$ and $\mathbf{v}$ as follows:

$$E[\mathbf{hv}]^{data} = P(\mathbf{h}|\mathbf{v})\mathbf{v}^T, \tag{7}$$

where $P(\mathbf{h}|\mathbf{v})$ stands for the probability of obtaining $\mathbf{h}$ given the visible vector (training data) $\mathbf{v}$:

$$P(h_j = 1|\mathbf{v}) = \sigma \left( \sum_{i=1}^{m} w_{ij} v_i + b_j \right), \tag{8}$$

where $\sigma(\cdot)$ stands for the logistic sigmoid function. Therefore, it is straightforward to compute $E[\mathbf{hv}]^{data}$: given a training data $\mathbf{x} \in \mathcal{X}$, where $\mathcal{X}$ stands for a training set, we just need to set $\mathbf{v} \leftarrow \mathbf{x}$ and then employ Eq. 8 to obtain $P(\mathbf{h}|\mathbf{v})$. Further, we use Eq. 7 to finally obtain $E[\mathbf{hv}]^{data}$.

The next question now is how to obtain $E[\mathbf{hv}]^{model}$, which is the model learned by the system. One possible strategy is to perform alternating Gibbs sampling starting at any random state of the visible units until a certain convergence criterion, such as $k$ steps, for instance. The Gibbs sampling consists of updating hidden units using Eq. 8 followed by updating the visible units using $P(\mathbf{v}|\mathbf{h})$, given by:

$$P(v_i = 1|\mathbf{h}) = \sigma \left( \sum_{j=1}^{n} w_{ij} h_j + a_i \right), \tag{9}$$

and then updating the hidden units once again using Eq. 8. In short, it is possible to obtain an estimative of $E[\mathbf{hv}]^{model}$ by initializing the visible unit with random values and then performing Gibbs sampling, in which $E[\mathbf{hv}]^{model}$ can be approximated after $k$ iterations. Notice a single iteration is defined by computing $P(\mathbf{h}|\mathbf{v})$, followed by computing $P(\mathbf{v}|\mathbf{h})$ and then computing $P(\mathbf{h}|\mathbf{v})$ once again.

For the sake of explanation, the Gibbs sampling employs $P(\mathbf{v}|\tilde{\mathbf{h}})$ instead of $P(\mathbf{v}|\mathbf{h})$, and $P(\tilde{\mathbf{h}}|\tilde{\mathbf{v}})$ instead of $P(\mathbf{h}|\mathbf{v})$. Essentially, they stand for the same meaning, but $P(\mathbf{v}|\mathbf{h})$ is used here to denote the visible unit $\mathbf{v}$ is going to be reconstructed using $\tilde{\mathbf{h}}$, which was obtained through $P(\mathbf{h}|\mathbf{v})$. The same takes place with $P(\tilde{\mathbf{h}}|\tilde{\mathbf{v}})$, that reconstructs $\tilde{\mathbf{h}}$ using $\tilde{\mathbf{v}}$, which was obtained through $P(\mathbf{v}|\tilde{\mathbf{h}})$.

However, the procedure depicted above is time-consuming, being also quite hard to establish suitable values for $k$. Fortunately, Hinton [6] introduced a faster methodology to compute $E[\mathbf{hv}]^{model}$ based on contrastive divergence. Basically, the idea is to initialize the visible units with a training sample, to compute the states of the hidden units using Eq. 8, and then to compute the states of the visible unit (reconstruction step) using Eq. 9. Roughly speaking, this is equivalent to perform Gibbs sampling using $k = 1$.

Based on the above assumption, we can now compute $E[\mathbf{hv}]^{model}$ as follows:

$$E[\mathbf{hv}]^{model} = P(\tilde{\mathbf{h}}|\tilde{\mathbf{v}})\tilde{\mathbf{v}}^T. \tag{10}$$

Therefore, the equation below leads to a simple learning rule for updating the weight matrix $\mathbf{W}$, as follows:

$$\begin{aligned}
\mathbf{W}^{t+1} &= \mathbf{W}^t + \eta(E[\mathbf{hv}]^{data} - E[\mathbf{hv}]^{model}) \\
&= \mathbf{W}^t + \eta(P(\mathbf{h}|\mathbf{v})\mathbf{v}^T - P(\tilde{\mathbf{h}}|\tilde{\mathbf{v}})\tilde{\mathbf{v}}^T),
\end{aligned} \tag{11}$$

where $\mathbf{W}^t$ stands for the weight matrix at time step $t$, and $\eta$ corresponds to the learning rate. Additionally, we have the following formulae to update the biases of the visible and hidden units:

$$\begin{aligned}
\mathbf{a}^{t+1} &= \mathbf{a}^t + \eta(\mathbf{v} - E[\mathbf{v}]^{model}) \\
&= \mathbf{a}^t + \eta(\mathbf{v} - \tilde{\mathbf{v}}),
\end{aligned} \tag{12}$$

and

$$\begin{aligned}
\mathbf{b}^{t+1} &= \mathbf{b}^t + \eta(E[\mathbf{h}]^{data} - E[\mathbf{h}]^{model}) \\
&= \mathbf{b}^t + \eta(P(\mathbf{h}|\mathbf{v}) - P(\tilde{\mathbf{h}}|\tilde{\mathbf{v}})),
\end{aligned} \tag{13}$$

where $\mathbf{a}^t$ and $\mathbf{b}^t$ stand for the visible and hidden units biases at time step $t$, respectively. In short, Eqs. 11, 12 and 13 are the vanilla formulation for updating the RBM parameters.

Later on, Hinton [7] introduced a weight decay parameter $\lambda$, which penalizes weights with large magnitude, as well as a momentum parameter $\alpha$ to control possible oscillations during the learning process. Therefore, we can rewrite Eqs. 11, 12 and 13 as follows:

$$\mathbf{W}^{t+1} = \mathbf{W}^t + \underbrace{\eta(P(\mathbf{h}|\mathbf{v})\mathbf{v}^T - P(\tilde{\mathbf{h}}|\tilde{\mathbf{v}})\tilde{\mathbf{v}}^T) - \lambda\mathbf{W}^t + \alpha\Delta\mathbf{W}^{t-1}}_{=\Delta\mathbf{W}^t}, \tag{14}$$

$$\mathbf{a}^{t+1} = \mathbf{a}^t + \underbrace{\eta(\mathbf{v} - \tilde{\mathbf{v}}) + \alpha\Delta\mathbf{a}^{t-1}}_{=\Delta\mathbf{a}^t} \tag{15}$$

and

$$\mathbf{b}^{t+1} = \mathbf{b}^t + \underbrace{\eta(P(\mathbf{h}|\mathbf{v}) - P(\tilde{\mathbf{h}}|\tilde{\mathbf{v}})) + \alpha\Delta\mathbf{b}^{t-1}}_{=\Delta\mathbf{b}^t}. \tag{16}$$

## 2.3  Stacked Restricted Boltzmann Machines

Truly speaking, DBNs are composed of a set of stacked RBMs, being each of them trained using the learning algorithm presented in Sect. 2.2 in a greedy fashion, which means an RBM at a certain layer does not consider others during its learning procedure. In this case, we have a DBN composed of $L$ layers, being $\mathbf{W}^i$ the weight matrix of RBM at layer $i$. Additionally, we can observe the hidden units at layer $i$ become the input units to the layer $i + 1$.

The approach proposed by Hinton et al. [8] for the training step of DBNs also considers a fine-tuning as a final step after the training of each RBM. Such procedure can be performed by means of a Backpropagation or Gradient descent algorithm, for instance, in order to adjust the matrices $\mathbf{W}^i$, $i = 1, 2, \ldots, L$. The optimization algorithm aims at minimizing some error measure considering the output of an additional layer placed on the top of the DBN after its former greedy training. Such layer is often composed of softmax or logistic units, or even some supervised pattern recognition technique.

# 3  Firefly Algorithm

The Firefly Algorithm is derived from the flash attractiveness of fireflies when mating partners (communication) and attracting potential preys [20]. Let $\mathbf{X}_{M \times N}$ be a firefly swarm, where $N$ is the number of variables to be optimized and $M$ stands for the number of fireflies. Also, each firefly $i$ is associated to a position $\boldsymbol{x}_i \in \Re^N$, $i = 1, 2, \ldots, M$, and its brightness is determined by the value of the objective function at that position, i.e. $f(\boldsymbol{x}_i)$.

The attractiveness of a firefly $\boldsymbol{x}_i$ with respect to another firefly $\boldsymbol{x}_j$, i.e. $\beta(\boldsymbol{x}_i, \boldsymbol{x}_j)$, varies with the distance $r_{ij}$ (i.e. Euclidean distance) between firefly $i$ and firefly $j$, which is modeled as follows:

$$\beta(\boldsymbol{x}_i, \boldsymbol{x}_j) = \beta_0 \exp(-\gamma * r_{i,j}^2), \tag{17}$$

where $\beta_0$ is the initial attractiveness and $\gamma$ the light absorption coefficient.

Each firefly at position $\boldsymbol{x}_i$ and time step $t$ is attracted towards every other brighter firefly at position $\boldsymbol{x}_j$ (i.e. $f(\boldsymbol{x}_j) > f(\boldsymbol{x}_i)$) as follows:

$$\boldsymbol{x}_i^{t+1} = \boldsymbol{x}_i^t + \beta(\boldsymbol{x}_i^t, \boldsymbol{x}_j^t)(\boldsymbol{x}_j^t - \boldsymbol{x}_i^t) + \alpha\left(\mu\frac{1}{2}\right), \tag{18}$$

where $\alpha$ is a randomization factor that controls the magnitude of the stochastic perturbation of $\boldsymbol{x}_i^t$, and $\mu \in \mathcal{U}(0, 1)$. Soon after all fireflies at time step $t$ had been modified, the best firefly $\boldsymbol{x}^*$ will perform a controlled random walk across the search space:

$$\boldsymbol{x}^* = \boldsymbol{x}_i^* + \alpha \left( \psi \frac{1}{2} \right), \tag{19}$$

where $\psi \in \mathcal{U}(0,1)$.

## 4    Methodology

In this section, we present the proposed approach for DBN parameter calibration, as well as we describe the employed datasets and the experimental setup.

### 4.1    Modelling DBN Parameter Optimization

We propose to model the problem of selecting suitable parameters considering DBN in the task of binary image reconstruction. As aforementioned in Sect. 2.2, the learning step has four parameters: the learning rate $\eta$, weight decay $\lambda$, penalty parameter $\alpha$, and the number of hidden units $n$. Therefore, we have a four-dimensional search space with three real-valued variables, as well as the integer-valued number of hidden units. Roughly speaking, the proposed approach aims at selecting the set of DBN parameters that minimizes the minimum squared error (MSE) of the reconstructed images from the training set. After that, the selected set of parameters is thus applied to reconstruct the images of the test set.

### 4.2    Datasets

We employed three datasets, as described below:

- MNIST dataset[1]: it is composed of images of handwritten digits. The original version contains a training set with $60,000$ images from digits '0'–'9', as well as a test set with $10,000$ images[2]. Due to the high computational burden for DBN model selection, we decided to employ the original test set together with a reduced version of the training set[3].
- CalTech 101 Silhouettes Data Set[4]: it is based on the former Caltech 101 dataset, and it comprises silhouettes of images from 101 classes with resolution of $28 \times 28$. We have used only the training and test sets, since our optimization model aims at minimizing the MSE error over the training set.

---

[1] http://yann.lecun.com/exdb/mnist/.
[2] The images are originally available in grayscale with resolution of $28 \times 28$.
[3] The original training set was reduced to 2 % of its former size, which corresponds to $1,200$ images.
[4] https://people.cs.umass.edu/~marlin/data.shtml.

- Semeion Handwritten Digit Data Set[5]: it is formed by 1,593 images from handwritten digits '0' –'9' written in two ways: the first time in a normal way (accurately) and the second time in a fast way (no accuracy). In the end, they were stretched with resolution of $16 \times 16$ in a grayscale of 256 values and then each pixel was binarized.

Figure 1 displays some training examples from the above datasets.

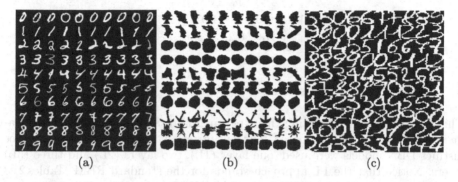

(a)                              (b)                              (c)

**Fig. 1.** Some training examples from (a) MNIST, (b) CalTech 101 Silhouettes and (c) Semeion datasets.

## 4.3  Experimental Setup

In this work, we compared the Firefly Algorithm against with the well-known PSO, Harmony Search [5] and its variant known as Improved Harmony Search (IHS) [11], which are meta-heuristic algorithms inspired in the improvisation process of music players.

In order to provide a statistical analysis by means of Wilcoxon signed-rank test [19], we conducted a cross-validation with 20 runnings. We employed 5 agents over 50 iterations for convergence considering all techniques. Table 1 presents the parameter configuration for each optimization technique[6].

Finally, we have set each DBN parameter according to the following ranges: $n \in [5, 100]$, $\eta \in [0.1, 0.9]$, $\lambda \in [0.1, 0.9]$ and $\alpha \in [0.00001, 0.01]$. Therefore, this means we have used such ranges to initialize the optimization techniques, as well as to conduct the baseline experiment by means of randomly set values. We also have employed $T = 10$ as the number of epochs for DBN learning weights procedure with mini-batches of size 20. In order to provide a more precise experimental validation, all DBNs were trained with three different algorithms[7]: Contrastive Divergence (CD) [6], Persistent Contrastive Divergence (PCD) [17] and Fast Persistent Contrastive Divergence (FPCD) [18].

---

[5] https://archive.ics.uci.edu/ml/datasets/Semeion+Handwritten+Digit.
[6] Notice these values have been empirically setup.
[7] One sampling iteration was used for all learning algorithms.

**Table 1.** Parameter configuration.

| Technique | Parameters |
|---|---|
| PSO | $c_1 = 1.7$, $c_2 = 1.7$, $w = 0.7$ |
| HS | $HMCR = 0.7$, $PAR = 0.7$, $\varrho = 10$ |
| IHS | $HMCR = 0.7$, $PAR_{MIN} = 0.1$ <br> $PAR_{MAX} = 0.7$, $\varrho_{MIN} = 1$ <br> $\varrho_{MAX} = 10$ |
| FFA | $\gamma = 1.0$, $\beta_0 = 1.0$ <br> $\alpha = 0.2$ |

## 5   Experimental Results

This section aims at presenting the experimental results concerning DBN parameter calibration. We compared four optimization methods, as well as three distinct DBN models were used: one layer (1L), two layers (2L) and three (3L) layers. Notice that the 1 L approach stands for the standard RBM. Tables 2, 3 and 4 present the average MSE results for MNIST, Caltech 101 Silhouettes and Semeion Handwritten Digit datasets, respectively. The most accurate results are in bold.

Clearly, one can observe FFA obtained the best results for all datasets, and it has been the sole technique in Caltech 101 Silhouettes and Semeion Handwritten datasets that obtained the most accurate result. In fact, these are the best results concerning the aforementioned datasets so far, since the results with HS and IHS were the very same ones reported by Papa et al. [15]. Interestingly, except for MNIST dataset, the best results were obtained by means of a CD-driven learning and one layer only.

One can realize the larger errors for Caltech 101 Silhouettes and Semeion Handwritten Digit datasets, since they pose a greater challenge, with more complex images, as well as with a greater diversity. Curiously, the other techniques seemed to obtain better results over these datasets by using more layers, which is somehow expected, since one can use models capable of describing better the images. However, such point can not be observed through FFA, which reported the best results with one single layer. Probably, FFA could obtain even better

**Table 2.** Average MSE over the test set considering MNIST dataset.

|  | 1L | | | 2L | | | 3L | | |
|---|---|---|---|---|---|---|---|---|---|
|  | CD | PCD | FPCD | CD | PCD | FPCD | CD | PCD | FPCD |
| PSO | 0.1057 | 0.1058 | 0.1057 | 0.1060 | 0.1059 | 0.1058 | 0.1058 | 0.1059 | 0.1058 |
| HS | 0.1059 | 0.1325 | 0.1324 | 0.1059 | 0.1061 | 0.1057 | 0.1059 | 0.1058 | 0.1057 |
| IHS | 0.0903 | **0.0879** | **0.0882** | 0.0885 | 0.0886 | 0.0886 | 0.0887 | 0.0885 | 0.0886 |
| FFA | **0.0876** | **0.0876** | **0.0882** | **0.0876** | **0.0876** | 0.0886 | **0.0876** | **0.0876** | 0.0885 |

**Table 3.** Average MSE over the test set considering Caltech 101 Silhouettes dataset.

|  | 1L | | | 2L | | | 3L | | |
|---|---|---|---|---|---|---|---|---|---|
|  | CD | PCD | FPCD | CD | PCD | FPCD | CD | PCD | FPCD |
| PSO | 0.1691 | 0.1690 | 0.1689 | 0.1689 | 0.1691 | 0.1688 | 0.1692 | 0.1692 | 0.1690 |
| HS | 0.1695 | 0.1696 | 0.1691 | 0.1695 | 0.1699 | 0.1693 | 0.1694 | 0.1696 | 0.1692 |
| IHS | 0.1696 | 0.1695 | 0.1693 | 0.1609 | 0.1607 | 0.1612 | 0.1611 | 0.1618 | 0.1606 |
| FFA | **0.1589** | 0.1598 | 0.1616 | 0.1606 | 0.1606 | 0.1635 | 0.1606 | 0.1606 | 0.1625 |

**Table 4.** Average MSE over the test set considering Semeion Handwritten Digit dataset.

|  | 1L | | | 2L | | | 3L | | |
|---|---|---|---|---|---|---|---|---|---|
|  | CD | PCD | FPCD | CD | PCD | FPCD | CD | PCD | FPCD |
| PSO | 0.2128 | 0.2128 | 0.2128 | 0.2128 | 0.2128 | 0.2128 | 0.2128 | 0.2128 | 0.2127 |
| HS | 0.2128 | 0.2128 | 0.2129 | 0.2202 | 0.2128 | 0.2128 | 0.2199 | 0.2128 | 0.2128 |
| IHS | 0.2131 | 0.2130 | 0.2128 | 0.2116 | 0.2114 | 0.2121 | 0.2103 | 0.2109 | 0.2119 |
| FFA | **0.2068** | 0.2078 | 0.2103 | 0.2097 | 0.2096 | 0.2125 | 0.2097 | 0.2096 | 0.2115 |

results if one allowed more iterations for convergence using 2 and 3 layers, for instance.

Although HS-based techniques are not swarm-oriented, which means they do not update all particles at a single iteration, they obtained very good results. Actually, it is expected that techniques based on swarm intelligence can obtain better results, since they interchange information among all possible solutions in the search space. On the other hand, HS-based techniques create only one new solution per time, but based on all possible solutions from the current iteration.

Previous works have highlighted IHS as one of the best techniques to learn parameters in DBNs and RBMs [13–15], but we have showed new and more accurate results obtained through FFA. As a matter of fact, Pal et al. [12] showed FFA is usually better than PSO when we have non-linear and noisy functions to be optimized, which seems to be the case addressed in this work.

One shortcoming of FFA concerns with its execution time, which is similar to PSO, since both are swarm-driven, but it is slower than HS and IHS. These latter techniques do not update all agents at each iteration, but they create a single new solution instead, which means they execute the fitness function only once per iteration. Such behaviour makes them much faster than swarm-based techniques, but having a slower convergence as well.

## 6   Conclusions

In this paper, we dealt with the problem of calibrating DBN parameters by means of the Firefly Algorithm. The experiments were carried out over three public datasets in the context of binary image reconstruction using DBNs with 1, 2 and 3 layers. Additionally, we also considered PSO, HS and IHS for comparison purposes.

The experiments have shown FFA obtained the best results for all datasets so far, as well as FFA required much less layers than the compared optimization techniques, which means it required less computational effort, since a single layer allowed to achieve the best results. We also believe one can obtain better results with FFA using more layers, but at the price of using more iterations for convergence. In regard to future works, we intend to investigate whether more iterations will improve FFA or not, as well as to employ other meta-heuristic-based techniques to fine-tune DBNs.

**Acknowledgment.** The authors would like to thank FAPESP grants #2014/24491-6, #2014/16250-9 and #2015/00801-9, and CNPq grants #303182/2011-3, #470571/2013-6 and #306166/2014-3.

# References

1. Ackley, D.H., Hinton, G.E., Sejnowski, T.J.: A learning algorithm for Boltzmann machines. In: Waltz, D., Feldman, J. (eds.) Connectionist Models and Their Implications: Readings from Cognitive Science, pp. 285–307. Ablex Publishing Corporation, Norwood (1988)
2. Bengio, Y.: Learning deep architectures for AI. Found. Trends Mach. Learn. **2**(1), 1–127 (2009)
3. Bengio, Y., Courville, A., Vincent, P.: Representation learning: a review and new perspectives. IEEE Trans. Pattern Anal. Mach. Intell. **35**(8), 1798–1828 (2013)
4. Bishop, C.: Neural Networks for Pattern Recognition. Oxford University Press, Oxford (1995)
5. Geem, Z.W.: Music-Inspired Harmony Search Algorithm: Theory and Applications, 1st edn. Springer Publishing Company, Incorporated, Berlin (2009)
6. Hinton, G.E.: Training products of experts by minimizing contrastive divergence. Neural Comput. **14**(8), 1771–1800 (2002)
7. Hinton, G.E.: A practical guide to training restricted Boltzmann machines. In: Montavon, G., Orr, G.B., Müller, K.-R. (eds.) Neural Networks: Tricks of the Trade. LNCS, vol. 7700, 2nd edn, pp. 599–619. Springer, Heidelberg (2012)
8. Hinton, G.E., Osindero, S., Teh, Y.W.: A fast learning algorithm for deep belief nets. Neural Comput. **18**(7), 1527–1554 (2006)
9. Kuremoto, T., Kimura, S., Kobayashi, K., Obayashi, M.: Time series forecasting using restricted Boltzmann machine. In: Huang, D.-S., Gupta, P., Zhang, X., Premaratne, P. (eds.) ICIC 2012. CCIS, vol. 304, pp. 17–22. Springer, Heidelberg (2012)
10. Liu, K., Zhang, L., Sun, Y.: Deep Boltzmann machines aided design based on genetic algorithms. In: Yarlagadda, P., Kim, Y.H. (eds.) Artificial Intelligence, Optimization Algorithms and Computational Mathematics. Applied Mechanics and Materials, pp. 848–851. Scientific.Net, Singapore (2014)
11. Mahdavi, M., Fesanghary, M., Damangir, E.: An improved harmony search algorithm for solving optimization problems. Appl. Math. Comput. **188**(2), 1567–1579 (2007)

12. Pal, S.K., Rai, C.S., Singh, A.P.: Comparative study of firefly algorithm and particle swarm optimization for noisy non-linear optimization problems. Int. J. Intell. Syst. Appl. **10**, 50–57 (2012)
13. Papa, J.P., Rosa, G.H., Costa, K.A.P., Marana, A.N., Scheirer, W., Cox, D.D.: On the model selection of Bernoulli restricted Boltzmann machines through harmony search. In: Proceedings of the Genetic and Evolutionary Computation Conference, pp. 1449–1450. ACM, New York (2015)
14. Papa, J.P., Rosa, G.H., Marana, A.N., Scheirer, W., Cox, D.D.: Model selection for discriminative restricted Boltzmann machines through meta-heuristic techniques. J. Comput. Sci. **9**, 14–18 (2015)
15. Papa, J.P., Scheirer, W., Cox, D.D.: Fine-tuning deep belief networks using harmony search. Appl. Soft Comput. **46**, 875–885 (2015)
16. Thornton, C., Hutter, F., Hoos, H.H., Leyton-Brown, K.: Auto-WEKA: combined selection and hyperparameter optimization of classification algorithms. In: Proceedings of the Proceedings of the 19th ACM SIGKDD International Conference on Knowledge Discovery, KDD 2013, pp. 847–855. ACM, New York (2013)
17. Tieleman, T.: Training restricted Boltzmann machines using approximations to the likelihood gradient. In: Proceedings of the 25th International Conference on Machine Learning, pp. 1064–1071. ACM, New York (2008)
18. Tieleman, T., Hinton, G.E.: Using fast weights to improve persistent contrastive divergence. In: Proceedings of the 26th Annual International Conference on Machine Learning, pp. 1033–1040. ACM, New York (2009)
19. Wilcoxon, F.: Individual comparisons by ranking methods. Biom. Bull. **1**(6), 80–83 (1945)
20. Yang, X.S.: Firefly algorithm, stochastic test functions and design optimisation. Int. J. Bio-Inspir. Comput. **2**(2), 78–84 (2010)

# Towards Effective Classification of Imbalanced Data with Convolutional Neural Networks

Vidwath Raj[(✉)], Sven Magg, and Stefan Wermter

Knowledge Technology, Department of Informatics, University of Hamburg,
Hamburg, Germany
vidwath0011@gmail.com

**Abstract.** Class imbalance in machine learning is a problem often found with real-world data, where data from one class clearly dominates the dataset. Most neural network classifiers fail to learn to classify such datasets correctly if class-to-class separability is poor due to a strong bias towards the majority class. In this paper we present an algorithmic solution, integrating different methods into a novel approach using a class-to-class separability score, to increase performance on poorly separable, imbalanced datasets using Cost Sensitive Neural Networks. We compare different cost functions and methods that can be used for training Convolutional Neural Networks on a highly imbalanced dataset of multi-channel time series data. Results show that, despite being imbalanced and poorly separable, performance metrics such as G-Mean as high as 92.8 % could be reached by using cost sensitive Convolutional Neural Networks to detect patterns and correctly classify time series from 3 different datasets.

## 1 Introduction

In supervised classification tasks, effective learning happens when there are sufficient examples for all the classes and class-to-class (C2C) separability is sufficiently large. However, real world datasets are often imbalanced and have poor C2C separability. A dataset is said to be imbalanced when a certain class is overrepresented compared to other classes in that dataset. In binary classification tasks, the class with too many examples is often referred to as the *majority class*, the other as the *minority class* respectively. Machine Learning algorithms performing classification on such datasets face the so-called '*class imbalance problem*', where learning is not as effective as it is with a balanced dataset [6,10,13], since it poses a bias in learning towards the majority class.

On the one hand, many of the real world datasets are imbalanced and on the other hand, most existing classification approaches assume that the underlying training set is evenly distributed. Furthermore, in many scenarios it is undesirable or dangerous to misclassify an example from a minority class. For example, in a continuous surveillance task, suspicious activity may occur as a rare event which is undesirable to go unnoticed by the monitoring system. In medical applications, the cost of erroneously classifying a sick person as healthy

© Springer International Publishing AG 2016
F. Schwenker et al. (Eds.): ANNPR 2016, LNAI 9896, pp. 150–162, 2016.
DOI: 10.1007/978-3-319-46182-3_13

can have larger risk (cost) than wrongly classifying a healthy person as sick. In these cases it is crucial for classification algorithms to have a higher identification rate for rare events, that means it is critical to not misclassify any minority examples while it is acceptable to misclassify few majority examples.

An extreme example for the imbalance problem would be a dataset where the area of the majority class overlaps that of the minority class completely and the overlapping region contains as many (or more) majority examples. Since the goal of learning is to minimise the overall cost of the cost function, such minimization can be obtained in this case by classifying all points to the majority class. Any other separation would result in misclassifying more data points of the majority class than correctly classifying data points of the minority class. This happens because a standard cost function, i.e. the cost due to erroneous classification, treats all individual errors as equally important.

Different methods such as Backpropagation Neural Networks (BPNN), Radial Basis Function (RBF) and Fuzzy ARTMAP when exposed to unbalanced datasets with different noise levels have this type of problem and it was shown that performance on imbalanced data greatly depends on how well the classes are separated [14]. When the classes are separated well enough, BPNN and Fuzzy ARTMAP could learn features well compared to RBF. Lately, Convolutional Neural Networks have gained a lot of interest in the pattern recognition community but are also subject to this problem as other neural network approaches using supervised learning [9].

Several possible solutions have been suggested already in the literature to tackle the class imbalance problem. There are two main approaches: Changing the training data or adjusting the algorithms. Training data can be changed by over-sampling the minority class (e.g. SMOTE [7]) or under-sampling the majority class (e.g. WWE [20]). Algorithmic approaches change the algorithm so that it favours the minority class. Examples are Threshold Moving [22], Snowball Learning for neural networks [19], ensemble methods [18], or other cost-sensitive learning methods for neural networks where different costs are associated with each class (e.g. [2]). For a comprehensive overview on different methods and an empirical study on cost-sensitive neural networks comparing resampling techniques to threshold-moving and hybrid ensembles, please see the work of Zhou et al. [22] where they report that Threshold Moving outperformed resampling. They also show results and discuss the effectiveness of the approaches for multi-class problems. Also hybrid methods have been used lately for neural networks and were shown to be effective [1,4]. Therefore, there is still a lack of addressing the imbalance problem, which we examine in this paper based on Convolutional Neural Networks.

## 2   Methods

The dataset mainly used for this work is highly imbalanced, which makes the use of resampling difficult since too many samples would be lost for training or have to be created artificially, leading to potential overfitting. We therefore investigated a cost-sensitive algorithmic approach with adaptable costs which is inspired by the recent works of Kahn et al. [9] and Castro et al. [5] for a binary class problem. As performance measure we use accuracy and the geometric mean (G-Mean)

$$\text{Accuracy} = \frac{TP+TN}{TP+TN+FP+FN}; \quad \text{G-Mean} = \sqrt{TPR*FPR} \qquad (1)$$

with TPR the true positive rate (FPR the false positive rate, respectively). Accuracy is the standard performance measure for classification, but leads to the class imbalance problem if optimised for, since all errors are treated equally. We therefore also use the G-Mean, which takes the performance of each class individually into account and thus leads to performance balancing between the classes.

## 2.1    Global Mean Square Error Separation

In the standard backpropagation algorithm, the weights are updated to minimise the overall error of the training data. However, if a certain class extensively dominates the training dataset and if C2C separability is low, the errors are contributed mainly from the majority class. Global Mean Square Error (GMSE) Separation [5] is an algorithmic method to differentiate the errors for different classes. The Mean Square Error (MSE) cost function minimises the error between predicted output from the network and the actual output. With GMSE, the cost to be minimised is the weighted sum of the cost of individual classes.

For a binary classification problem, let $T$ be the training set of size $n$ and $T^+$ and $T^-$ be the set of positive and negative examples respectively. Let us assume $T^+$ represents the minority class and $w$ represents the weight parameters of the entire network. The error of the $p^{th}$ training example $\varepsilon_p(w)$ is given by

$$\varepsilon_p(w) = \frac{1}{2}(d_p - y_p)^2 \qquad (2)$$

where $y_p$ is the predicted output and $d_p$ is the desired output of the $p^{th}$ training example. The global mean squared error is given by

$$E(w) = \frac{1}{n} \sum_{p=1}^{n} \varepsilon_p(w). \qquad (3)$$

By separating the global mean squared error into a weighted sum of errors of each class, we have

$$\varepsilon_p^{(k)}(w) = \frac{1}{2}\left(d_p^{(k)} - y_p^{(k)}\right)^2 \text{ for } k = \{+, -\}. \qquad (4)$$

Therefore the mean squared error of each class is given by

$$E^{(k)}(w) = \frac{1}{n^{(k)}} \sum_{p=1}^{n^{(k)}} \varepsilon_p^{(k)}(w). \qquad (5)$$

Optimization is achieved by minimizing the sum of errors for all $k$.

**Weight Update:** In standard backpropagation the weights are updated as follows

$$w^{(m+1)} = w^{(m)} - \eta \nabla E(w^{(m)})\tag{6}$$

where $m$ is the epoch, $\eta$ is the learning rate, and $\nabla E(w^{(m)})$ is the gradient from the $m^{th}$ epoch. To separate the GMSE, gradients are calculated separately and a weighted sum is calculated as follows

$$\nabla E(w) = \frac{1}{\lambda^+} \nabla E^+(w) + \frac{1}{\lambda^-} \nabla E^-(w).\tag{7}$$

$\lambda^+$ and $\lambda^-$ can be any positive constant to penalise the classes for misclassification [5]. This also gives flexibility to achieve desired specificity and sensitivity. In standard backpropagation, both $\lambda^+$ and $\lambda^-$ are equal to 1. Values generally used for $\lambda^+$ and $\lambda^-$ are the sizes of the respective classes $n^+$ and $n^-$, or a heuristically determined value suitable for the problem. However, using $n^+$ and $n^-$ with a very large dataset reduces the learning rate significantly. Since the XING dataset used for training in our experiments had 1 million examples and the ratio between the classes was 114:1, using the class size was not suitable and the ratio of 114 was used for $\lambda^+$ (with $\lambda^- = 1$). This means the cost for misclassifying a minority example was 114 times higher than the cost of misclassifying a majority example.

The core idea of this method is to have a weighted sum of the gradients of each class so as to achieve the desired specificity and sensitivity. In other words, each class has a different learning rate since the rate at which the weights are updated is different. One problem is that many libraries for GPU programming do not give easy access to the error gradients, but similar results can be achieved by having different costs of misclassification for each class, thus rewriting the error function as

$$E = \frac{1}{n} \sum_{p=1}^{n} (d_p - \kappa * y_p)^2 \quad \text{where} \quad \kappa = \begin{cases} 1, & \text{if } p \in \text{majority} \\ \dfrac{max(n^+, n^-)}{min(n^+, n^-)}, & \text{otherwise} \end{cases}\tag{8}$$

where $\kappa = 1$ if $p$ is from the majority class and the ratio of class sizes if not. $n^+$ and $n^-$ are again the size of $T^+$ and $T^-$ respectively. Our implementation (see Algorithm 1) is similar to Kukar et al. [11] and simplifies the implementation when using the python GPU library Theano[1] [3]. It can be shown mathematically that the implementation has the same effect of GMSE separation [16].

One of the drawbacks of this method is to find the optimum value for $\kappa$. There could exist a dataset dependent value $\kappa^*$ with which better performance could be achieved. We now make $\kappa$ an adaptable parameter and show empirically that it can achieve better results compared to a static value.

## 2.2 Learnable $\kappa$ in Cost Function

In the previous approach we used a fixed $\kappa$ to penalise more for misclassifying minority examples. Since the overall goal is to obtain a maximum G-Mean, $\kappa$ can

---

[1] http://deeplearning.net/software/theano/.

---

**Algorithm 1.** Optimised Implementation of Global Mean Square Error Separation

---

**Input**: Training data
**Output**: Learned weights

1  set maxepoch;
2  initialise weights of the network randomly;
3  **while** *epoch not equal to maxepoch* **do**
4      **for** *minibatch* **do**
5          Forward pass;
6          Calculate error as per Eq. 8;
7          Calculate gradients for the error;
8          Update weights;
9      **end**
10 **end**

---

be optimised or learned to maximise G-Mean for a specific dataset. The ideas and methods are originally motivated by the work of Khan et al. [9].

$$E = \frac{1}{n} \sum_{p=1}^{n} (d_p - \kappa * y_p)^2 \text{ where } \kappa = \begin{cases} 1, & \text{if } p \in \text{majority} \\ \kappa^*, & \text{otherwise} \end{cases} \tag{9}$$

$\kappa$ is initialised to 1 and is updated to optimise G-Mean or the combination of both G-Mean and accuracy. The effect of optimising different metrics are discussed in the subsequent sections. In stochastic gradient learning, the weight parameters are updated for each minibatch. But $\kappa$ is updated at the end of each epoch and has its own learning rate. The implementation can be seen in Algorithm 2.

**$\kappa$ Optimization:** The goal of the learning is to jointly learn weight parameters as well as $\kappa$ alternatively, keeping one of these parameters constant and minimizing the cost with respect to the other [9]. Some modifications on optimising $\kappa$ were made compared to the original approach. In our work, $\kappa$ is optimised as follows

$$\kappa^* = argminF(\kappa); \quad F(\kappa) = ||T - \kappa||^2 \tag{10}$$

and using a gradient decent algorithm

$$\nabla F(\kappa) = \nabla ||T - \kappa||^2 = -(T - \kappa). \tag{11}$$

For $T$ we have used different methods to gain more insights into the effect of optimising for different metrics. $H$ is the maximum cost applicable to the minority class which in this case is the ratio of imbalance:

– Optimising for G-Mean and Accuracy

$$T_1 = H * exp\left(-\frac{GMean}{2}\right) * exp\left(-\frac{Accuracy}{2}\right) \tag{12}$$

---

**Algorithm 2.** Learning Optimal Parameters

---
**Input**: Training data, Validation Data, maxepoch, Learning Rate for $w$,
         learning rate for $\kappa$
**Output**: Learned parameters, $w^*$ and $\kappa^*$
1  Initialise Network;
2  Initialise Network weights randomly;
3  Initialise $\kappa$ to 1;
4  **while** *epoch not equal to maxepoch* **do**
5     **for** *minibatch* **do**
6        Forward pass;
7        Calculate error as per Eq. 9;
8        Calculate gradients for the error;
9        Update weights;
10    **end**
11    Compute gradients for $\kappa$ using Eq. 10 with $T$ either $T_1$, $T_2$, or $T_3$;
12    Update $\kappa$;
13 **end**

---

- Optimizing only for G-Mean

$$T_2 = H * exp\left( - \frac{GMean}{2} \right) \tag{13}$$

- Optimizing for G-Mean and validation errors (1 - accuracy). The motivation behind this equation is to see if bringing down accuracy would help improve G-Mean.

$$T_3 = H * exp\left( - \frac{GMean}{2} \right) * exp\left( - \frac{(1 - Accuracy)}{2} \right) \tag{14}$$

This explores a $\kappa$ between 1 and $H$, which can be set to the ratio from minority to majority class. For brevity, let us call the adaptable $\kappa$ using Eqs. 12, 13 and 14 as $\kappa_1$, $\kappa_2$ and $\kappa_3$ respectively.

**Class Separability:** Learning in imbalance depends on how well the classes are separated [14]. However, $\kappa$ is never affected by C2C separability so far. Thus it is meaningful, to introduce an effect on $\kappa$ with respect to class-to-class separability. The idea is that when C2C separability is good, i.e. when the classification is easier, errors should cost more and, respectively, errors on hard classification problems should be punished less.

For this, the Silhouette score [17] was used as a C2C separability measure. This technique quantifies how well each data point lies within its own cluster and the value ranges from $-1$ to $+1$. For individual points, $+1$ means that the point lies perfectly within the own cluster, 0 means the point is on the boundary between both clusters, and $-1$ means the point is perfectly within the opposite cluster. The sum over all points gives a measure on how tightly formed and separated the clusters are. Given classes A and B,

$$s(i) = \frac{b(i) - a(i)}{max\{a(i), b(i)\}} \tag{15}$$

where, $b(i) = $ minimum $d(i, B)$ and $d(i, B)$ is the average dissimilarity of $i$ from Class A to all other objects of Class B ($a(i)$ respectively) [17]. The average of $s(i)$ over all data points is the measure of how well the classes form clusters or how well they can be separated.

Let us denote the Silhouette score as $S$ and with the imbalance ratio $IR = max(n^+, n^-)/min(n^+, n^-)$, $H$ is now given by

$$H_{adjusted} = IR(1 + |S|). \tag{16}$$

This adjusts the maximum reachable cost for the minority class based on its separability. If two classes are well separable, the maximum cost is twice $IR$. Notice that for perfectly balanced classes, if the classes are clearly separable ($|S| = 1$), we incur twice the cost on one of the classes. This may seem wrong, but we will show this case has no practical effects on learning.

## 3  Experiments and Results

In this section experiments are reported to show the classification performance with the different methods we have introduced above. We have used three real datasets (XING, Earthquake, ECG) and some specifically selected subsets to enforce specific IR or C2C scores. The *XING dataset* was the main target and includes activity timelines of one million users of the XING social network, tracking 14 different activities. The likelihood of a job change had to be predicted on the basis of the recent past. Since job changes are very rare compared to normal activity, the IR was 114:1. The Silhouette separability score $S$ for XING was calculated on a small random sample of 10,000 examples, since it is a computationally expensive function and a good estimate was sufficient. The mean $S$ for five random samples was $-0.184 \pm 0.037$. The negative sign signifies that many majority examples are within the area of the minority class. This was expected, because a lot of users' activities is similar to the activities of a job changer, except that they did not change their job in the end.

The other datasets are taken from [8], also include timeline data, and have been used mainly for comparison at the end: The *Earthquake dataset* contains hourly averages of sensory readings. The positive instances are major events greater than 5.0 on the Richter scale and not preceded by another major event for at least 512 hours. Negative examples are readings with Richter scale less than 4.0. This dataset has a very low class separability score $S$ of 0.0005 and has an imbalance ratio of 3.95.

The *ECG dataset* contains electrocardiography recordings from two different days on the same patient. The positive instances had an anomalous heartbeat.

The data exhibits linear drift, which in medical terms is called a wandering baseline. Globally, both positive and negative instances are very similar with separability score $S$ being only 0.09. The raw dataset was perfectly balanced, therefore imbalance was created artificially by removing positive examples to reach an imbalance ratio of 20.

The primary set of experiments include a comparison of methods in terms of performance on the XING dataset as well as other datasets. The secondary goal was to analyse the behaviour of different adaptable $\kappa$ and the effects of adjusting the maximum applicable cost $H$ by incorporating the silhouette score of the corresponding datasets.

A multi-channel CNN with one-dimensional kernels for each feature was used similar to [21]. 90 % of the datasets was used for training and 10 % for validation in each epoch. Since training CNNs takes a long time, all experiments on the XING data were stopped after the 100th epoch since little improvement was found beyond that point. The size of each minibatch was set to 5000 and for training Stochastic Gradient Descent was used due to its faster convergence [12]. Smaller minibatches were not used to make sure a minority example was contained with high chance. Also, $L2$ regularization was used to avoid overfitting [15]. All reported values are means ($\pm$ standard deviation) of five runs for each experiment to account for variability in the random training/validation splits of the datasets.

## 3.1 Comparison of Methods

Figure 1 shows a typical example of performance over epochs of CNNs with different cost methods. With no cost sensitivity, G-Mean remains 0 due to the class imbalance problem. Further experiments revealed, no matter the number of iterations, there was no further improvement in the G-Mean score. Figure 2 shows how accuracy and G-Mean vary with cost. With just a little loss in overall accuracy, the classifier's performance on both classes together can be increased significantly.

Table 1 summarises the performance of different methods on a smaller random subset of the XING data with 200,000 examples. Using the imbalance ratio as $\kappa$ shows significant improvement in comparison with having no cost at all. It also shows that learning $\kappa$ gave better performance in comparison to constant $\kappa$ (t-test: p = 0.0192). The adaptable $\kappa$ starts by incrementally increasing the cost, and these increments get smaller as the G-Mean increases. Therefore a gradual increase in G-Mean can be noticed and at the end a slight improvement in comparison to a constant $\kappa$, thus an improvement by tuning $\kappa$ to the dataset was possible as hypothesised.

## 3.2 Closer Look at Adaptable $\kappa$

We have proposed three different optimisation approaches, $\kappa_1$, $\kappa_2$, and $\kappa_3$. Figure 3 shows G-Mean and accuracy scores of all three methods over iterations. As can be seen, $\kappa_1$ improves the G-Mean score over time while trying to

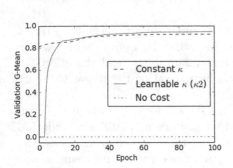

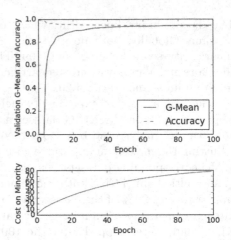

**Fig. 1.** Results of example runs of different costs functions on G-Mean metric over learning time.

**Fig. 2.** Example of how Accuracy and G-Mean (a) are affected by cost ($\kappa_2$) over epochs (b)

**Table 1.** Performance comparison of different cost functions.

| Method | G-Mean Validation | Accuracy Validation |
|---|---|---|
| Const. $\kappa$ | 86.4% ± 2.8% | 77.2% ± 5.6% |
| $\kappa 1$ | 90.1% ± 1.5% | 92.5% ± 1.3% |
| $\kappa 2$ | 90.3% ± 1.0% | 87.7% ± 1.3% |
| $\kappa 3$ | 90.8% ± 1.1% | 88.5% ± 1.0% |
| No Cost | 0% ± 0% | 99.1% ± 0% |

keep a high accuracy. The other two focus on optimisation of G-Mean with $\kappa_3$ being more aggressive than $\kappa_2$. Nevertheless, in the end, $\kappa_3$ achieves a slightly higher score for both performance measures (compare Table 1).

### 3.3   Combined Effect of Separability and Imbalance

We also introduced the usage of the class separability score $S$ to adjust the maximum applicable cost for minorities. Table 2 compares the performance outcome of using a maximum cost on minority to be $H$ and $H_{adjusted}$. Results clearly show an advantage of having adjusted cost (t-test: p = 0.00018). However, when classes are balanced and fairly separable, imposing greater cost on one of the classes appears improper both theoretically and intuitively. Nevertheless, experiments were conducted with $\kappa_2$ to evaluate practical outcomes of such scenario. Table 3 shows the performance of regular learning and cost-sensitive learning on an artificial subset of $XING$ with different $S$ and imbalance ratios.

**Table 2.** Effect of adjusting maximum cost $H$ to impose on minority using C2C separability score compared to using $H$ as imbalance ratio.

|  | Max Cost on Minority | G-Mean using Cost Sensitive CNN ($\kappa2$) |
|---|---|---|
| $H$ | 114 | 92.8% $\pm$ 0.24% |
| $H_{adjusted}$ | 142 | **94.1% $\pm$ 0.25%** |

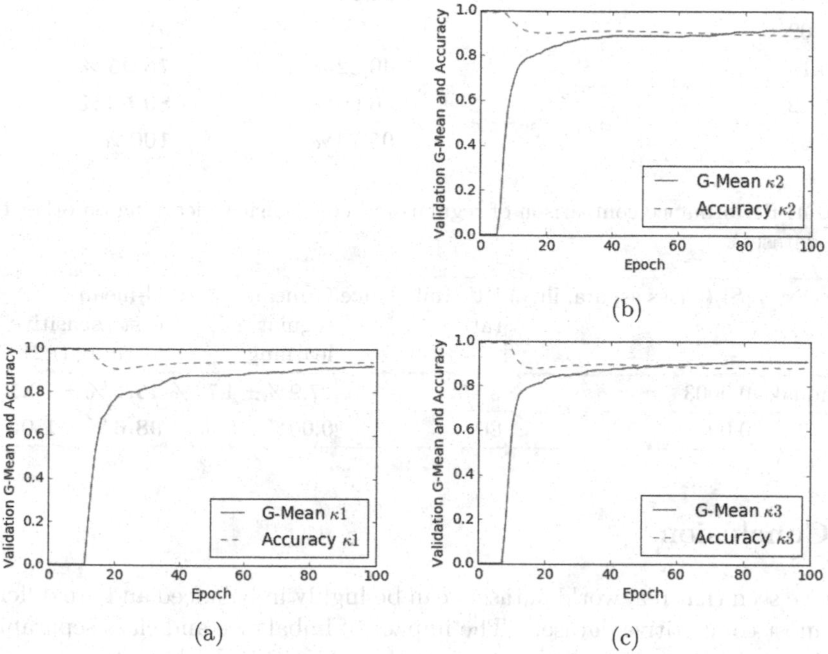

**Fig. 3.** Rate of change of G-Mean and Accuracy with (a) Learnable $\kappa1$, (b) Learnable $\kappa2$ and (c) Learnable $\kappa3$ in example runs on XING data

These results show again a clear benefit when using $\kappa_2$ with adjusting $H$ to the respective class separability when there exists an imbalance in class distribution. Punishing one class in the balanced case has a small but negligible effect compared to regular learning. For the case of imbalance with $|S| = 0.02$ the learning failed to distinguish the classes. Table 4 shows results of running experiments 5 times on each of the other real datasets. In both cases cost sensitive learning has clearly outperformed regular learning. In the case of ECG, only the method with adaptable $\kappa$ and adjusted $H$ was successful.

**Table 3.** Effect of S and IR on learning of a small artificially adjusted dataset.

| |S| (Class separability) | IR (Imbalance ratio) | G-mean regular learning | G-mean cost sensitive learning ($\kappa2$) |
|---|---|---|---|
| 0.0005 | 1 | **50.42 %** | 49.80 % |
| 0.17 | 1 | **77.46 %** | 72.43 % |
| 0.42 | 1 | **91.49 %** | 91.39 % |
| 0.69 | 1 | **100 %** | 98.99 % |
| 0.02 | 5 | 0 % | 0 % |
| 0.16 | 5 | 40.82 % | **73.95 %** |
| 0.42 | 5 | 86.60 % | **89.56 %** |
| 0.69 | 5 | 95.74 % | **100 %** |

**Table 4.** Performance comparison of regular and cost sensitive learning on other time series datasets.

| Dataset | |S| (Class separability) | IR (Imbalance ratio) | G-mean regular learning | G-mean cost sensitive learning ($\kappa2$) |
|---|---|---|---|---|
| Earthquake | 0.0003 | 3.95 | 27.9 % ± 4.72 % | **45.4 % ± 10.1 %** |
| ECG | 0.09 | 20 | 0.00 % ± 0 % | **98.6 % ± 1.03 %** |

## 4    Conclusion

We have seen that real world datasets can be highly imbalanced and are different from most competitive datasets. The impact of imbalance and class separability would make most machine learning algorithms biased towards one single majority class, leading to the class imbalance problem.

We have combined several methods and for the first time used them for learning time series with multi-channel CNNs. Using Global Mean Error Separation from Castro et al. [5] to achieve cost-sensitive learning and combining them with a method to learn the parameter $\kappa$ for the specific dataset introduced by Kahn et al. [9], we were able to show successful learning on a highly imbalanced dataset from XING with low C2C separability. The chosen approach also permits a simpler implementation with GPU libraries like Theano. Further improvements could be achieved with the novel use of the Silhouette score to adjust the maximum applicable cost to the C2C separability of the given dataset, and thus defining a better range for $\kappa$. Although the used adjustment function can be further improved to not punish a class when using balanced data, we have shown that this seemed to lead to no significant negative effect on learning. These results have also been confirmed for two other imbalanced time series datasets (Earthquake and ECG data) with different C2C separability, showing significantly improved results to regular learning without cost sensitivity.

Further work to generalise and evaluate the approach on multi-class problems is necessary. Although the single methods and metrics can be extended to several classes, it was already shown that this does not automatically mean a solution for multi-class imbalances [22]. Also a more comprehensive comparison with other approaches in the literature that could be used for CNNs is needed. Overall, the proposed method was shown to be an effective algorithmic approach to solve the class imbalance problem for binary classes when using convolutional deep neural networks that can easily be integrated into different neural network architectures.

# References

1. Alejo, R., Valdovinos, R.M., García, V., Pacheco-Sanchez, J.: A hybrid method to face class overlap and class imbalance on neural networks and multi-class scenarios. Pattern Recogn. Lett. **34**(4), 380–388 (2013)
2. Berardi, V.L., Zhang, G.P.: The effect of misclassification costs on neural network classifiers. Decis. Sci. **30**(3), 659–682 (1999)
3. Bergstra, J., Breuleux, O., Bastien, F., Lamblin, P., Pascanu, R., Desjardins, G., Turian, J., Warde-Farley, D., Bengio, Y.: Theano: a CPU and GPU math expression compiler. In: Proceedings of the Python for Scientific Computing Conference (SciPy), vol. 4, p. 3, Austin, TX (2010)
4. Cao, P., Zhao, D., Zaïane, O.R.: A PSO-based cost-sensitive neural network for imbalanced data classification. In: Li, J., Cao, L., Wang, C., Tan, K.C., Liu, B., Pei, J., Tseng, V.S. (eds.) PAKDD 2013. LNCS, vol. 7867, pp. 452–463. Springer, Heidelberg (2013)
5. Castro, C.L., de Pádua Braga, A.: Artificial neural networks learning in ROC space. In: IJCCI, pp. 484–489 (2009)
6. Chan, P.K., Stolfo, S.J.: Toward scalable learning with non-uniform class and cost distributions: a case study in credit card fraud detection. In: KDD, vol. 1998, pp. 164–168 (1998)
7. Chawla, N.V., Bowyer, K.W., Hall, L.O., Kegelmeyer, W.P.: Smote: synthetic minority over-sampling technique. J. Artif. Intell. Res. **16**, 321–357 (2002)
8. Chen, Y., Keogh, E., Hu, B., Begum, N., Bagnall, A., Mueen, A., Batista, G.: The UCR time series classification archive, July 2015. www.cs.ucr.edu/~eamonn/time_series_data/
9. Khan, S.H., Bennamoun, M., Sohel, F., Togneri, R.: Cost sensitive learning of deep feature representations from imbalanced data (2015). arXiv preprint arXiv:1508.03422
10. Khoshgoftaar, T.M., Van Hulse, J., Napolitano, A.: Supervised neural network modeling: an empirical investigation into learning from imbalanced data with labeling errors. IEEE Trans. Neural Netw. **21**(5), 813–830 (2010)
11. Kukar, M., Kononenko, I., et al.: Cost-sensitive learning with neural networks. In: ECAI, pp. 445–449. Citeseer (1998)
12. LeCun, Y.A., Bottou, L., Orr, G.B., Müller, K.-R.: Efficient backprop. In: Orr, G.B., Müller, K.-R. (eds.) Neural Networks: Tricks of the Trade. LNCS, vol. 1524, pp. 9–48. Springer, Heidelberg (2012)
13. Liu, X.Y., Zhou, Z.H.: The influence of class imbalance on cost-sensitive learning: an empirical study. In: Sixth International Conference on Data Mining, 2006, ICDM 2006, pp. 970–974. IEEE (2006)

14. Murphey, Y.L., Guo, H., Feldkamp, L.A.: Neural learning from unbalanced data. Appl. Intell. **21**(2), 117–128 (2004)
15. Ng, A.Y.: Feature selection, l 1 vs. l 2 regularization, and rotational invariance. In: Proceedings of the Twenty-First International Conference on Machine Learning, p. 78. ACM (2004)
16. Raj, V.: Towards effective classification of imbalanced data with convolutional neural networks. Master's thesis, Department of Informatics, University of Hamburg, Vogt-Koelln-Str. 22527 Hamburg, Germany, April 2016
17. Rousseeuw, P.J.: Silhouettes: a graphical aid to the interpretation and validation of cluster analysis. J. Comput. Appl. Math. **20**, 53–65 (1987)
18. Sun, Y., Kamel, M.S., Wong, A.K., Wang, Y.: Cost-sensitive boosting for classification of imbalanced data. Pattern Recogn. **40**(12), 3358–3378 (2007)
19. Wang, J., Jean, J.: Resolving multifont character confusion with neural networks. Pattern Recogn. **26**(1), 175–187 (1993)
20. Wilson, D.L.: Asymptotic properties of nearest neighbor rules using edited data. IEEE Trans. Syst. Man Cybern. **3**, 408–421 (1972)
21. Zheng, Y., Liu, Q., Chen, E., Ge, Y., Zhao, J.L.: Time series classification using multi-channels deep convolutional neural networks. In: Li, F., Li, G., Hwang, S., Yao, B., Zhang, Z. (eds.) WAIM 2014. LNCS, vol. 8485, pp. 298–310. Springer, Heidelberg (2014)
22. Zhou, Z.H., Liu, X.Y.: Training cost-sensitive neural networks with methods addressing the class imbalance problem. IEEE Trans. Knowl. Data Eng. **18**(1), 63–77 (2006)

# On CPU Performance Optimization
# of Restricted Boltzmann Machine
# and Convolutional RBM

Baptiste Wicht[1,2](✉), Andreas Fischer[1,2], and Jean Hennebert[1,2]

[1] University of Applied Science of Western Switzerland, Delémont, Switzerland
{baptiste.wicht,jean.hennebert}@hefr.ch, andreas.fischer@unifr.ch
[2] University of Fribourg, Fribourg, Switzerland

**Abstract.** Although Graphics Processing Units (GPUs) seem to currently be the best platform to train machine learning models, most research laboratories are still only equipped with standard CPU systems. In this paper, we investigate multiple techniques to speedup the training of Restricted Boltzmann Machine (RBM) models and Convolutional RBM (CRBM) models on CPU with the Contrastive Divergence (CD) algorithm. Experimentally, we show that the proposed techniques can reduce the training time by up to 30 times for RBM and up to 12 times for CRBM, on a data set of handwritten digits.

## 1 Introduction

Although most of the recent research has shown that learning on Graphics Processing Units (GPUs) is generally more efficient than training on Central Processing Units (CPUs) [13,14,20], especially for Convolutional Neural Networks (CNNs) [7,9,16], GPUs are not accessible everywhere. Some researchers may not have access to them and some laboratories may not want to upgrade their CPU clusters to GPU clusters. Therefore, it remains important to be able to train neural networks in reasonable time on machines equipped only with CPUs.

Restricted Boltzmann Machines (RBMs) are old models [19], that resurged recently to initialize the weights of an Artificial Neural Network (ANN) [4] or to extract features from samples [2]. Later on, the model was extended with the Convolutional RBM (CRBM) [11]. Performance optimization of these models was investigated on GPUs only [8,15].

In the present paper, we present several techniques to reduce the training time of RBM and CRBM models. Techniques such as CPU vectorization, usage of BLAS kernels and reduction of convolutions to other functions are explored. To evaluate the performance, several networks are trained on 60,000 images of handwritten digits from the MNIST data set.

The rest of this paper is organized as follows. The system setup for the experiments is presented in Sect. 2. Section 3 presents techniques to speed up an RBM while optimizations for CRBM are detailed in Sect. 4. Section 5 covers the training of a DBN. Finally, conclusions are drawn in Sect. 6.

© Springer International Publishing AG 2016
F. Schwenker et al. (Eds.): ANNPR 2016, LNAI 9896, pp. 163–174, 2016.
DOI: 10.1007/978-3-319-46182-3_14

## 2    System Setup

The experiments have been computed on a Gentoo Linux machine with 12Go of RAM running an Intel® Core™ i7-2600 with a frequency of 3.40 GHz. The tests were written in C++ using our own Deep Learning Library (DLL)[1] and Expression Templates Library (ETL)[2] libraries. The programs were compiled with GNU Compiler Collection (GCC) 4.9. Vector operations are vectorized using AVX. Intel® Math Kernel Library (MKL) is used as the BLAS implementation.

The experiments are conducted on the MNIST data set [10]. It contains grayscale images of handwritten digits, normalized to a size of $28 \times 28$ pixels. Each experiment is done on the 60,000 training images for 5 epochs and the average time per epoch is used as the final result.

## 3    Restricted Boltzmann Machine

A Restricted Boltzmann Machine (RBM) [19] is a generative stochastic Artificial Neural Network (ANN), developed to learn the probability distribution of some input. Training an RBM using the algorithm for general Boltzmann Machines [3] is very slow. Hinton et al. proposed a new technique, Contrastive Divergence (CD) [4], depicted in Fig. 1. It is quite similar to the Stochastic Gradient Descent method, used to train regular ANNs. It approximates the Log-Likelihood gradients by minimizing the reconstruction error, thus training the model into an autoencoder. The algorithm performs a certain number of steps of Gibbs sampling (*CD-n*). When the RBM is used as a feature extractor or as a way of pretraining a Deep Belief Network [6], CD-1 is generally sufficient [5].

The original RBM model was designed with binary visible and binary hidden units (also called a Bernoulli RBM). Several different types of units were since developed (for instance Gaussian, ReLu or Softmax) [5]. This research focuses on binary units, but the conclusions stand for all general types of units. Indeed,

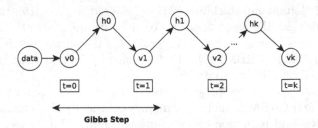

**Fig. 1.** Graphical representation of the contrastive divergence algorithm. The algorithm CD-k stops at $t = k$. Each iteration performs a full Gibbs step.

---

[1] https://github.com/wichtounet/dll/.
[2] https://github.com/wichtounet/etl/.

only the activation functions would change. The probability activations of visible
and hidden units can be computed as follows:

$$p(h_j = 1|v) = \sigma(c_j + \sum_i^m v_i W_{i,j}) \tag{1}$$

$$p(v_i = 1|h) = \sigma(b_i + \sum_j^n h_j W_{i,j}) \tag{2}$$

The states of the units are obtained by sampling the activation probabilities.
For binary units, Bernoulli sampling is performed to obtain the states:

$$s_j = \begin{cases} 1 & \text{if } p_j > \text{Unif}(0,1) \\ 0 & \text{otherwise} \end{cases} \tag{3}$$

$$s_i = \begin{cases} 1 & \text{if } p_i > \text{Unif}(0,1) \\ 0 & \text{otherwise} \end{cases} \tag{4}$$

From an implementation point of view, an RBM is made of a vector $v$ of $m$
visible units, a vector $h$ of $n$ hidden units, a matrix $W$ of weights connecting the
visible and the hidden units, a vector $b$ of $m$ visible biases and a vector $c$ of $n$
hidden biases. In practice, the weights are represented as single-precision floating
point numbers rather than double-precision. Indeed, some single-precision computations can be as much as twice faster than their double-precision counterparts. Moreover, the precision is generally more than sufficient for CD training.

---

**Algorithm 1.** Standard CD-1 algorithm (one sample)

---

$v_0$ = training sample
$h_0$ = sample hidden activations from $v_0$
$v_1$ = sample visible activations from $h_0$
$h_1$ = sample hidden activations from $v_1$
$W_{pos} = v_0 \otimes h_0$
$W_{neg} = v_1 \otimes h_1$
$\nabla W = \epsilon(W_{pos} - W_{neg})$
$\nabla b = \epsilon(v_0 - v_1)$
$\nabla c = \epsilon(h_0 - h_1)$

---

Algorithm 1 describes the CD-1 algorithm for one sample. The same procedure is done for each sample of the data set and is repeated for as many epochs as
necessary. In practice, it is important to note that the hidden activations should
be computed directly from the visible activation probabilities rather than from
the states [5]. Therefore, it is never necessary to compute the states of the visible
units during training. Moreover, the last update of the hidden unit is only used

to compute the positive gradients, in which case the probabilities are used rather than the states. Therefore, it is not necessary to sample the states of the hidden units for the last update.

In the algorithm and activation formulas, several computation routines are well-known and can be optimized. The Basic Linear Algebra Subprogams (BLAS) are a collection of small and highly optimized linear algebra routines. In the activation formulas, the sums are simply vector-matrix multiplication and matrix-vector multiplication. They can be implemented using the SGEMV operation from BLAS. The outer products to compute the positive and negative gradients can be implemented using the SGER routine. Finally, the computation of the visible and hidden biases gradients can be done with the SAXPY operation. For evaluation, the following networks are trained:

- A: 784 visible units, 500 hidden units
- B: 500 visible units, 500 hidden units
- C: 500 visible units, 2000 hidden units
- D: 2000 visible units, 10 hidden units

**Table 1.** Training time for an epoch of RBM training, in seconds. The speedup is the improvement gained by using BLAS kernels for linear algebra operations.

|              | A      | B     | C      | D     |
|--------------|--------|-------|--------|-------|
| Base         | 161.99 | 47.00 | 167.20 | 70.78 |
| Base + BLAS  | 141.91 | 43.03 | 114.18 | 36.12 |
| Speedup      | 1.14   | 1.09  | 1.46   | 1.95  |

Table 1 shows the time, in seconds, necessary to train an epoch of the networks. Even if the computations are simple, BLAS operations can bring an important speedup to CD training, compared to standard implementations of these operations. For the tested networks, the speedup ranges from 1.09 to 1.95. The MKL BLAS implementation is highly tuned for Intel processor and each routine is especially optimized for cache and maximum throughput.

Experimentally, we find that more than 75 % of the training time is spent inside the BLAS library, 8 % in the sigmoid function and around 7 % in random number generation. The sigmoid time could be optimized further by using an approximation of the sigmoid function or a vectorized version of the exponential function. Since this represents only a fraction of the total time, it would only slightly improve the general training time.

### 3.1 Mini-Batch Training

In practice, CD is rarely performed one element at a time, but rather on a mini-batch. The data set is split into several mini-batches of the same size. The

**Algorithm 2.** Mini-batch CD-1 algorithm (one mini-batch)

---

**for all** $v_0 \in$ mini-batch **do**

    $h_0 =$ sample hidden activations from $v_0$

    $v_1 =$ sample visible activations from $h_0$

    $h_1 =$ sample hidden activations from $v_1$

    $W_{pos} \stackrel{\pm}{=} v_0 \otimes h_0$

    $W_{neg} \stackrel{\pm}{=} v_1 \otimes h_1$

**end for**

$\nabla W = \frac{\epsilon}{B}(W_{pos} - W_{neg})$

$\nabla b = \frac{\epsilon}{B}(v_0 - v_1)$

$\nabla c = \frac{\epsilon}{B}(h_0 - h_1)$

---

gradients are computed for a complete batch before the weights are updated. Algorithm 2 shows the updated version of CD-1 for mini-batch training.

In practice, this could be implemented by accumulating the gradients element after element. However, it is better to compute the gradients independently for each element of the mini-batch. This needs more memory to store the intermediary results for the complete mini-batch. However, this is only for a small portion of the data set and has the advantage of allowing higher level optimizations of the loop body. Each iteration being completely independent, this could seem like an excellent candidate for parallelization. However, this is not the case. Depending on the dimensions of the matrices, a small speedup can be obtained by computing each iteration in parallel before aggregating the results sequentially. However, since most of the time will be spent in memory-bound operations (matrix-vector multiplication and outer product), there won't be enough bandwidth for many cores to process the data in parallel. A better optimization is to compute the activations and states of the units for a complete mini-batch at once instead of one sample at time. If we consider $h$ as a $[B, n]$ matrix and $v$ as a $[B, m]$ matrix, they can be computed directly as follows[3]:

$$h = \sigma(repmat(c, B) + v * W) \tag{5}$$

$$v = \sigma(repmat(b, B) + (W * h^T)^T) \tag{6}$$

This has the great advantage of performing a single large matrix-matrix multiplication instead of multiple small vector-matrix multiplication. In practice, this is much more efficient. In that case, the SGEMM operation of the BLAS library is used to compute the activation probabilities. Moreover, if the matrices are big enough, it is also possible to use a parallel version of the matrix-matrix multiplication algorithm. Figure 2 shows the time necessary to train each network with different batch sizes. It compares the base version with a hand-crafted matrix multiplication and the version using BLAS. The parallel BLAS version is also included in the results. On average the BLAS version is twice faster than the standard version and the parallel version of BLAS reduces the time by another factor of two.

---

[3] *repmat* vertically stacks the array $B$ times.

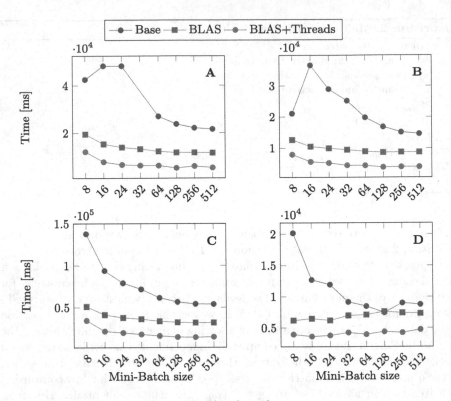

**Fig. 2.** Mini-batch performance

Generally, increasing the mini-batch size reduces the training time. However, due to the small output dimension of the network D, the possible speedup is greatly reduced and larger mini-batch do not provide any substantial improvements. However, a too large mini-batch size may have negative impact on the classification performance of the network since many gradients will be averaged. On the other hand, a small batch size is generally leading to a more stable convergence. The batch size must be chosen as a trade-off between training time and classification performance. Moreover, a large mini-batch also increases the need for the inputs to be shuffled prior to each epoch. For MNIST, mini-batch size of up to 128 samples are still reasonable, but higher mini-batch are increasing the

**Table 2.** Final results for standard RBM training, in seconds.

|  | A | B | C | D |
|---|---|---|---|---|
| Base | 161.99 | 47.00 | 167.20 | 70.78 |
| Mini-Batch + BLAS + Threads | 5.35 | 3.94 | 14.57 | 4.39 |
| Speedup | 30.27 | 11.92 | 11.47 | 16.12 |

overall training time, by decreasing the learning each epoch does. To conclude this section, Table 2 compares the basic implementation and the final optimized version with mini-batch (128 samples) and a threaded BLAS library. Depending on the network, the optimized version is between 11 and 30 times faster.

## 4    Convolutional Restricted Boltzmann Machine

The original RBM model can be extended to the Convolutional Restricted Boltzmann Machine (CRBM) [11]. The visible and hidden layers are connected together by convolution, allowing the model to learn features shared among all locations of the input, thus improving translation invariance of the model. While this research focuses on two-dimensional CRBM, one-dimensional CRBM are also possible, for instance for audio [12] and the model can be adapted for three-dimensional inputs. Only square inputs and filters are described here for the sake of simplicity, but the model is able to handle rectangular inputs and filters.

A CRBM model has a matrix $V$ of $C \times N_V \times N_V$ visible units. It has $K$ groups of $N_H \times N_H$ hidden units. There are $C \times K$ convolutional filters of dimension $N_W \times N_W$ (by convolutional properties $N_W \triangleq N_V - N_H + 1$). There is a single visible bias $c$ and a vector $b$ of $K$ hidden biases. The notation $\bullet_v$ is used to denote a valid convolution and $\bullet_f$ is for a full convolution. A tilde over a matrix ($\tilde{A}$) is used to indicate that the matrix is flipped horizontally and vertically. For a network with binary units, the probability activation are computed as follows:

$$P(h_{ij}^k = 1 | v_c) = \sigma((\sum_c \tilde{W}_c^k \bullet_v v_c)_{ij} + b_k) \tag{7}$$

$$P(v_{cij}^k = 1 | h) = \sigma((\sum_k W^k \bullet_f h^k)_{ij} + c) \tag{8}$$

A CRBM is trained similarly to an RBM, with an adapted version of formulas to compute the positive and negative gradients:

$$W_{ck}^{pos} = v_c^0 \bullet_v \tilde{h}_k^0 \tag{9}$$

$$W_{ck}^{neg} = v_c^1 \bullet_v \tilde{h}_k^1 \tag{10}$$

Training a CRBM requires a large number of convolutions for each epoch. Indeed, for each sample, there would be $2KC$ valid convolutions for the gradients, $2KC$ valid convolutions to compute the hidden activation probabilities (done twice in CD) and $KC$ full convolutions for the visible units. Contrary to matrix multiplication, there is no general convolution reference implementation. The first optimization that can be applied is to vectorize the convolution implementations. Modern processors are able to process several floating point operations in one instruction. For instance, AVX instructions process 8 floats at once, while SSE instructions process 4 floats once. While modern compilers are able to vectorize simple code, vectorizing complex program must be done by hand. We vectorized the inner loop of the convolutions, with AVX for large kernels and SSE for small kernels (smaller than 8 pixels). For evaluation, the following networks are trained:

- A: $1 \times 28 \times 28$ visible units, 40 $9 \times 9$ filters
- B: $40 \times 20 \times 20$ visible units, 40 $5 \times 5$ filters
- C: $40 \times 16 \times 16$ visible units, 96 $5 \times 5$ filters
- D: $96 \times 12 \times 12$ visible units, 8 $5 \times 5$ filters

Table 3 shows the time necessary to train the different networks and the obtained speedup. Due to the small images and filters in the measured networks, the speedups are only very interesting for the first layer of the network, with larger kernel and images. Moreover, the two-dimensional property of the algorithms adds overhead to the vectorized version, reducing the possible speedups.

**Table 3.** Results for Convolutional RBM training, in seconds.

|                        | A      | B       | C       | D      |
| ---------------------- | ------ | ------- | ------- | ------ |
| Base                   | 380.37 | 3013.82 | 3947.46 | 338.16 |
| Base + Vectorization   | 198.21 | 2174.66 | 3358.76 | 295.83 |
| Speedup                | 1.91   | 1.38    | 1.17    | 1.14   |

When training the model using mini-batch, it becomes interesting to compute the gradients of each sample concurrently. For convolutions, there is no simple technique to compute the gradients of a complete batch, therefore parallelization inside batches is the best option. Figure 3 shows the results with different number of threads, with a mini-batch of 64. The performance increases almost linearly with the number of threads until four threads are used and then only slightly improves with more threads, exhibiting memory-bound computation behaviour. Since threads on the same core share the same cache, having more threads than cores does not improve the performance substantially in this case.

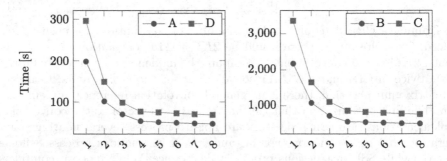

**Fig. 3.** Parallel performance

## 4.1 Valid Convolution

As seen previously, training a CRBM requires four times more valid convolutions than full convolutions. Thus, it is extremely important to make it as fast as possible. By rearranging the image to be convolved, it is possible to reduce a valid convolution to a vector-matrix multiplication [18]. The general algorithm is presented in Algorithm 3.

---

**Algorithm 3.** Convolution $C = I \bullet_v K$ with Matrix Multiplication

$W' = \text{reshape}(\tilde{W}, [1, k_1 k_2])$
$I' = \text{matrix}(k_1 k_2, c_1 c_2)$
$I' = \text{im2col}(K, [k_1 k_2])$
$C = W' * I'$

---

However, because of the memory-inefficient im2col operation, this is experimentally slower than the vectorized version. Nevertheless, since the same image is convolved with $K$ filters, the overhead of im2col can be greatly mitigated, by doing it only once for $K$ convolutions. Moreover, the multiple vector-matrix operations become a single matrix-matrix multiplication. Finally, since the computation of the activation probabilities and the gradients operates on flipped weights and that flipping is an involution, the computation can be done directly on the original weights, saving several flipping operations. Table 4 presents the results obtained when using this optimization for all the valid convolutions on the parallel version. On average, the training time is divided by two.

Experimentally, the difference in precision is found to be very small between the different versions and the reference. On average, the average difference between the vectorized version and the reference is in the order of $1e^{-5}\%$ and in the order of $5e^{-5}\%$ for the reduction with matrix multiplication. No difference has been observed when training a CRBM with different versions of the valid convolution. This difference may vary between different BLAS implementations.

**Table 4.** Results for Convolutional RBM training, in seconds.

|                     | A      | B      | C      | D     |
|---------------------|--------|--------|--------|-------|
| Parallel            | 46.69  | 494.52 | 756.70 | 68.47 |
| Parallel + Reduction | 28.45  | 241.79 | 336.56 | 40.12 |
| Speedup             | 1.64   | 2.04   | 2.24   | 1.70  |

## 4.2 Full Convolution

While there are no standard implementation of the full convolution, it is possible to reduce it to a another algorithm for which there exists efficient implementations. Using the convolution theorem, a full convolution can be reduced to a

Fourier transform. Indeed, the convolution in the time domain is equal to the pointwise multiplication in the frequency domain [1]. Since the image and the kernel may not be of the same size, it is necessary to pad them with zeroes before computing their transforms. Algorithm 4 shows the steps used to compute a full convolution using a Fourier transform.

---

**Algorithm 4.** Convolution $C = I *_f K$ with Fourier Transform

$I' = \text{pad}(I)$
$K' = \text{pad}(K)$
$C' = \mathcal{F}(I') \cdot \mathcal{F}(I')$
$C = \mathcal{F}^{-1}(C')$

---

In practice, this can be implemented using the Fast Fourier Transform (FFT). There exists some standard and very efficient implementations of the FFT. While it is not a part of BLAS, the MKL library provides an FFT implementation.

Unfortunately, this is not always faster than a properly vectorized convolution. Table 5 shows the performance for different image and kernel sizes. The FFT convolution is around 3 times slower for an $16 \times 16$ image and a kernel of $5 \times 5$, while it is almost 12 times faster for an image of $256 \times 256$ and a kernel of $31 \times 31$. This shows that using an FFT algorithm to perform the full convolution can brings very large speedup to the training of a CRBM. However, it is only really interesting for large models. Another optimization that can be done when computing the full convolution by FFT is to precompute the Fourier transforms of the images. Indeed, each image will be convolved several times with different kernels, therefore only one transform per image is necessary. On the evaluated networks, this does not bring any substantial performance improvements. Only the network $A$ has big enough images and kernels to profit from this, but this only result in a speedup of an epoch by less than 1%.

Again, the difference in precision is found to be very small. On average, the average difference for the vectorized version is found to be in the order of $1e^{-4}\%$ and in the order of $3e^{-5}\%$ for the FFT reduction. No difference has been observed when training a CRBM with different versions of the full convolution. The difference may vary between different FFT implementations.

**Table 5.** Performance of full convolution by FFT, in milliseconds

| Image | $12 \times 12$ | $16 \times 16$ | $16 \times 16$ | $28 \times 28$ | $50 \times 50$ | $128 \times 128$ | $128 \times 128$ | $256 \times 256$ |
|---|---|---|---|---|---|---|---|---|
| Kernel | $5 \times 5$ | $5 \times 5$ | $9 \times 9$ | $9 \times 9$ | $17 \times 17$ | $17 \times 17$ | $31 \times 31$ | $31 \times 31$ |
| Vectorized | 4.98 | 8.16 | 20.89 | 49.72 | 367.78 | 2010 | 7139 | 30787 |
| FFT | 11.59 | 24.49 | 25.8 | 46.38 | 122.43 | 368.83 | 1700 | 2598 |
| Speedup | 0.42 | 0.33 | 0.83 | 1.07 | 3.00 | 5.45 | 4.19 | 11.85 |

# 5    Deep Belief Network

A Deep Belief Network (DBN) is a network formed by stacking RBMs on top of each other. It is pretrained by training each layer with Contrastive Divergence. Once the first layer has been trained, its activation probabilities are computed for each input sample and these values are taken as the input of the next layer, and so on until the last layer of the network. A Convolutional DBN (CDBN) is similar, except that it stacks CRBMs.

Since pretraining a DBN consists in training RBMs with CD, the same optimizations discussed in previous sections apply. If there is enough memory, it is important to keep the entire data set in memory as well as the intermediate results (the activation probabilities of the previous layer) during training to maximize the performance. When this is not possible, the best course of action is to keep a multiple of the mini-batch size of samples in memory (and their intermediate output) for training. Ideally, computing the activation probabilities of the previous layer should be done in a separate thread so that CD always has data ready for training.

If the network is to be used for classification, it can then be fine-tuned using standard algorithms like for instance Stochastic Gradient Descent, Conjugate Gradient or Limited-Memory BFGS. Performance of these algorithms is out of the scope of this paper and has already been studied [17].

# 6    Conclusion and Future Work

Several techniques were presented to speedup training of RBM and CRBM models, on a single-CPU system. By using these techniques, RBM's training time has been reduced by up to 30 times and CRBM's training time has been reduced by up to 12 times. This demonstrates that even on CPU, many techniques can be used to substantially speedup the training of RBM models and train large models within reasonable time.

Future work could go in several directions. Combining several full convolutions together and using the FFT reduction could reduce its overhead and allow better performance even for small kernels. The performance of the vectorized convolution versions can also be improved further by vectorizing the convolution at the image level rather than just at the kernel level. Finally, once the large operations are fully optimized, operations such as sigmoid or Bernoulli sampling could also be considered for optimization.

# References

1. Bracewell, R.: The Fourier Transform and Its Applications. McGraw-Hill Electrical and Electronic Engineering Series, vol. 1. McGraw-Hill, New York (1965)
2. Coates, A., Ng, A.Y., Lee, H.: An analysis of single-layer networks in unsupervised feature learning. In: Proceedings of the International Conference on Artificial Intelligence and Statistics, pp. 215–223 (2011)

3. Hinton, G.E., Sejnowski, T.J.: Learning and relearning in Boltzmann machines. In: Parallel Distributed Processing: Explorations in the Microstructure of Cognition, vol. 1, pp. 282–317. MIT Press, Cambridge (1986). http://dl.acm.org/citation.cfm?id=104279.104291

4. Hinton, G.E.: Training products of experts by minimizing contrastive divergence. Neural Comput. **14**, 1771–1800 (2002)

5. Hinton, G.E.: A practical guide to training restricted Boltzmann machines. In: Montavon, G., Orr, G.B., Müller, K.-R. (eds.) Neural Networks: Tricks of the Trade. LNCS, vol. 7700, 2nd edn, pp. 599–619. Springer, Heidelberg (2012)

6. Hinton, G.E., Salakhutdinov, R.R.: Reducing the dimensionality of data with neural networks. Science **313**(5786), 504–507 (2006)

7. Jia, Y., Shelhamer, E., Donahue, J., Karayev, S., Long, J., Girshick, R., Guadarrama, S., Darrell, T.: Caffe: convolutional architecture for fast feature embedding. In: Proceedings of the ACM International Conference on Multimedia, pp. 675–678. ACM (2014)

8. Krizhevsky, A., Hinton, G.: Convolutional Deep belief Networks on CIFAR-10. Unpublished manuscript 40 (2010)

9. Krizhevsky, A., Sutskever, I., Hinton, G.E.: Imagenet classification with deep convolutional neural networks. In: Advances in Neural Information Processing Systems, pp. 1097–1105 (2012)

10. LeCun, Y., Bottou, L., Bengio, Y., Haffner, P.: Gradient-based learning applied to document recognition. Proc. IEEE **86**(11), 2278–2324 (1998)

11. Lee, H., Grosse, R., Ranganath, R., Ng, A.Y.: Convolutional deep belief networks for scalable unsupervised learning of hierarchical representations. In: Proceedings of the International Conference on Machine Learning, pp. 609–616. ACM (2009)

12. Lee, H., Pham, P., Largman, Y., Ng, A.Y.: Unsupervised feature learning for audio classification using CDBNs. In: Proceedings of the Advances in Neural Information Processing Systems, pp. 1096–1104 (2009)

13. Lee, V.W., Kim, C., Chhugani, J., Deisher, M., Kim, D., Nguyen, A.D., Satish, N., Smelyanskiy, M., Chennupaty, S., Hammarlund, P., et al.: Debunking the 100x GPU vs. CPU myth: an evaluation of throughput computing on CPU and GPU. In: ACM SIGARCH Computer Architecture News, vol. 38, pp. 451–460. ACM (2010)

14. Lopes, N., Ribeiro, B.: Towards adaptive learning with improved convergence of deep belief networks on graphics processing units. Pattern Recogn. **47**(1), 114–127 (2014)

15. Ly, D.L., Paprotski, V., Yen, D.: Neural networks on GPUs: Restricted Boltzmann Machines (2008). http://www.eecg.toronto.edu/~moshovos/CUDA08/doku.php

16. Mathieu, M., Henaff, M., LeCun, Y.: Fast training of convolutional networks through FFTs (2013). arXiv preprint arXiv:1312.5851

17. Ngiam, J., Coates, A., Lahiri, A., Prochnow, B., Le, Q.V., Ng, A.Y.: On optimization methods for deep learning. In: Proceedings of the 28th International Conference on Machine Learning (ICML 2011), pp. 265–272 (2011)

18. Ren, J.S., Xu, L.: On vectorization of deep convolutional neural networks for vision tasks (2015). arXiv preprint arXiv:1501.07338

19. Smolensky, P.: Information processing in dynamical systems: foundations of harmony theory. Parallel Distrib. Process. **1**, 194–281 (1986)

20. Upadhyaya, S.R.: Parallel approaches to machine learning: a comprehensive survey. J. Parallel Distrib. Comput. **73**(3), 284–292 (2013)

# Comparing Incremental Learning Strategies
# for Convolutional Neural Networks

Vincenzo Lomonaco$^{(\boxtimes)}$ and Davide Maltoni

DISI - University of Bologna, Bologna, Italy
{vincenzo.lomonaco,davide.maltoni}@unibo.it

**Abstract.** In the last decade, Convolutional Neural Networks (CNNs) have shown to perform incredibly well in many computer vision tasks such as object recognition and object detection, being able to extract meaningful high-level invariant features. However, partly because of their complex training and tricky hyper-parameters tuning, CNNs have been scarcely studied in the context of incremental learning where data are available in consecutive batches and retraining the model from scratch is unfeasible. In this work we compare different incremental learning strategies for CNN based architectures, targeting real-word applications.

**Keywords:** Deep learning · Incremental learning · Convolutional neural networks

## 1 Introduction

In recent years, deep learning has established itself as a real game changer in many domains like speech recognition [1], natural language processing [2] and computer vision [3]. The real power of deep convolutional neural networks resides in their ability of directly processing high-dimensional raw data and produce meaningful high-level representations regardless of the specific task at hand [4]. So far, computer vision has been one of the most benefited fields and tasks like object recognition and object detection have seen significant advances in terms of accuracy (ref. to *ImageNet LSVRC* [5] and *PascalVOC* [6]). However, most of the recent successes are based on benchmarks providing an enormous quantity of data with a fixed training and test set. This is not always the case of real world applications where training data are often only partially available at the beginning and new data keep coming while the system is already deployed and working, like for recommendation systems and anomaly detection where learning sudden behavioral changes becomes quite critical. One possible approach to deal with this incremental scenario is to store all the previously seen past data, and retrain the model from scratch as soon as a new batch of data is available (in the following we refer to this approach as *cumulative* approach). However, this solution is often impractical for many real world systems where memory and computational resources are subject to stiff constraints. As a matter of fact, employing convolutional neural networks is expensive and training them on large datasets could take days or weeks, even on modern GPUs [3]. A different approach to address this issue, is to *update* the model based *only* on the new available batch of data. This is computationally

© Springer International Publishing AG 2016
F. Schwenker et al. (Eds.): ANNPR 2016, LNAI 9896, pp. 175–184, 2016.
DOI: 10.1007/978-3-319-46182-3_15

cheaper and storing past data is not necessary. Of course, nothing comes for free, and this approach may lead to a substantial loss of accuracy with respect to the cumulative strategy. Moreover, the stability-plasticity dilemma may arise and dangerous shifts of the model are always possible because of catastrophic forgetting [7–10]. Forgetting previously learned patterns can be conveniently seen as overfitting the new available training data. Nevertheless, in the incremental scenario, the issue is much more puzzling than overfitting a single, fixed-size training set, since incremental batches are usually smaller, highly variable and class biased. It is worth noting that this is also different and much more complicated than *transfer learning* where the previously used dataset and task are no longer of interest and forgetting previously learned patterns is not a concern.

In this work we compare and evaluate different incremental learning strategies for CNN-based architectures, targeting real-word applications. The aim is understanding how to minimize the impact of forgetting thus reducing as much as possible the gap with respect to the cumulative strategy.

In Sect. 2, different incremental tuning strategies for CNNs are introduced. In Sect. 3, we provide details of the two (very different) real-world datasets used for the comparative evaluation, while, in Sect. 4, we describe the experimentations carried out together with their results. Finally, in Sect. 5, some conclusions are drawn.

## 2   Incremental Learning Strategies

In this section we present different incremental tuning strategies employing CNNs, which are sufficiently general to be applied to any dataset with a number of possible variations. Then, we illustrate the specific implementations used in our experiments on the two selected datasets.

The different possibilities we explored to deal with an incremental tuning/learning scenario, can be conveniently framed in three main strategies:

1. Training/tuning an ad hoc CNN architecture suitable for the problem.
2. Using an already trained CNN as a fixed feature extractor in conjunction with an incremental classifier.
3. Fine-tuning an already trained CNN.

Of course, all the above strategies can be used also outside the context of incremental learning, but very little is known about their applicability and relevance if such is the case. In our experiments we tested three instantiations of the aforementioned strategies:

- **LeNet7:** consists of the classical "LeNet7" proposed by Yan LeCun in 2004 [11]. Its architecture is based on seven layers (much less than current state-of-the-art CNNs designed for large-scale datasets). However, it has been successfully applied to many object recognition datasets (NORB, COIL, CIFAR, etc.) with colorful or grayscale images of size varying from $32 \times 32$ to $96 \times 96$, and is still competitive on low/medium scale problems.
- **CaffeNet + SVM**: In this strategy we employ a pre-trained CNN provided in the Caffe library [12] "*Model Zoo*", *BVLC Reference CaffeNet*, which is based on the

well-known *"AlexNet"* architecture proposed in [3] and trained on ImageNet. This model is used *off-the-shelf* to extract high-level features from the second-last hidden layer following the strategy proposed in [13–15]. Then a linear and incremental SVM[1] is used (instead of the native soft-max output layer) for the final classification.

- **CaffeNet + FT**: Even in this case the *BVLC Reference CaffeNet* is employed. However, instead of using it as a fixed feature extractor the network is fine-tuned to suit the new task. Even if for fine-tuning it is generally recommended to diversify the learning rate of the last layer (which is re-initialized to suit the novel number of output neurons) from the others, we found no significant difference during our exploratory analysis and therefore we kept the hyper-parametrization as homogeneous as possible.

Furthermore, for the BigBrother dataset (see following section) we decided to test an additional pair of strategies: **VGG_Face + SVM** and **VGG_Face + FT**. They are identical respectively to **CaffeNet + SVM** and **CaffeNet + FT** with exception of the pre-trained model used. The VGG_Face is a very deep architecture (16-levels) that has been trained directly on a very large dataset of faces (2,622 Subjects and 2.6 M images) [17].

## 3   Datasets

In the literature, there are few labeled image datasets suitable for incremental learning. In particular, we are interested in datasets where the objects of interest have been acquired in a number of successive sessions and the environmental condition can change among the sessions.

We focused on two applicative fields where incremental learning is very relevant: robotics and biometrics. The two datasets selected, *iCubWorld28* [18] and *BigBrother* [19], are briefly described below.

### 3.1   iCubWorld28

The *iCubWorld28* dataset [18] consists of 28 distinct domestic objects evenly organized into 7 categories (see Fig. 1). Images are 128 × 128 pixels in RGB format. The acquisition session of a single object consists in a video recording of about 20 s where the object is slowly moved/rotated in front of the camera. Each acquisition session results in about 200 train and 200 test images for each of the 28 objects. Being designed to assess the incremental learning performance of the iCub robot visual recognition subsystem, the same acquisition approach has been repeated for 4 consecutive days, ending up with four subsets (Day 1, to 4) of around 8 K images each (39,693 in total).

To better assess the capabilities of our incremental learning strategies we split each training set of Day 1, 2 and 3 in three parts of equal size. On the contrary, Day 4 was

---

[1] We used incremental SVM from LibLinear implementation [16].

Laundry     Plate     Dishwashing  Sponge     Cup     Soap     Sprayer
detergent             detergent

**Fig. 1.** Example images of the 28 objects (7 categories) from one of the 4 subsets constituting icubWorld28.

left unchanged and entirely used as test set (as in [18]). In Table 1, we report the full details about the size of the training and test set used for our experiments.

**Table 1.** *iCubWorld28* batches size and membership to the original Day.

| Partition name | Images count | Original day |
| --- | --- | --- |
| $Batch_1$ | 1341 | $Day_1$ |
| $Batch_2$ | 1341 | $Day_1$ |
| $Batch_3$ | 1341 | $Day_1$ |
| $Batch_4$ | 1789 | $Day_2$ |
| $Batch_5$ | 1788 | $Day_2$ |
| $Batch_6$ | 1788 | $Day_2$ |
| $Batch_7$ | 1836 | $Day_3$ |
| $Batch_8$ | 1836 | $Day_3$ |
| $Batch_9$ | 1836 | $Day_3$ |
| Test | 5550 | $Day_4$ |

## 3.2   BigBrother

The *BigBrother* dataset [19] has been created starting from 2 DVDs made commercially available at the end of the 2006 edition of the "Big Brother" reality show produced for the Italian TV and documenting the 99 days of permanence of 20 participants in a closed environment. It consists of 23,842 70 × 70 gray-scale images of faces belonging to 19 subjects (one participant was immediately eliminated at the beginning of the reality show). In addition to the typical training and test sets, an additional large set of images (called "updating set") is provided for incremental learning/tuning purposes. Details about the composition of each set can be found in [19], together with the number of days the person lived in the house. However, some subjects lived in the house for a short period and too few images are thus available for an in-depth evaluation. For this reason, a subset of the whole database, referred to as SETB, has been defined by the authors of [19]. It includes the images of the 7 subjects who lived in the house for a longer period (such number of users seems realistic for a home environment application).

In this work, we compare our incremental tuning strategies on the SETB of the Big-Brother dataset consisting of a total of 54 incremental batches. In Fig. 2, an example image for each of the different seven subjects of the SETB is shown. It is worth noting that images have been automatically extracted from the video frames by Viola and Jones detector [20] and are often characterized by bad lighting, poor focus, occlusions, and non-frontal pose.

**Fig. 2.** Example images of the seven subjects contained in the SETB of the BigBrother Dataset.

## 4 Experiments and Results

For all the strategies employed in this work, we trained the models until full convergence on the first batch of data and tuned them on the successive incremental batches, trying to balance the trade-off between accuracy gain and forgetting. This protocol fits the requirements of many real-world applications where a reasonable initial accuracy is demanded and the first batch is large enough to reach that accuracy.

For the *iCubWorld28* dataset the three strategies have been validated over 10 runs where we randomly shuffled the position of the batches $TrainB_2, \ldots, TrainB_9$.

To control forgetting during the incremental learning phase we relied on early stopping and for each batch a fixed number of iterations were performed depending on the specific strategy. For example, for the **LeNet7**, trained with stochastic gradient descent (SGD), we chose a learning rate of 0.01, a mini-batch size of 100 and a number of iterations of 50 for all the eight incremental batches. We found that tuning these hyper-parameters can have a significant impact on forgetting (see an example on the *BigBrother* dataset in Fig. 4).

To better understand the efficacy of the proposed incremental strategies and to quantify the impact of forgetting, we also tested each model on the corresponding *cumulative* strategies. In Fig. 3, the average accuracy over the ten runs is reported for each strategy. We note that:

- The **CaffeNet + SVM** has a quite good recognition rate increment along the 9 batches, moving from an accuracy of 41,63 % to 66,97 %. The standard deviation is initially higher with respect to the other strategies, but it rapidly decreases as new batches of data are available and the SVM model is updated. Furthermore, the small gap with respect to its cumulative counterpart proves that a fixed features extractor favors stability and reduces forgetting.
- The **CaffeNet + FT** is the most effective strategy for this dataset. This is probably because the features originally learned on the ImageNet dataset are very general and the *iCubWorld28* dataset can be thought as a specific sub-domain where feature fine-tuning can help pattern classification. Moreover, even if splitting the dataset in 9 batches makes the task harder, we managed to achieve an averaged accuracy of 78.40 % that outperforms previously proposed methods on the same datasets [18]. Even if in this case the gap with respect to the cumulative approach is slightly higher, the proper adjustment of early stopping and learning rate during the incremental phase allows to effectively control forgetting.
- The **LeNet7** on this dataset is probably not able to learn (being the number of patterns too limited) complex invariant features that are necessary to deal with the multi-axes rotations, partial occlusions and the complex backgrounds which characterize this problem. The gap with respect to the cumulative approach is here high. This is in line with previous studies [7, 10] showing that smaller models without pre-training are much more susceptible to forgetting.

For the *SETB* of the *BigBrother* dataset, in order to make reproducible and comparable results, we decided to keep fixed (i.e., no shuffling) the order of the 54 updating batches as contained in the original dataset [19]. In Fig. 5, accuracy results are reported for each of the 5 tested strategies. It is worth pointing out that in this case the 54 incremental batches used for updating the model have a very high variance in terms of number of patterns they contain: in particular, it can vary from few dozens to many hundreds. This is typical in many real-world systems, where the hypothesis of collecting uniform and equally sized batches is often unrealistic.

Controlling forgetting was here more complex than for the iCubWorld28 dataset. In fact, in this case, due to the aforementioned high variation in the number of patterns in the different incremental batches, we found that adapting the tuning strength[2] to the batch size can lead to relevant improvements.

In Fig. 4, an exemplifying parameterization for the **CaffeNet + FT** strategy is reported where we compare the learning trend by using (i) a low learning rate, (ii) a high learning rate, (iii) an adjustable learning rate depending on the size of the batch. Results show that using an adjustable learning rate leads to better results. Therefore, in the rest of the experiments on the *BigBrother* dataset, an adjustable learning rate[3] is used.

---

[2] In terms of number of iterations and/or learning rate.

[3] By now, we used a simple thresholding approach where the learning rate was varied among three fixed values, since in these experiments we did not found any significant difference using a continuous approach.

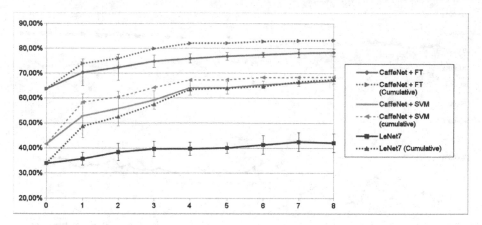

**Fig. 3.** IcubWorld28 dataset: average accuracy during incremental training (8 batches). The bars indicate the standard deviation of the ten runs performed for each strategy. The dotted lines denote the cumulative strategies.

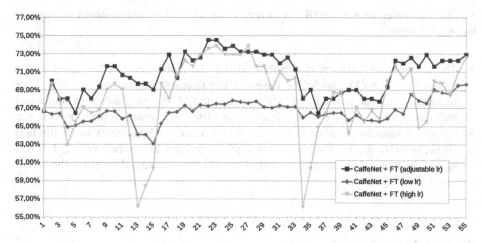

**Fig. 4.** Accuracy results of different parameterizations for the *CaffeNet + FT* strategy: an example of the impact of the learning rate on the *BigBrother* dataset, in our incremental scenario (54 batches).

In Fig. 5 the accuracy on the *BigBrother* dataset for all the strategies introduced in Sect. 2 is reported. For this dataset we note that:

- Unlike the previous experiment, here **LeNet7** model performs slightly better than **CaffeNet + SVM** or **CaffeNet + FT**. This is probably because of the high peculiar features (and invariance) requested for face recognition. Hence, learning the features from scratch for this dataset seems more appropriate than adapting general features by fine-tuning.

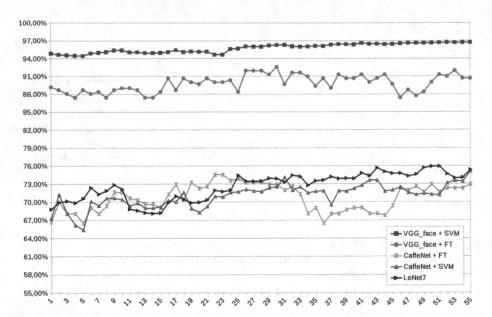

**Fig. 5.** *BigBrother* dataset (SETB): accuracy of the different strategies during incremental training (54 batches).

- The previous observation is corroborated by the really good performance of the **VGG_Face + SVM** and **VGG_Face + FT** strategies. In fact, since VGG_Face features have been learned in a face recognition task by using a dataset containing millions of faces, they are pretty effective for a transfer learning in the same domain.
- Since the features are already optimal, **VGG_Face + SVM** seems to be the best choice both for the accuracy and the stability. It reaches an accuracy of 96,73 % that is 24,1 % better than accuracy reported in [21] for the same dataset (in the supervised learning scenario).

## 5   Conclusions

Incremental and on-line learning is still scarcely studied in the field of deep learning (especially for CNNs) but it is essential for many real-world applications. In this work we explored different strategies to train Convolutional Neural Networks incrementally. We recognize that the empirical evaluations carried out in this work are still limited, and further studies are necessary to better understand the advantages and weakness of each strategy. However, it seems that the lesson learned in classical transfer learning [13, 14, 22] holds here too:

- Forgetting can be a very detrimental issue: hence, when possible (i.e., transfer learning from the same domain), it is preferable to use CNN as a fixed feature extractor to feed an incremental classifier. In general, this results in better stability and often in improved efficiency (i.e., tuning all CNN layers can be computationally expensive).

- If the features are not optimized (transfer learning from a different domain), the tuning of low level layers may be preferable and the learning strength (i.e., learning rate, number of iteration, etc.) can be used to control forgetting.
- Training a CNN from scratch can be advantageous if the problem patterns (and feature invariances) are highly specific and a sufficient number of samples are available.

In the near future, we plan to extend this work with a more extensive experimental evaluation, finding a more principled way to control forgetting and adapting the tuning parameters to the size (and bias) of each incremental batch. Lastly, we are interested in studying real-world applications of semi-supervised incremental learning strategies for CNNs, with approaches similar to [23].

# References

1. Hinton, G., Deng, L., Yu, D., Dahl, G.E., Mohamed, A., Jaitly, N., Senior, A., Vanhoucke, V., Nguyen, P., Sainath, T.N., Kingsbury, B.: Deep neural networks for acoustic modeling in speech recognition. IEEE Signal Process. Mag. 82–97 (2012)
2. Mikolov, T., Chen, K., Corrado, G., Dean, J.: Distributed representations of words and phrases and their compositionality. Adv. Neural Inf. Process. Syst. (NIPS) 1–9 (2013)
3. Krizhevsky, A., Sulskever, I., Hinton, G.E.: ImageNet classification with deep convolutional neural networks. Adv. Neural Inf. Process. Syst. (NIPS) 1–9 (2012)
4. Bengio, Y., Courville, A., Vincent, P.: Unsupervised feature learning and deep learning: a review and new perspectives. IEEE Trans. Pattern Anal. Mach. Intell. 35, 1798–1828
5. Russakovsky, O., Deng, J., Su, H., Krause, J., Satheesh, S., Ma, S., Huang, Z., Karpathy, A., Khosla, A., Bernstein, M., Berg, A.C., Fei-Fei, L.: ImageNet large scale visual recognition challenge. Int. J. Comput. Vis. 115, 211–252 (2015)
6. Everingham, M., Van Gool, L., Williams, C.K.I., Winn, J., Zisserman, A.: The PASCAL visual object classes (voc) challenge. Int J Comput Vis. 88, 303–338 (2010)
7. Mermillod, M., Bugaiska, A., Bonin, P.: The stability-plasticity dilemma: investigating the continuum from catastrophic forgetting to age-limited learning effects. Front. Psychol. 4, 504 (2013)
8. Mccloskey, M., Cohen, N.J.: Catastrophic interference in connectionist networks: the sequential learning problem. Psychol. Learn. Motiv. 109–165 (1989)
9. French, R.M.: Catastrophic forgetting in connectionist networks. Encycl. Cogn. Sci. Nadel/Cogn. (2006)
10. Goodfellow, I.J., Mirza, M., Courville, A., Bengio, Y.: An empirical investigation of catastrophic forgetting in gradient-based neural networks (2012). arXiv preprint arXiv:1312. 6211v3 (2015)
11. LeCun, Y., Huang, F.J., Bottou, L.: Learning methods for generic object recognition with invariance to pose and lighting. In: Proceedings of the 2004 IEEE Computer Society Conference on Computer Vision Pattern Recognition, 2004, CVPR 2004, vol. 2, pp. 97–104 (2004)

12. Jia, Y., Shelhamer, E., Donahue, J., Karayev, S., Long, J., Girshick, R., Guadarrama, S., Darrell, T.: Caffe: convolutional architecture for fast feature embedding. In: Proceedings of the ACM International Conference Multimedia, pp. 675–678 (2014)

13. Donahue, J., Jia, Y., Vinyals, O., Hoffman, J., Zhang, N., Tzeng, E., Darrell, T.: DeCAF: a deep convolutional activation feature for generic visual recognition. In: International Conference on Machine Learning (ICML), vol. 32, pp. 647–655 (2014)

14. Razavian, A.S., Azizpour, H., Sullivan, J., Carlsson, S.: CNN features off-the-shelf: an astounding baseline for recognition. In: IEEE Computer Society Conference Computer Vision Pattern Recognition Workshops, pp. 512–519 (2014)

15. Chatfield, K., Simonyan, K., Vedaldi, A., Zisserman, A.: Return of the devil in the details: delving deep into convolutional nets. In: Proceedings of the British Machine Vision Conference pp. 1–11 (2014)

16. Tsai, C.-H., Lin, C.-Y., Lin, C.-J.: Incremental and decremental training for linear classification. In: 20th ACM SIGKDD International Conference Knowledge Discovery Data Mining, KDD 2014, pp. 343–352 (2014)

17. Parkhi, O.M., Vedaldi, A., Zisserman, A.: Deep face recognition. In: Proceedings of the British Machine Vision Conference 2015, pp. 41.1–41.12 (2015)

18. Pasquale, G., Ciliberto, C., Odone, F., Rosasco, L., Natale, L.: Real-world object recognition with off-the-shelf deep conv nets: how many objects can iCub learn? arXiv:1504.03154 [cs] (2015)

19. Franco, A., Maio, D., Maltoni, D.: The big brother database: evaluating face recognition in smart home environments. In: Tistarelli, M., Nixon, M.S. (eds.) ICB 2009. LNCS, vol. 5558, pp. 142–150. Springer, Heidelberg (1999)

20. Viola, P., Jones, M.: Rapid object detection using a boosted cascade of simple features. In: Computer Vision Pattern Recognition, vol. 1, pp. I–511–I–518 (2001)

21. Franco, A., Maio, D., Maltoni, D.: Incremental template updating for face recognition in home environments. Pattern Recognit. **43**, 2891–2903 (2010)

22. Yosinski, J., Clune, J., Bengio, Y., Lipson, H.: How transferable are features in deep neural networks? In: Advances Neural Information Processing Systems (NIPS), vol. 27, pp. 1–9 (2014)

23. Maltoni, D., Lomonaco, V.: Semi-supervised tuning from temporal coherence. Technical report, DISI – University of Bologna, pp. 1–14 (2015). http://arXiv.org/pdf/1511.03163v3.pdf

# Approximation of Graph Edit Distance by Means of a Utility Matrix

Kaspar Riesen[1,3]($\boxtimes$), Andreas Fischer[2], and Horst Bunke[3]

[1] Institute for Information Systems, University of Applied Sciences FHNW,
Riggenbachstrasse 16, 4600 Olten, Switzerland
kaspar.riesen@fhnw.ch
[2] Department of Informatics, University of Fribourg and HES-SO,
1700 Fribourg, Switzerland
andreas.fischer@unifr.ch
[3] Institute of Computer Science and Applied Mathematics, University of Bern,
Neubrückstrasse 10, 3012 Bern, Switzerland
bunke@iam.ch

**Abstract.** Graph edit distance is one of the most popular graph matching paradigms available. By means of a reformulation of graph edit distance to an instance of a linear sum assignment problem, the major drawback of this dissimilarity model, viz. the exponential time complexity, has been invalidated recently. Yet, the substantial decrease of the computation time is at the expense of an approximation error. The present paper introduces a novel transformation that processes the underlying cost model into a utility model. The benefit of this transformation is that it enables the integration of additional information in the assignment process. We empirically confirm the positive effects of this transformation on three standard graph data sets. That is, we show that the accuracy of a distance based classifier can be improved with the proposed transformation while the run time remains nearly unaffected.

## 1 Introduction

Graphs are recognized as versatile alternative to feature vectors and thus, they found widespread application in pattern recognition and related fields [1,2]. However, one drawback of graphs, when compared to feature vectors, is the significant increase of the complexity of many algorithms. Regard, for instance, the algorithmic comparison of two patterns (which is actually a basic requirement for pattern recognition). Due to the homogeneous nature of feature vectors, pairwise comparisons is straightforward and can be accomplished in linear time with respect to the length of the two vectors. Yet, the same task for graphs, commonly referred to as *graph matching*, is much more complex, as one has to identify common parts of the graphs by considering all of their subsets of nodes. Regarding that there are $O(2^n)$ subsets of nodes in a graph with $n$ nodes, the inherent difficulty of graph matching becomes obvious.

In the last four decades a huge number of procedures for graph matching have been proposed in the literature [1,2]. They range from *spectral methods* [3,4], over

© Springer International Publishing AG 2016
F. Schwenker et al. (Eds.): ANNPR 2016, LNAI 9896, pp. 185–194, 2016.
DOI: 10.1007/978-3-319-46182-3_16

*graph kernels* [5,6], to reformulations of the discrete graph matching problem to an instance of a *continuous optimization problem* (basically by relaxing some constraints) [7]. *Graph edit distance* [8,9], introduced about 30 years ago, is still one of the most flexible graph distance models available and topic of various recent research projects.

In order to compute the graph edit distance often A* based search techniques using some heuristics are employed (e.g. [10]). Yet, exact graph edit distance computation based on a tree search algorithm is exponential in the number of nodes of the involved graphs. Formally, for two graphs with $m$ and $n$ nodes we observe a time complexity of $O(m^n)$. This means that for large graphs the computation of the exact edit distance is intractable.

In [11] authors of the present paper introduced an algorithmic framework for the approximation of graph edit distance. The basic idea of this approach is to reduce the difficult problem of graph edit distance to a *linear sum assignment problem* (LSAP), for which an arsenal of efficient (i.e. cubic time) algorithms exist [12]. In two recent papers [13,14] the optimal algorithm for the LSAP has been replaced with a suboptimal greedy algorithm which runs in quadratic time. Due to the lower complexity of this suboptimal assignment process, a substantial speed up of the complete approximation procedure has been observed. However, it was also reported that the distance accuracy of this extension is slightly worse than with the original algorithm. Major contribution of the present paper is to improve the overall distance accuracy of this recent procedure by means of an elaborated transformation of the underlying cost model.

The remainder of this paper is organized as follows. Next, in Sect. 2, the computation of graph edit distance is thoroughly reviewed. In particular, it is shown how the graph edit distance problem can be reduced to a linear sum assignment problem. In Sect. 3, the transformation of the cost model into a utility model is outlined. Eventually, in Sect. 4, we empirically confirm the benefit of this transformation in a classification experiment on three graph data sets. Finally, in Sect. 5, we conclude the paper.

## 2   Graph Edit Distance (GED)

### 2.1   Exact Computation of GED

A graph $g$ is a four-tuple $g = (V, E, \mu, \nu)$, where $V$ is the finite set of nodes, $E \subseteq V \times V$ is the set of edges, $\mu : V \to L_V$ is the node labeling function, and $\nu : E \to L_E$ is the edge labeling function. The labels for both nodes and edges can be given by the set of integers $L = \{1, 2, 3, \ldots\}$, the vector space $L = \mathbb{R}^n$, a set of symbolic labels $L = \{\alpha, \beta, \gamma, \ldots\}$, or a combination of various label alphabets from different domains. Unlabeled graphs are obtained by assigning the same (empty) label $\varnothing$ to all nodes and edges, i.e. $L_V = L_E = \{\varnothing\}$.

Given two graphs, $g_1 = (V_1, E_1, \mu_1, \nu_1)$ and $g_2 = (V_2, E_2, \mu_2, \nu_2)$, the basic idea of *graph edit distance* (*GED*) [8,9] is to transform $g_1$ into $g_2$ using edit operations, viz. *insertions*, *deletions*, and *substitutions* of both nodes and edges. The substitution of two nodes $u$ and $v$ is denoted by $(u \to v)$, the deletion of node $u$ by $(u \rightarrow \varepsilon)$, and the insertion of node $v$ by $(\varepsilon \rightarrow v)$[1]. A set of edit operations $\lambda(g_1, g_2) = \{o_1, \ldots, o_k\}$ that transform $g_1$ completely into $g_2$ is called an edit path between $g_1$ and $g_2$.

Note that edit operations on edges are uniquely defined by the edit operations on their adjacent nodes. That is, whether an edge $(u, v)$ is substituted with an existing edge from the other graph, deleted, or inserted actually depends on the operations performed on both adjacent nodes $u$ and $v$. Thus, we define that an edit path $\lambda(g_1, g_2)$ explicitly contains the edit operations between the graphs' nodes $V_1$ and $V_2$, while the edge edit operations are implicitly given by these node edit operations.

A cost function that measures the strength of an edit operation is commonly introduced for graph edit distance. The edit distance between two graphs $g_1$ and $g_2$ is then defined by the sum of cost of the minimum cost edit path $\lambda_{\min}$ between $g_1$ and $g_2$. In fact, the problem of finding the minimum cost edit path $\lambda_{\min}$ between $g_1$ and $g_2$ can be reformulated as a *quadratic assignment problem* (QAP). Roughly speaking, QAPs deal with the problem of assigning $n$ entities of a first set $S = \{s_1, \ldots, s_n\}$ to $n$ entities of a second set $Q = \{q_1, \ldots, q_n\}$ under some (computationally demanding) side constraints. A common way to formally represent assignments between the entities of $S$ and $Q$ is given by means of permutations $(\varphi_1, \ldots, \varphi_n)$ of the integers $(1, 2, \ldots, n)$. A permutation $(\varphi_1, \ldots, \varphi_n)$ refers to the assignment where the first entity $s_1 \in S$ is mapped to entity $q_{\varphi_1} \in Q$, the second entity $s_2 \in S$ is assigned to entity $q_{\varphi_2} \in Q$, and so on.

By reformulating the graph edit distance problem to an instance of a QAP, two major issues have to be resolved. First, QAPs are generally stated on sets with equal cardinality. Yet, in case of graph edit distance the elements to be assigned to each other are given by the sets of nodes (and edges) with unequal cardinality in general. Second, solutions to QAPs refer to assignments of elements in which every element of the first set is assigned to exactly one element of the second set and vice versa (i.e. a solution to a QAP corresponds to a bijective assignment of the underlying entities). Yet, GED is a more general assignment problem as it explicitly allows both deletions and insertions to occur on the basic entities (rather than only substitutions).

These two issues can be simultaneously resolved by adding an appropriate number of empty "nodes" $\varepsilon$ to both graphs $g_1$ and $g_2$. Formally, assume that $|V_1| = n$ and $|V_2| = m$, we extend $V_1$ and $V_2$ according to

$$V_1^+ = V_1 \cup \overbrace{\{\varepsilon_1, \ldots, \varepsilon_m\}}^{m \; empty \; nodes} \quad \text{and} \quad V_2^+ = V_2 \cup \underbrace{\{\varepsilon_1, \ldots, \varepsilon_n\}}_{n \; empty \; nodes}.$$

Since both graphs $g_1$ and $g_2$ have now an equal number of nodes, viz. $(n + m)$, their corresponding adjacency matrices $\mathbf{A}$ and $\mathbf{B}$ offer also equal dimension.

---

[1] A similar notation is used for edges.

These adjacency matrices of $g_1$ and $g_2$ are defined by

$$
\mathbf{A} = \begin{array}{c} \\ 1 \\ \vdots \\ n \\ 1 \\ \vdots \\ m \end{array}
\begin{array}{c} \overset{1 \quad \cdots \quad n}{\phantom{x}} \; \overset{1 \cdots m}{\phantom{x}} \\
\left[ \begin{array}{ccc|ccc}
a_{11} & \cdots & a_{1n} & \varepsilon & \cdots & \varepsilon \\
\vdots & \ddots & \vdots & \vdots & \ddots & \vdots \\
a_{n1} & \cdots & a_{nn} & \varepsilon & \cdots & \varepsilon \\
\varepsilon & \cdots & \varepsilon & \varepsilon & \cdots & \varepsilon \\
\vdots & \ddots & \vdots & \vdots & \ddots & \vdots \\
\varepsilon & \cdots & \varepsilon & \varepsilon & \cdots & \varepsilon
\end{array} \right] \end{array}
\quad \text{and} \quad
\mathbf{B} = \begin{array}{c} \\ 1 \\ \vdots \\ m \\ 1 \\ \vdots \\ n \end{array}
\begin{array}{c} \overset{1 \quad \cdots \quad m}{\phantom{x}} \; \overset{1 \cdots n}{\phantom{x}} \\
\left[ \begin{array}{ccc|ccc}
b_{11} & \cdots & b_{1m} & \varepsilon & \cdots & \varepsilon \\
\vdots & \ddots & \vdots & \vdots & \ddots & \vdots \\
b_{m1} & \cdots & b_{mm} & \varepsilon & \cdots & \varepsilon \\
\varepsilon & \cdots & \varepsilon & \varepsilon & \cdots & \varepsilon \\
\vdots & \ddots & \vdots & \vdots & \ddots & \vdots \\
\varepsilon & \cdots & \varepsilon & \varepsilon & \cdots & \varepsilon
\end{array} \right] \end{array}
\tag{1}
$$

If there actually is an edge between node $u_i \in V_1$ and $v_j \in V_1$, entry $a_{ij}$ refers to this edge $(u_i, v_j) \in E_1$, and otherwise to the empty "edge" $\varepsilon$. Note that there cannot be any edge from an existing node in $V_1$ to an empty node $\varepsilon$ and thus the corresponding entries $a_{ij} \in \mathbf{A}$ with $i > n$ and/or $j > n$ are also empty. The same observations account for entries $b_{ij}$ in $\mathbf{B}$.

Next, based on the extended node sets $V_1^+$ and $V_2^+$ of $g_1$ and $g_2$, respectively, a *cost matrix* $\mathbf{C}$ can be established as follows.

$$
\mathbf{C} = \begin{array}{c} \\ u_1 \\ u_2 \\ \vdots \\ u_n \\ \varepsilon_1 \\ \varepsilon_2 \\ \vdots \\ \varepsilon_m \end{array}
\begin{array}{c} \overset{v_1 \quad v_2 \quad \cdots \quad v_m}{\phantom{x}} \; \overset{\varepsilon_1 \; \varepsilon_2 \; \cdots \; \varepsilon_n}{\phantom{x}} \\
\left[ \begin{array}{cccc|cccc}
c_{11} & c_{12} & \cdots & c_{1m} & c_{1\varepsilon} & c_{1\varepsilon} & \cdots & c_{1\varepsilon} \\
c_{21} & c_{22} & \cdots & c_{2m} & c_{2\varepsilon} & c_{2\varepsilon} & & \vdots \\
\vdots & \vdots & \ddots & \vdots & \vdots & & \ddots & \vdots \\
c_{n1} & c_{n2} & \cdots & c_{nm} & c_{n\varepsilon} & \cdots & c_{n\varepsilon} & c_{n\varepsilon} \\
c_{\varepsilon 1} & c_{\varepsilon 1} & \cdots & c_{\varepsilon 1} & 0 & 0 & \cdots & 0 \\
c_{\varepsilon 2} & c_{\varepsilon 2} & & \vdots & 0 & 0 & & \vdots \\
\vdots & & \ddots & \vdots & \vdots & & \ddots & 0 \\
c_{\varepsilon m} & \cdots & c_{\varepsilon m} & c_{\varepsilon m} & 0 & \cdots & 0 & 0
\end{array} \right] \end{array}
\tag{2}
$$

Entry $c_{ij}$ thereby denotes the cost $c(u_i \to v_j)$ of the node substitution $(u_i \to v_j)$, $c_{i\varepsilon}$ denotes the cost $c(u_i \to \varepsilon)$ of the node deletion $(u_i \to \varepsilon)$, and $c_{\varepsilon j}$ denotes the cost $c(\varepsilon \to v_j)$ of the node insertion $(\varepsilon \to v_j)$. Obviously, the left upper part of the cost matrix represents the costs of all possible node substitutions, the right upper part the costs of all possible node deletions, and the bottom left part the costs of all possible node insertions. The bottom right part of the cost matrix is set to zero since substitutions of the form $(\varepsilon \to \varepsilon)$ should not cause any cost.

Given the adjacency matrices $\mathbf{A}$ and $\mathbf{B}$ as well as the cost matrix $\mathbf{C}$ (Eqs. 1 and 2), the following optimization problem can now be stated.

$$
(\varphi_1, \ldots, \varphi_{(n+m)}) = \underset{(\varphi_1, \ldots, \varphi_{(n+m)}) \in \mathcal{S}_{(n+m)}}{\arg \min} \left[ \sum_{i=1}^{n+m} c_{i\varphi_i} + \sum_{i=1}^{n+m} \sum_{j=1}^{n+m} c(a_{ij} \to b_{\varphi_i \varphi_j}) \right],
$$

where $\mathcal{S}_{(n+m)}$ refers to the set of all $(n+m)!$ possible permutations of the integers $1, 2, \ldots, (n + m)$. Note that this optimal permutation $(\varphi_1, \ldots, \varphi_{(n+m)})$ (as well as any other valid permutation) corresponds to a bijective assignment

$$\lambda = \{(u_1 \rightarrow v_{\varphi_1}), (u_2 \rightarrow v_{\varphi_2}), \ldots, (u_{m+n} \rightarrow v_{\varphi_{m+n}})\}$$

of the extended node set $V_1^+$ of $g_1$ to the extended node set $V_2^+$ of $g_2$. That is, assignment $\lambda$ includes node edit operations of the form $(u_i \rightarrow v_j)$, $(u_i \rightarrow \varepsilon)$, $(\varepsilon \rightarrow v_j)$, and $(\varepsilon \rightarrow \varepsilon)$ (the latter can be dismissed, of course). In other words, an arbitrary permutation $(\varphi_1, \ldots, \varphi_{(n+m)})$ perfectly corresponds to a valid edit path $\lambda$ between two graphs.

The optimization problem stated in Eq. 2.1 exactly corresponds to a standard QAP. Note that the linear term $\sum_{i=1}^{n+m} c_{i\varphi_i}$ refers to the sum of cost of all node edit operations, which are defined by the permutation $(\varphi_1, \ldots, \varphi_{n+m})$. The quadratic term $\sum_{i=1}^{n+m} \sum_{j=1}^{n+m} c(a_{ij} \rightarrow b_{\varphi_i\varphi_j})$ refers to the implied edge edit cost defined by the node edit operations. That is, since node $u_i \in V_1^+$ is assigned to a node $v_{\varphi_i} \in V_2^+$ and node $u_j \in V_1^+$ is assigned to a node $v_{\varphi_j} \in V_2^+$, the edge $(u_i, u_j) \in E_1 \cup \{\varepsilon\}$ (stored in $a_{ij} \in \mathbf{A}$) has to be assigned to the edge $(v_{\varphi_i}, v_{\varphi_j}) \in E_2 \cup \{\varepsilon\}$ (stored in $b_{\varphi_i\varphi_j} \in \mathbf{B}$).

## 2.2 Approximate Computation of GED

In fact, QAPs are very hard to solve as they belong to the class of NP-hard problems. Authors of the present paper introduced an algorithmic framework which allows the approximation of graph edit distance in a substantially faster way than traditional methods [11]. The basic idea of this approach is to reduce the QAP of graph edit distance computation to an instance of a *Linear Sum Assignment Problem* (*LSAP*). LSAPs are similar to QAPs in the sense of also formulating an assignment problem of entities. Yet, in contrast with QAPs, LSAPs are able to optimize the permutation $(\varphi_1, \ldots, \varphi_{(n+m)})$ with respect to the linear term $\sum_{i=1}^{n+m} c_{i\varphi_i}$ only. That is, LSAPs consider a single cost matrix $\mathbf{C}$ without any side constraints. For solving LSAPs a large number of efficient (i.e. polynomial) algorithms exist (see [12] for an exhaustive survey on LSAP solvers).

Yet, by omitting the quadratic term $\sum_{i=1}^{n+m} \sum_{j=1}^{n+m} c(a_{ij} \rightarrow b_{\varphi_i\varphi_j})$ during the optimization process, we neglect the structural relationships between the nodes (i.e. the edges between the nodes). In order to integrate knowledge about the graph structure, to each entry $c_{ij} \in \mathbf{C}$, i.e. to each cost of a node edit operation $(u_i \rightarrow v_j)$, the minimum sum of edge edit operation costs, implied by the corresponding node operation, can be added. Formally, for every entry $c_{ij}$ in the cost matrix $\mathbf{C}$ one might solve an LSAP on the ingoing and outgoing edges of node $u_i$ and $v_j$ and add the resulting cost to $c_{ij}$. That is, we define

$$c_{ij}^* = c_{ij} + \min_{(\varphi_1, \ldots, \varphi_{(n+m)}) \in \mathcal{S}_{(n+m)}} \sum_{k=1}^{n+m} c(a_{ik} \rightarrow b_{j\varphi_k}) + c(a_{ki} \rightarrow b_{\varphi_k j}),$$

where $\mathcal{S}_{(n+m)}$ refers to the set of all $(n+m)!$ possible permutations of the integers $1, \ldots, (n+m)$. To entry $c_{i\varepsilon}$, which denotes the cost of a node deletion, the cost of the deletion of all incident edges of $u_i$ can be added, and to the entry $c_{\varepsilon j}$, which denotes the cost of a node insertion, the cost of all insertions of the incident edges of $v_j$ can be added. We denote the cost matrix which is enriched with structural information with $\mathbf{C}^* = (c_{ij}^*)$ from now on.

In [11] the cost matrix $\mathbf{C}^* = (c_{ij}^*)$ as defined above is employed in order to optimally solve the LSAP by means of *Munkres Algorithm* [15][2]. The LSAP optimization consists in finding a permutation $(\varphi_1^*, \ldots, \varphi_{n+m}^*)$ of the integers $(1, 2, \ldots, (n+m))$ that minimizes the overall assignment cost $\sum_{i=1}^{(n+m)} c_{i\varphi_i^*}^*$. Similar to the permutation $(\varphi_1, \ldots, \varphi_{n+m})$ obtained on the QAP, the permutation $(\varphi_1^*, \ldots, \varphi_{n+m}^*)$ corresponds to a bijective assignment of the entities in $V_1^+$ to the entities in $V_2^+$. In other words, the permutation $(\varphi_1^*, \ldots, \varphi_{(n+m)}^*)$ refers to an admissible and complete (yet not necessarily minimal cost) edit path between the graphs under consideration. We denote this approximation framework with *BP-GED* from now on.

Recently, it has been proposed to solve the LSAP stated on $\mathbf{C}^*$ with an approximation rather than with an exact algorithm [13, 14]. This algorithm iterates through $\mathbf{C}^*$ from top to bottom through all rows and assigns every element to the minimum unused element in a greedy manner. Clearly, the complexity of this suboptimal assignment algorithm is $O((n+m)^2)$. For the remainder of this paper we denote the graph edit distance approximation where the LSAP on $\mathbf{C}^*$ is solved by means of this greedy procedure with *GR-GED*.

## 3    Building the Utility Matrix

Similar to [13, 14] we aim at solving the basic LSAP in $O(n^2)$ time in order to approximate the graph edit distance. Yet, in contrast with this previous approach, which considers the cost matrix $\mathbf{C}^* = (c_{ij}^*)$ directly as its basis, we transform the given cost matrix into a *utility matrix* with equal dimension as $\mathbf{C}^*$ and work with this matrix instead.

The rationale behind this transformation is based on the following observation. When picking the minimum element $c_{ij}$ from cost matrix $\mathbf{C}^*$, i.e. when assigning node $u_i$ to $v_j$, we exclude both nodes $u_i$ and $v_j$ from any future assignment. However, it may happen that node $v_j$ is not only the best choice for $u_i$ but also for another node $u_k$. Because $v_j$ is no longer available, we may be forced to map $u_k$ to another, very expensive node $v_l$, such that the total assignment cost becomes higher than mapping node $u_i$ to some node that is (slightly) more expensive than $v_j$. In order to take such situations into account, we incorporate additional information in the utility matrix about the the minimum and maximum value in each row, and each column.

Let us consider the $i$-th row of the cost matrix $\mathbf{C}^*$ and let *row-min$_i$* and *row-max$_i$* denote the minimum and maximum value occurring in this row, respectively. Formally, we have

$$row\text{-}min_i = \min_{j=1,\ldots,(n+m)} c_{ij}^* \quad \text{and} \quad row\text{-}max_i = \max_{j=1,\ldots,(n+m)} c_{ij}^*.$$

---

[2] The time complexity of this particular algorithm is cubic in the size of the problem, i.e. $O((n+m)^3)$.

If the node edit operation $(u_i \rightarrow v_j)$ is selected, one might interpret the quantity

$$row\text{-}win_{ij} = \frac{row\text{-}max_i - c_{ij}^*}{row\text{-}max_i - row\text{-}min_i}$$

as a *win* for $(u_i \rightarrow v_j)$, when compared to the locally worst case situation where $v_k$ with $k = \arg\max_{j=1,\ldots,(n+m)} c_{ij}^*$ is chosen as target node for $u_i$. Likewise, we might interpret

$$row\text{-}loss_{ij} = \frac{c_{ij}^* - row\text{-}min_i}{row\text{-}max_i - row\text{-}min_i}$$

as a *loss* for $(u_i \rightarrow v_j)$, when compared to selecting the minimum cost assignment which would be possible in this row. Note that both $row\text{-}win_{ij}$ and $row\text{-}loss_{ij}$ are normalized to the interval $[0, 1]$. That is, when $c_{ij}^* = row\text{-}min_i$ we have a maximum win of 1 and a minimum loss of 0. Likewise, when $c_{ij}^* = row\text{-}max_i$ we observe a minimum win of 0 and a maximum loss of 1.

Overall we define the *utility* of the node edit operation $(u_i \rightarrow v_j)$ with respect to row $i$ as

$$row\text{-}utility_{ij} = row\text{-}win_{ij} - row\text{-}loss_{ij} = \frac{row\text{-}max_i + row\text{-}min_i - 2c_{ij}^*}{row\text{-}max_i - row\text{-}min_i}.$$

Clearly, when $c_{ij} = row\text{-}min_i$ we observe a row utility of $+1$, and vice versa, when $c_{ij} = row\text{-}max_i$ we have a row utility of $-1$.

So far the utility of a node edit operation $(u_i \rightarrow v_j)$ is quantified with respect to the $i$-th row only. In order to take into account information about the $j$-th column, we seek for the minimum and maximum values that occur in column $j$ by

$$col\text{-}min_j = \min_{i=1,\ldots,(n+m)} c_{ij}^* \quad \text{and} \quad col\text{-}max_j = \max_{i=1,\ldots,(n+m)} c_{ij}^*.$$

Eventually, we define

$$col\text{-}win_{ij} = \frac{col\text{-}max_j - c_{ij}^*}{col\text{-}max_j - col\text{-}min_j} \quad \text{and} \quad col\text{-}loss_{ij} = \frac{c_{ij}^* - col\text{-}min_j}{col\text{-}max_j - col\text{-}min_j}.$$

Similarly to the utility of the node edit operation $(u_i \rightarrow v_j)$ with respect to row $i$ we may define the utility of the same edit operation with respect to column $j$ as

$$col\text{-}utility_{ij} = col\text{-}win_{ij} - col\text{-}loss_{ij} = \frac{col\text{-}max_j + col\text{-}min_j - 2c_{ij}^*}{col\text{-}max_j - col\text{-}min_j}.$$

To finally estimate the utility $u_{ij}$ of a node edit operation $(u_i \rightarrow v_j)$ with respect to both row $i$ and column $j$ we compute the sum

$$u_{ij} = row\text{-}utility_{ij} + col\text{-}utility_{ij}.$$

Since both $row\text{-}utility_{ij}$ and $col\text{-}utility_{ij}$ lie in the interval $[-1, 1]$, we have $u_{ij} \in [-2, 2]$ for $i, j = 1, \ldots, (n + m)$. We denote the final utility matrix by $\mathbf{U} = (u_{ij})$.

## 4    Experimental Evaluation

In the experimental evaluation we aim at investigating the benefit of using the utility matrix $\mathbf{U}$ instead of the cost matrix $\mathbf{C}^*$ in the framework GR-GED. In particular, we aim at assessing the quality of the different distance approximations by means of comparisons of the sum of distances and by means of a distance based classifier. Actually, a nearest-neighbor classifier (NN) is employed. Note that there are various other approaches to graph classification that make use of graph edit distance in some form. Yet, the nearest neighbor paradigm is particularly interesting for the present evaluation because it directly uses the distances without any additional classifier training.

We use three real world data sets from the IAM graph database repository [16][3]. Two graph data sets involve graphs that represent molecular compounds (AIDS and MUTA). These data set consists of two classes, which represent molecules with activity against HIV or not (AIDS), and molecules with and without the *mutagen* property (MUTA), respectively. The third data set consists of graphs representing proteins stemming from six different classes (PROT).

**Table 1.** The mean run time for one matching ($\varnothing t$), the relative increase of the sum of distances compared with BP-GED, and the recognition rate (rr) of a nearest-neighbor classifier using a specific graph edit distance algorithm.

| Data Set | Algorithm | | | | | | | | |
|---|---|---|---|---|---|---|---|---|---|
| | BP-GED($C^*$) | | | GR-GED($C^*$) | | | GR-GED($U$) | | |
| | $\varnothing t$ | sod | rr | $\varnothing t$ | sod | rr | $\varnothing t$ | sod | rr |
| AIDS | 3.61 | - | 99.07 | 1.21 | 1.92 | 98.93 | 1.34 | 2.40 | 99.00 |
| MUTA | 33.89 | - | 70.20 | 4.56 | 1.50 | 70.10 | 5.06 | 0.68 | 71.60 |
| PROT | 25.54 | - | 67.50 | 13.31 | 10.86 | 64.50 | 14.11 | 2.71 | 66.00 |

In Table 1 the results obtained with three different graph edit distance approximations are shown. The first algorithm is BP-GED($\mathbf{C}^*$), which solves the LSAP on $\mathbf{C}^*$ in an optimal manner in cubic time [11]. The second algorithm is GR-GED($\mathbf{C}^*$), which solves the LSAP on $\mathbf{C}^*$ in a greedy manner in quadratic time [13,14]. Finally, the third algorithm is GR-GED($\mathbf{U}$), which operates on the utility matrix $\mathbf{U}$ instead of $\mathbf{C}^*$ (also using the greedy assignment algorithm).

We first focus on the mean run time for one matching in ms ($\varnothing t$) and compare BP-GED with GR-GED that operates on the original cost matrix $\mathbf{C}^*$. On all data sets substantial speed-ups of GR-GED($\mathbf{C}^*$) can be observed. On the AIDS data set, for instance, the greedy approach GR-GED($\mathbf{C}^*$) is approximately three times faster than BP-GED. On the MUTA data set the mean matching time is decreased from 33.89 ms to 4.56 ms (seven times faster) and on the PROT data the greedy approach approximately halves the matching time (25.43 ms

---

[3] www.iam.unibe.ch/fki/databases/iam-graph-database.

vs. 13.31 ms). Comparing GR-GED($\mathbf{C}^*$) with GR-GED($\mathbf{U}$) we observe only a small increase of the matching time when the latter approach is used. The slight increase of the run time, which is actually observable on all data sets, is due to the computational overhead that is necessary for transforming the cost matrix $\mathbf{C}^*$ to the utility matrix $\mathbf{U}$.

Next, we focus on the distance quality of the greedy approximation algorithms. Note that all of the employed algorithms return an upper bound on the true edit distance, and thus, the lower the sum of distances of a specific algorithm is, the better is its approximation quality. For our evaluation we take the sum of distances returned by BP-GED as reference point and measure the relative increase of the sum of distances when compared with BP-GED (*sod*). We observe that GR-GED($\mathbf{C}^*$) increases the sum of distances by 1.92 % on the AIDS data when compared with BP-GED. On the other two data sets the sum of distances is also increased (by 1.50 % and 10.86 %, respectively). By using the utility matrix $\mathbf{U}$ rather than the cost matrix $\mathbf{C}$ in the greedy assignment algorithm, we observe smaller sums of distances on the MUTA and PROT data sets. Hence, we conclude that GR-GED($\mathbf{U}$) is able to produce more accurate approximations than GR-GED($\mathbf{C}$) in general.

Finally, we focus on the recognition rate (*rr*) of a NN-classifier that uses the different distance approximations. We observe that the NN-classifier that is based on the distances returned by GR-GED($\mathbf{C}^*$) achieves lower recognition rates than the same classifier that uses distances from BP-GED (on all data sets). This loss in recognition accuracy may be attributed to the fact that the approximations in GR-GED are coarser than those in BP-GED. Yet, our novel procedure, i.e. GR-GED($\mathbf{U}$), improves the recognition accuracy on all data sets when compared to GR-GED($\mathbf{C}^*$). Moreover, we observe that GR-GED($\mathbf{U}$) is inferior to BP-GED in two out of three cases only.

## 5   Conclusions and Future Work

In the present paper we propose to use a utility matrix instead of a cost matrix for the assignment of local substructures in a graph. The motivation for this transformation is based on the greedy behavior of the basic assignment algorithm. More formally, with the transformation of the cost matrix into a utility matrix we aim at increasing the probability of selecting a correct node edit operation during the optimization process. With an experimental evaluation on three real world data sets, we empirically confirm that our novel approach is able to increase the accuracy of a distance based classifier, while the run time is nearly not affected.

In future work we aim at testing other (greedy) assignment algorithms on the utility matrix $\mathbf{U}$. Moreover, there seems to be room for developing and researching variants of the utility matrix with the aim of integrating additional information about the trade-off between wins and losses of individual assignments.

**Acknowledgements.** This work has been supported by the *Hasler Foundation* Switzerland.

# References

1. Conte, D., Foggia, P., Sansone, C., Vento, M.: Thirty years of graph matching in pattern recognition. Int. J. Pattern Recognit. Art Intell. **18**(3), 265–298 (2004)
2. Foggia, P., Percannella, G., Vento, M.: Graph matching and learning in pattern recognition in the last 10 years. Int. J. Pattern Recognit. Art Intell. **28**(1), 1450001 (2014)
3. Luo, B., Wilson, R.C., Hancock, E.R.: Spectral feature vectors for graph clustering. In: Caelli, T.M., Amin, A., Duin, R.P.W., Kamel, M.S., de Ridder, D. (eds.) SPR 2002 and SSPR 2002. LNCS, vol. 2396, p. 83. Springer, Heidelberg (2002)
4. Wilson, R.C., Hancock, E.R., Luo, B.: Pattern vectors from algebraic graph theory. IEEE Trans. Pattern Anal. Mach. Intell. **27**(7), 1112–1124 (2005)
5. Gaüzère, B., Brun, L., Villemin, D.: Two new graphs kernels in chemoinformatics. Pattern Recognit. Lett. **33**(15), 2038–2047 (2012)
6. Borgwardt, K., Kriegel, H.-P.: Graph kernels for disease outcome prediction from protein-protein interaction networks. Pac. Symp. Biocomput. **2007**, 4–15 (2007)
7. Torsello, A., Hancock, E.: Computing approximate tree edit distance using relaxation labeling. Pattern Recognit. Lett. **24**(8), 1089–1097 (2003)
8. Bunke, H., Allermann, G.: Inexact graph matching for structural pattern recognition. Pattern Recognit. Lett. **1**, 245–253 (1983)
9. Sanfeliu, A., Fu, K.S.: A distance measure between attributed relational graphs for pattern recognition. IEEE Trans. Syst. Man Cybern. (Part B) **13**(3), 353–363 (1983)
10. Fischer, A., Plamondon, R., Savaria, Y., Riesen, K., Bunke, H.: A hausdorff heuristic for efficient computation of graph edit distance. In: Fränti, P., Brown, G., Loog, M., Escolano, F., Pelillo, M. (eds.) S+SSPR 2014. LNCS, vol. 8621, pp. 83–92. Springer, Heidelberg (2014)
11. Riesen, K., Bunke, H.: Approximate graph edit distance computation by means of bipartite graph matching. Image Vis. Comput. **27**(4), 950–959 (2009)
12. Burkard, R., Dell'Amico, M., Martello, S.: Assignment Problems. Society for Industrial and Applied Mathematics, Philadelphia (2009)
13. Riesen, K., Ferrer, M., Dornberger, R., Bunke, H.: Greedy graph edit distance. In: Perner, P. (ed.) MLDM 2015. LNCS, vol. 9166, pp. 3–16. Springer, Heidelberg (2015)
14. Riesen, K., Ferrer, M., Fischer, A., Bunke, H.: Approximation of graph edit distance in quadratic time. In: Liu, C.-L., Luo, B., Kropatsch, W.G., Cheng, J. (eds.) GbRPR 2015. LNCS, vol. 9069, pp. 3–12. Springer, Heidelberg (2015)
15. Munkres, J.: Algorithms for the assignment and transportation problems. J. Soci. Ind. Appl. Math. **5**(1), 32–38 (1957)
16. Riesen, K., Bunke, H.: IAM graph database repository for graph based pattern recognition and machine learning. In: da Vitoria Lobo, N., Kasparis, T., Roli, F., Kwok, J.T., Georgiopoulos, M., Anagnostopoulos, G.C., Loog, M. (eds.) Structural, Syntactic, and Statistical Pattern Recognition. LNCS, vol. 5342, pp. 287–297. Springer, Heidelberg (2008)

# Applications

# Time Series Classification in Reservoir- and Model-Space: A Comparison

Witali Aswolinskiy[1]([✉]), René Felix Reinhart[2], and Jochen Steil[1]

[1] Research Institute for Cognition and Robotics - CoR-Lab, Universitässtraße 25, 33615 Bielefeld, Germany
waswolinskiy@cor-lab.uni-bielefeld.de
[2] Fraunhofer Research Institution for Mechatronic Systems Design IEM, Zukunftsmeile 1, 33102 Paderborn, Germany

**Abstract.** Learning in the space of Echo State Network (ESN) output weights, i.e. model space, has achieved excellent results in time series classification, visualization and modelling. This work presents a systematic comparison of time series classification in the model space and the classical, discriminative approach with ESNs. We evaluate the approaches on 43 univariate and 18 multivariate time series. It turns out that classification in the model space achieves often better classification rates, especially for high-dimensional motion datasets.

**Keywords:** Time series classification · Echo State Network · Model space

## 1 Introduction

The idea of learning in the model space [6] is to train models on parts of the data and then use the model parameters for further processing. Recently, this approach was applied to learn parameterized skills in robotics [16,19,20], to model parameterized processes [2] and to classify [7,8] and to visualize [12,14] time series.

While the idea to use parameters of linear models for time series classification appeared as early as in 1997 [10], only the more recent usage of non-linear, reservoir-based models allowed the method to achieve results similar or better as state-of-the-art methods [1,7,8]. The models used in the latter publications are variants of Echo State Networks (ESN, [13]). For each time series, an ESN-model is trained and the model parameters are used as features in a consecutive classification stage [8]. Typically, the models are trained to minimize the one-step-ahead prediction error on the time series. The number of model parameters is independent of the length of the time series, which allows for the deployment of any feature-based classifier in the model space.

Time series classification in the model space of ESNs was evaluated on a smaller number of datasets in [7,8], however, a systematic comparison to the classical, discriminative approach to time series classification with ESNs [15] is

© Springer International Publishing AG 2016
F. Schwenker et al. (Eds.): ANNPR 2016, LNAI 9896, pp. 197–208, 2016.
DOI: 10.1007/978-3-319-46182-3_17

missing. Our contribution is the evaluation of a large number of datasets and the comparison of the classification performance in the model space with that of two discriminative ESN architectures. The results show that model space learning (MSL) is often superior to discriminative ESN training and on par with other state-of-the-art methods.

The remainder of this paper is structured as follows. In Sect. 2, work related to learning in the model space is presented. In the following three sections, feature based time series classification is defined and the reservoir- and model-based feature creation presented. Section 6 describes the datasets and our approach to parameter search and evaluation. Finally, the results are presented and discussed.

## 2    Related Work

While the idea of time series classification in the model space of reservoir networks was first explicitly formulated in [6,8], its roots reach far back. For example, the mapping of a time series with the Fourier transform into the frequency spectrum can be seen as the most rudimentary form of a model space transformation. To our knowledge, the first approach to use a predictive model was employed in [10], where for each time series an Autoregressive Moving Average model was fitted and the estimated parameters were used for classification. The models can also be domain-specific as was shown in [5], where fMRI measurements were represented by parameters of biophysically motivated, generative models. Other types of models include frequencies of local polynomials [5], covariance matrices [3] and Hidden Markov Models [11].

Time series classification in the model space of reservoir networks was first introduced in [8]. The authors used a cycle reservoir with regular jumps (CRJ, [17]). The classification was carried out via kernels: The input sequences were transformed into models and three different kernels, each defined by a different model distance measure, were evaluated. One of the kernels, 'RV', used an adapted Euclidean distance between the readout weights. Another kernel ('Sampling RV') compared the models via their prediction error. An evaluation of these and other kernels on nine univariate and three multivariate datasets showed that Sampling RV outperformed the other kernels. The parameters of the CRJ-reservoir were determined via cross-validation. In [7] another approach was taken: The reservoir parameters and the readout weights were learned simultaneously to minimize the prediction error, the distance between models of the same class and to maximize the distance between models of different classes. The authors named this approach Model Metric Co-Learning (MMCL). The evaluation of MMCL on ten datasets (subset of those from [8]) showed, that MMCL had lower error rates than RV on most of the datasets, but was also slower.

In our work we classify directly in the model space using a linear classifier, which decreases the complexity of the approach and increases it's speed. This is a form of feature-based time series classification, which is formally introduced in the next section.

## 3    Feature-Based Time Series Classification with a Linear Classifier

Let a dataset consist of $S$ time series $\boldsymbol{u}_i, i = 1, \ldots, S$ with varying lengths $K_i$: $\boldsymbol{u}_i(1), \ldots, \boldsymbol{u}_i(K_i)$. $K$ is the total number of time steps: $K = \sum_{i=1}^{S} K_i$. The goal of time series classification is to assign to each time series a class label $c \in \{1, \ldots, C\}$. In feature-based time series classification, each sequence is transformed into a feature vector and then classified [22]. Alternatively, the features can be computed and classified stepwise, resulting in sequences of features and class predictions, respectively. Then, the class predictions have to be combined to have a prediction per sequence.

For classification we use a linear classifier: $\boldsymbol{y} = \boldsymbol{W}\boldsymbol{h}$, where $\boldsymbol{y}$ is the prediction, $\boldsymbol{W}$ the classifier weights and $\boldsymbol{h}$ the features. The classifier weights are efficiently learned with ridge regression: $\boldsymbol{W} = (\boldsymbol{H}^T \boldsymbol{H} + \alpha \boldsymbol{I})^{-1} \boldsymbol{H}^T \boldsymbol{T}$, where $\boldsymbol{H}$ are the features, collected either per sequence or per step, $\boldsymbol{T}$ are the target values, $\alpha$ is the regularization strength and $\boldsymbol{I}$ is the identity matrix. The target values are the class labels encoded in the so-called 1-of-$K$ scheme and have the dimensionality $C$.

If the features are collected per sequence, the class of a sequence is decided via winner-takes-all: $\hat{c}_i = \arg\max_c(y_{i,c})$, where $y_{i,c}$ is the prediction for the class $c$ of the $i$-th sequence. If the features and predictions are computed step-wise, the step-wise class predictions are averaged over the length of the time series, before winner-takes-all is applied: $\hat{c}_i = \arg\max_c(1/K_i * \sum_{j=1}^{K_i} y_{i,c}(j))$, where $y_{i,c}(j)$ is the prediction for the class $c$ of the $j$-th step in the $i$-th sequence.

## 4    Classification in the Space of Reservoir Activations

An ESN consists of two parts: A reservoir of recurrently connected neurons and a linear readout. The reservoir provides a non-linear fading memory of the inputs. The reservoir states $\mathbf{x}$ and the readouts $\boldsymbol{y}$ are updated according to

$$\boldsymbol{x}(k) = (1 - \lambda)\boldsymbol{x}(k-1) + \lambda f(\boldsymbol{W}^{rec}\boldsymbol{x}(k-1) + \boldsymbol{W}^{in}\boldsymbol{u}(k)) \tag{1}$$

$$\boldsymbol{y}(k) = \boldsymbol{W}^{out}\boldsymbol{x}(k), \tag{2}$$

where $\lambda$ is the leak rate, $f$ the activation function, e.g. $tanh$, $\mathbf{W}^{rec}$ the recurrent weight matrix, $\mathbf{W}^{in}$ the weight matrix from the inputs to the reservoir neurons and $\mathbf{W}^{out}$ the weight matrix from the reservoir neurons to the readouts $\boldsymbol{y}$. $\mathbf{W}^{in}$ and $\mathbf{W}^{rec}$ are initialized randomly, scaled and remain fixed. $\mathbf{W}^{rec}$ is scaled to fulfill the Echo State Property (ESP, [13]). The necessary condition for the ESP is typically achieved by scaling the spectral radius of $\mathbf{W}^{rec}$ to be smaller than one. The scaling of $\mathbf{W}^{in}$ is task-dependent and has strong influence on the network performance [15].

The input time series are fed into the reservoir and the readout is typically trained to predict the class label for each step of the time series. The readout weights are then: $\boldsymbol{W}^{out} = (\boldsymbol{H}^T \boldsymbol{H} + \alpha \boldsymbol{I})^{-1} \boldsymbol{H}^T \boldsymbol{T}$, where $\boldsymbol{H} \in \mathbb{R}^{K \times N}$ is the matrix of the row-wise collected N reservoir neuron activations for all input time steps $x(k), k = 1, \ldots, K$. During testing, the step-wise class predictions are averaged over the length of the time series and the class is decided via winner-takes-all (see Sect. 3).

Alternatively, during training and testing, the reservoir states can be averaged over parts of the time series [15]. We consider here the case, where the reservoir activations are averaged over the complete length of the time series and only then are fitted to the class labels. Each sequence will be represented by a single, averaged reservoir activations vector independent of the length of the sequence. The readout weights are then: $W^{out} = (\bar{H}^T \bar{H} + \alpha I)^{-1} \bar{H}^T T$, where the i-th row of $\bar{H} \in \mathbb{R}^{S \times N}$ is $h_i = 1/K_i \sum_{j=1}^{K_i} x_i(j)$ and represents the i-th sequence.

We explore these two discriminative ESN architectures for time series classification using the following notation:

- "ESN": Reservoir activations for all time steps of the input sequence are used to learn the class labels.
- "ESN": The averaged reservoir activations per input sequence are used to learn the class labels.

## 5    Classification in the Model Space

To enable learning in the model space, the ESNs are an intermediate step to create a time-independent representation of the time series. For each time series $u_i$, an ESN is trained to predict from the previous step $u_i(k)$ the next step $u_i(k+1)$ in the time series. The ESNs are trained independently, but share the same reservoir parameters $\mathbf{W}^{in}$ and $\mathbf{W}^{rec}$ in order to create a coherent model space. Thus, for each $u_i$ a $W_i^{out}$ is trained by minimizing the one-step-ahead prediction error

$$E(W_i^{out}) = \frac{1}{K_i - 1} \sum_{k=2}^{K_i} (u_i(k) - W_i^{out} x_i(k-1))^2 + \alpha \|W_i^{out}\|^2, \qquad (3)$$

$$W_i^{out} = (X_i^T X_i + \alpha I)^{-1} X_i^T U_i, \qquad (4)$$

where $X_i \in \mathbb{R}^{(K_i-1) \cdot N}$ are the row-wise collected reservoir activations for the input $i$ and $\alpha$ is the regularization strength. This results in $S$ readout weight matrices $W_i^{out}$. The classifier weights are then: $W^{clf} = (\Omega^T \Omega + \gamma I)^{-1} \Omega^T T$, where the i-th row of $\Omega \in \mathbb{R}^{S \times (D \cdot N)}$ is $\omega_i$, the vectorized version of $W_i^{out}$. The dimensionality of this model space for signals with $D$ dimensions is $D \cdot N$, or $D \cdot (N + 1)$ if a regression intercept is used. $\gamma$ is a regularization parameter.

We denote this approach as model space learning (MSL). The idea of MSL is sketched in Fig. 1. The example in Fig. 1 shows the classification of noisy sums of sine waves. In the example, the time series and hence the reservoir activations are 100 steps long. For comparison, in the lower part of Fig. 1, each time series

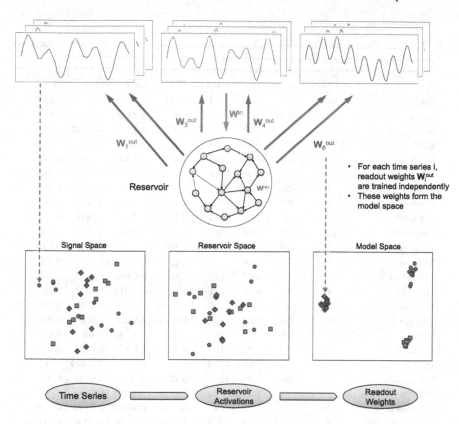

**Fig. 1.** Learning in the Model Space of ESNs. In the upper part of the diagram, noisy sums of sine waves are shown. The goal is to classify the time series as indicated by the colors. For this purpose, the time series are fed into the reservoir. For each time series a readout is trained to predict the next input. In the lower part of the diagram, the time series, the reservoir activations and the readout weights are visualized with matching colors. The signal, reservoir activation and readout weight matrices were projected to a plane using PCA. (Color figure online)

is represented as a point in signal space, reservoir space (the space of reservoir neuron activations) and model space. For visualization purposes, the data was projected to two dimensions via principal component analysis. In this example, time series from different classes can be easily separated in the model space.

## 6    Evaluation

### 6.1    Datasets and Other State-of-the-Art Approaches

We evaluated ESN, $\overline{\text{ESN}}$ and MSL on 43 univariate datasets from the UCR archive [9], 13 multivariate datasets gathered in [4] and five Kinect skeleton datasets from [21]. The Tables 1, 2 and 3 provide information on the datasets.

**Table 1.** Univariate datasets from UCR [9]. Properties and classification error rates. The lowest error rate for a task is marked bold.

| Task | Classes | Train | Test | Length | ESN | $\overline{\text{ESN}}$ | MSL | LPS |
|------|---------|-------|------|--------|------|------|------|------|
| 50words | 50 | 450 | 455 | 270 | 0.244 | **0.2** | 0.2725 | 0.213 |
| Adiac | 37 | 390 | 391 | 176 | 0.3478 | 0.3248 | 0.335 | **0.211** |
| Beef | 5 | 30 | 30 | 470 | 0.4667 | 0.5 | **0.1333** | 0.367 |
| CBF | 3 | 30 | 900 | 128 | 0.04 | 0.0022 | 0.14 | **0.002** |
| ChlorineConc. | 3 | 467 | 3840 | 166 | 0.4674 | **0.1549** | 0.1589 | 0.352 |
| CinC_ECG_torso | 4 | 40 | 1380 | 1639 | 0.5123 | 0.4746 | 0.2942 | **0.064** |
| Coffee | 2 | 28 | 28 | 286 | 0.1786 | 0.1429 | 0.2143 | **0.071** |
| Cricket_X | 12 | 390 | 390 | 300 | 0.4513 | **0.2256** | 0.2718 | 0.282 |
| Cricket_Y | 12 | 390 | 390 | 300 | **0.1795** | 0.2051 | 0.1846 | 0.208 |
| Cricket_Z | 12 | 390 | 390 | 300 | **0.1744** | 0.2231 | 0.2026 | 0.305 |
| DiatomSizeRed. | 4 | 16 | 306 | 345 | 0.3072 | 0.2222 | 0.0948 | **0.049** |
| ECG200 | 2 | 100 | 100 | 96 | 0.13 | 0.11 | 0.11 | **0.061** |
| ECGFiveDays | 2 | 23 | 861 | 136 | **0.0** | 0.0046 | 0.0488 | 0.155 |
| FaceAll | 14 | 560 | 1690 | 131 | 0.1592 | **0.1018** | 0.1314 | 0.242 |
| FaceFour | 4 | 24 | 88 | 350 | 0.0455 | 0.1591 | 0.1818 | **0.04** |
| FacesUCR | 14 | 200 | 2050 | 131 | **0.0405** | 0.0517 | 0.0561 | 0.098 |
| Fish | 7 | 175 | 175 | 463 | **0.04** | 0.0686 | **0.04** | 0.094 |
| Gun_Point | 2 | 50 | 150 | 150 | 0.02 | 0.02 | 0.0867 | **0.0** |
| Haptics | 5 | 155 | 308 | 1092 | 0.737 | 0.5649 | **0.5422** | 0.562 |
| InlineSkate | 7 | 100 | 550 | 1882 | 0.7727 | 0.6636 | 0.5473 | **0.494** |
| ItalyPowerDemand | 2 | 67 | 1029 | 24 | 0.0525 | **0.0292** | 0.0534 | 0.053 |
| Lighting2 | 2 | 60 | 61 | 637 | 0.459 | 0.3607 | 0.2459 | **0.197** |
| Lighting7 | 7 | 70 | 73 | 319 | 0.3288 | 0.3425 | **0.274** | 0.411 |
| MALLAT | 8 | 55 | 2345 | 1024 | 0.1706 | 0.0631 | 0.0627 | **0.093** |
| MedicalImages | 10 | 381 | 760 | 99 | 0.3961 | **0.2461** | 0.2605 | 0.297 |
| MoteStrain | 2 | 20 | 1252 | 84 | 0.1893 | 0.1589 | 0.2939 | **0.114** |
| OSULeaf | 6 | 200 | 242 | 427 | 0.1736 | 0.1901 | 0.1446 | **0.134** |
| OliveOil | 4 | 30 | 30 | 570 | 0.6 | 0.1333 | **0.1** | 0.133 |
| SonyAIBORobot | 2 | 20 | 601 | 70 | **0.1531** | 0.3311 | 0.3494 | 0.225 |
| SonyAIBORobot II | 2 | 27 | 953 | 65 | **0.0661** | 0.0766 | 0.1312 | 0.123 |
| StarLightCurves | 3 | 1000 | 8236 | 1024 | 0.1663 | 0.1463 | 0.0389 | **0.033** |
| SwedishLeaf | 15 | 500 | 625 | 128 | 0.1248 | 0.0896 | 0.1248 | **0.072** |
| Symbols | 6 | 25 | 995 | 398 | 0.1658 | 0.1206 | 0.0533 | **0.03** |
| Synthetic Control | 6 | 300 | 300 | 60 | 0.0233 | **0.0067** | 0.03 | 0.027 |
| Trace | 4 | 100 | 100 | 275 | 0.02 | **0.0** | **0.0** | 0.02 |
| TwoLeadECG | 2 | 23 | 1139 | 82 | 0.0123 | 0.0483 | **0.0018** | 0.061 |
| Two_Patterns | 4 | 1000 | 4000 | 128 | 0.0053 | **0.001** | 0.0015 | 0.014 |
| Wafer | 2 | 1000 | 6164 | 152 | 0.1079 | 0.0045 | 0.012 | **0.001** |
| WordsSynonyms | 25 | 267 | 638 | 270 | 0.3339 | 0.326 | 0.4169 | **0.27** |
| Yoga | 2 | 300 | 3000 | 426 | 0.2687 | 0.2347 | 0.2177 | **0.136** |
| uWaveGestureX | 8 | 896 | 3582 | 315 | 0.2596 | **0.1812** | 0.2192 | 0.189 |
| uWaveGestureY | 8 | 896 | 3582 | 315 | 0.3364 | 0.2901 | 0.3199 | **0.263** |
| uWaveGestureZ | 8 | 896 | 3582 | 315 | 0.3063 | 0.2543 | 0.2775 | **0.253** |

**Table 2.** Multivariate datasets from [4]. Properties and classification error rates. The lowest error rate for a task is marked bold.

| Task | Classes | Dim | Train | Test | Length | ESN | $\overline{\text{ESN}}$ | MSL | LPS |
|---|---|---|---|---|---|---|---|---|---|
| AUSLAN* | 95 | 22 | 1140 | 1425 | 45–136 | 0.047 | 0.0533 | **0.0253** | 0.246 |
| ArabicDigits | 10 | 13 | 6600 | 2200 | 4–93 | 0.0132 | 0.0077 | **0.0055** | 0.029 |
| CMUsubject16* | 2 | 62 | 29 | 29 | 127–580 | **0.0** | 0.069 | **0.0** | **0.0** |
| CharacterTraj.* | 20 | 3 | 300 | 2558 | 109–205 | 0.0117 | **0.009** | 0.0141 | 0.035 |
| ECG | 2 | 2 | 100 | 100 | 39–152 | 0.17 | **0.11** | 0.16 | 0.18 |
| JapaneseVowels | 9 | 12 | 270 | 370 | 7–29 | 0.0189 | **0.0054** | 0.0081 | 0.049 |
| KickvsPunch* | 2 | 62 | 16 | 10 | 274–841 | 0.1 | 0.4 | **0.0** | 0.1 |
| Libras* | 15 | 2 | 180 | 180 | 45 | 0.1667 | 0.1556 | **0.0944** | 0.097 |
| NetFlow | 2 | 4 | 803 | 534 | 50–997 | 0.0899 | 0.0562 | 0.0749 | **0.032** |
| PEMS | 7 | 165 | 267 | 173 | 144 | 0.2948 | 0.3121 | 0.2775 | **0.156** |
| UWave* | 8 | 3 | 200 | 4278 | 315 | 0.0657 | 0.0683 | 0.0736 | **0.02** |
| Wafer | 2 | 6 | 298 | 896 | 104–198 | 0.0179 | 0.0134 | **0.01** | 0.038 |
| WalkvsRun* | 2 | 62 | 28 | 16 | 128–1918 | **0.0** | **0.0** | **0.0** | **0.0** |

*Motion datasets

**Table 3.** Kinect skeleton datasets from [21]. Properties and classification error rates. The lowest error rate for a task is marked bold.

| Task | Classes | Dim | Samples | Length | ESN | $\overline{\text{ESN}}$ | MSL | PLG |
|---|---|---|---|---|---|---|---|---|
| Florence3d | 9 | 42 | 215 | 35 | 0.1455 | 0.1727 | 0.1213 | **0.0912** |
| MSR-Action3d AS1 | 8 | 57 | 219 | 76 | 0.2188 | 0.2868 | **0.0191** | 0.0471 |
| MSR-Action3d AS2 | 8 | 57 | 228 | 76 | 0.3673 | 0.2888 | **0.0731** | 0.1613 |
| MSR-Action3d AS3 | 8 | 57 | 222 | 76 | 0.0383 | 0.0375 | **0.0124** | 0.0178 |
| UTKinect | 10 | 57 | 199 | 74 | 0.1066 | 0.1275 | 0.0845 | **0.0292** |

The datasets in these collections have predefined training and test splits which allows to compare the reservoir approaches additionally to the methods in [4,21], respectively. In Learned Pattern Similarity (LPS, [4]) each time series is represented as a matrix of segments and a tree-based learning strategy is used to learn the dependency structure of the segments. Classification utilizes a similarity measure based on this representation. In [21], a representation of skeleton motions as points in a Lie group (here abbreviated as PLG) was proposed. Classification is carried out on the covariance matrices for skeleton joint locations over time.

## 6.2 Parameter Search

ESNs have several parameters with strong influence on performance: The number of neurons, the scaling of the input and recurrent weights, the regularization strength and others. An exhaustive grid search of the high-dimensional parameter space is not possible - it would be necessary to evaluate between 2.000 and 100.000 combinations or more, depending on the level of detail. Instead we employed a heuristic, iterative approach resembling hill climbing: From a starting parameter combination, neighbor combinations are evaluated and the combination with the lowest error is selected as the next starting point. Thus, the error decreases until a local minimum is reached or the allotted computation time is exceeded. Since the number of neighboring points in a high-dimensional space is large, we optimize only one parameter at a time. When the parameter change (increase or decrease) does not lead to a better performance, the next parameter is selected. The order of the parameters was: input weight scaling, regularization factor(s), leak rate, recurrent weight scaling, bias scaling and number of neurons.

The parameter search was performed on training data only - the test data was used only for testing. For the Kinect datasets, ten train/test splits were provided. The first split was used for parameter search.

The parameter search was run approximately for five hours on a standard workstation with four cores. The maximal number of neurons was restricted to 1000 for the ESN-Architectures and to 500 for MSL. Since the input and recurrent ESN weight matrices are random, the best error rate over five network initializations is reported.

### 6.3  Data Preprocessing and Evaluation

In order to evaluate the classification approaches and not the preprocessing methods, only scaling was applied. All datasets were scaled to the range $[-1/d, 1/d]$, where $d$ is the number of dimensions in the dataset. This makes the scaling of the input weights independent of the dimension of the dataset, which allows identical initial conditions for the parameter search and eases manual experimentation. During testing, the scaling factors computed from the training data were applied, hereby allowing the test data to exceed the range.

The 963-dimensional PEMS dataset was reduced to 165 dimensions via PCA to reduce the amount of computation time.

## 7  Classification Errors

The classification test error rates for ESN, $\overline{\text{ESN}}$, MSL and either LPS [4] or PLG [21] are listed in the Tables 1, 2 and 3. The lowest error rate for a task is marked bold. The average error rates with standard deviations are shown in Fig. 2. For univariate and multivariate datasets together, the difference between the approaches is small. The comparison on the multivariate and especially the multivariate motion datasets reveals a stronger difference in the achieved classification error rates. MSL reduces the test classification rate roughly by half compared to the ESN-Architectures. In five of the seven motion datasets in Table 2 MSL performs on par or better than the other approaches. For the five Kinect datasets in Table 3, MSL was far better than ESN and $\overline{\text{ESN}}$ in all tasks and better than PLG in three tasks. Notably, PLG [21] was specifically developed for Kinect skeleton motion classification, while MSL has remarkably good results in other areas as well, e.g. the ArabicDigits speech recognition task.

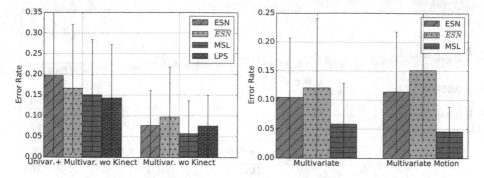

**Fig. 2.** Average test error rates for different selections of the datasets. From the left: 43 univariate and 13 multivariate without Kinect, 13 multivariate without Kinect, 18 multivariate datasets including Kinect and 12 motion datasets (7 multivariate + 5 Kinect). Error bars indicate standard deviation.

To test the significance of the results we performed a Friedman+Nemenyi test with [18]. Over all datasets, MSL is significantly different from ESN ($p = 0.01$), but not from $\overline{ESN}$. The ranking is: MSL: 1.81, $\overline{ESN}$: 1.84, ESN: 2.34. The difference between MSL and $\overline{ESN}$ is only significant over the 18 multivariate datasets (Friedman Aligned Ranks $p = 0.0025$) with the ranking MSL: 16.3, ESN: 32.3.8, $\overline{ESN}$: 33.9. There is no significant difference between MSL and LPS and between MSL and PLG for the Kinect datasets.

# 8 Classification in Model Space Invariant Against 'Standstills' in the AUSLAN Data

The Australian Sign Language Signs (AUSLAN) dataset consists of 2565 samples (27 repetitions × 95 signs), recorded from a native signer using hand position trackers. The sequences are between 45 and 136 steps long and have 22 input dimensions. MSL achieves an error rate approximately half as large as the ESN-Architectures. The investigation of the dataset structure revealed a factor responsible for the difference in performance. Figure 3 shows three randomly selected samples from different classes. The first half of the time series

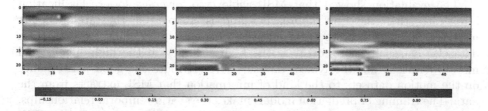

**Fig. 3.** Heatmaps of three samples from different classes in the AUSLAN dataset.

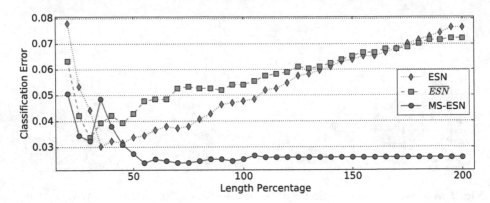

**Fig. 4.** AUSLAN test classification performance in dependence to the percentage of the sequences relative to the original sequence length. The sequences were either shortened or lengthened by padding with the last value.

contains the information necessary to differentiate the samples. The second half is comparatively constant and not distinctive.

We investigated, how these 'standstills' affect classification performance in the AUSLAN dataset. To that effect we varied the length of the sequences by either shortening or padding them with their last step. Figure 4 shows the test classification error rates of MSL and $\overline{ESN}$ in dependence to the percentage of the sequence relative to their original length.

ESN and $\overline{ESN}$ achieve their lowest classification error rate at the 30 % and 35 % mark, respectively. Then, the error increases with the sequence length. In contrast, after the 50 % mark MSL maintains throughout a low classification error. Thus, time series classification by MSL appears to be invariant against the addition of sequence segments with constant feature values.

## 9   Discussion

In our work we use standard, random reservoirs as basis for the predictive models. Reservoir networks are especially suited for model space learning, since the read-out weights are trained with ridge regression, which has an unique solution. This reduces the probability that two similar sequences have very different models.

The evaluation showed that MSL achieves on average lower classification error rates than the ESN approaches and is comparable to state-of-the-art methods. Significant, however, is only the difference on the multivariate motion datasets. Especially noteworthy is the high accuracy in the high-dimensional motion datasets, like the Kinect datasets. We attribute the good performance on the motion datasets to the kind of information that MSL extracts from the data: The training of prediction models makes the spatio-temporal relationships, which are prevalent in motion data, explicit. On the other hand, in the ESN architectures, the classificator operates on the reservoir activations, which reflect spatio-temporal relations without explicitly integrating the predictive aspect of the data. For the motion data, learning predictors results in a well-separable models. For time series with less spatio-temporal coherence, the learning of predictors might be less meaningful and results in less separable models.

# 10   Conclusion

In this paper we presented a comparison between time series classification in reservoir space and model space. In the classical, discriminative classification with ESNs, the reservoir activations are mapped directly to the class labels. In model space learning, for each time series, an ESN is trained to predict the next step in the time series. This step yields a time series specific readout weight matrix, which represents the spatio-temporal structure of the time series. This representation is then mapped to class labels in a second step. The evaluation on numerous datasets showed that learning in the model space outperforms on average the ESN-approaches and is on par with other state-of-the-art methods.

The datasets used here are preprocessed and contain a collection of sequences with the implication that each sequence belongs only to one class and does not contain (parts of) patterns from other classes. In real-world applications, the classification will probably be carried out on streams of data, which makes it necessary to either pre-segment the incoming data and classify the segments or apply the classification step by step. Learning in the model space offers structural benefits for such applications, e.g. for time series clustering and segmentation, which are investigated in future work.

**Acknowledgments.** This project is funded by the German Federal Ministry of Education and Research (BMBF) within the Leading-Edge Cluster Competition "it's OWL" (intelligent technical systems OstWestfalenLippe) and managed by the Project Management Agency Karlsruhe (PTKA). The authors are responsible for the contents of this publication.

# References

1. Aswolinskiy, W., Reinhart, R., Steil, J.: Impact of regularization on the model space for time series classification. In: Machine Learning Reports, pp. 49–56 (2015)
2. Aswolinskiy, W., Reinhart, F., Steil, J.: Modelling parameterized processes via regression in the model space. In: European Symposium on Artificial Neural Networks (2016)
3. Barachant, A., Bonnet, S., Congedo, M., Jutten, C.: Classification of covariance matrices using a Riemannian-based kernel for BCI applications. Neurocomputing **112**, 172–178 (2013)
4. Baydogan, M.G., Runger, G.: Time series representation and similarity based on local autopatterns. Data Min. Knowl. Discov. 1–34 (2015). www.mustafabaydogan. com/files/viewcategory/20-data-sets.html
5. Brodersen, K.H., Schofield, T.M., Leff, A.P., Ong, C.S., Lomakina, E.I., Buhmann, J.M., Stephan, K.E.: Generative embedding for model-based classification of fMRI data. PLoS Comput. Biol. **7**(6), e1002079 (2011)

6. Chen, H., Tino, P., Rodan, A., Yao, X.: Learning in the model space for cognitive fault diagnosis. IEEE Trans. Neural Netw. Learn. Syst. **25**(1), 124–136 (2014)
7. Chen, H., Tang, F., Tino, P., Cohn, A.G., Yao, X.: Model metric co-learning for time series classification. In: International Joint Conference on Artificial Intelligence, pp. 3387–3394 (2015)
8. Chen, H., Tang, F., Tino, P., Yao, X.: Model-based kernel for efficient time series analysis. In: ACM SIGKDD International Conference on Knowledge Discovery and Data Mining, pp. 392–400 (2013)
9. Chen, Y., Keogh, E., Hu, B., Begum, N., Bagnall, A., Mueen, A., Batista, G.: The UCR Time Series Classification Archive, July 2015. www.cs.ucr.edu/~eamonn/time_series_data/
10. Deng, K., Moore, A.W., Nechyba, M.C.: Learning to recognize time series: combining ARMA models with memory-based learning. In: IEEE International Symposium on Computational Intelligence in Robotics and Automation, pp. 246–251 (1997)
11. Ghassempour, S., Girosi, F., Maeder, A.: Clustering multivariate time series using hidden Markov models. Int. J. Environ. Res. Public Health **11**(3), 2741–2763 (2014)
12. Gianniotis, N., Kügler, S., Tino, P., Polsterer, K., Misra, R.: Autoencoding time series for visualisation. In: European Symposium on Artificial Neural Networks (2015)
13. Jaeger, H.: The "echo state" approach to analysing and training recurrent neural networks-with an erratum note. GMD Technical report 148, p. 34 (2001)
14. Kügler, S., Gianniotis, N., Polsterer, K.: An explorative approach for inspecting kepler data. Mon. Not. R. Astron. Soc. **455**(4), 4399–4405 (2016)
15. Lukoševičius, M.: A practical guide to applying echo state networks. In: Montavon, G., Orr, G.B., Müller, K.-R. (eds.) Neural Networks: Tricks of the Trade, 2nd edn. LNCS, vol. 7700, pp. 659–686. Springer, Heidelberg (2012)
16. Reinhart, R., Steil, J.: Efficient policy search in low-dimensional embedding spaces by generalizing motion primitives with a parameterized skill memory. Auton. Robots **38**(4), 331–348 (2015)
17. Rodan, A., Tiňo, P.: Simple deterministically constructed cycle reservoirs with regular jumps. Neural Comput. **24**(7), 1822–1852 (2012)
18. Rodríguez-Fdez, I., Canosa, A., Mucientes, M., Bugarín, A.: STAC: a web platform for the comparison of algorithms using statistical tests. In: IEEE International Conference on Fuzzy Systems (2015)
19. Stulp, F., Raiola, G., Hoarau, A., Ivaldi, S., Sigaud, O.: Learning compact parameterized skills with a single regression. In: IEEE-RAS International Conference on Humanoid Robots (Humanoids), pp. 417–422 (2013)
20. Ude, A., Riley, M., Nemec, B., Kos, A., Asfour, T., Cheng, G.: Synthesizing goal-directed actions from a library of example movements. In: IEEE-RAS International Conference on Humanoid Robots, pp. 115–121 (2007)
21. Vemulapalli, R., Arrate, F., Chellappa, R.: Human action recognition by representing 3D skeletons as points in a Lie group. In: IEEE Conference on Computer Vision and Pattern Recognition, pp. 588–595 (2014). http://ravitejav.weebly.com/kbac.html
22. Xing, Z., Pei, J., Keogh, E.: A brief survey on sequence classification. ACM SIGKDD Explor. Newsl. **12**(1), 40–48 (2010)

# Objectness Scoring and Detection Proposals in Forward-Looking Sonar Images with Convolutional Neural Networks

Matias Valdenegro-Toro[✉]

Ocean Systems Laboratory, School of Engineering and Physical Sciences,
Heriot-Watt University, Edinburgh EH14 4AS, UK
m.valdenegro@hw.ac.uk

**Abstract.** Forward-looking sonar can capture high resolution images of underwater scenes, but their interpretation is complex. Generic object detection in such images has not been solved, specially in cases of small and unknown objects. In comparison, detection proposal algorithms have produced top performing object detectors in real-world color images. In this work we develop a Convolutional Neural Network that can reliably score objectness of image windows in forward-looking sonar images and by thresholding objectness, we generate detection proposals. In our dataset of marine garbage objects, we obtain 94 % recall, generating around 60 proposals per image. The biggest strength of our method is that it can generalize to previously unseen objects. We show this by detecting chain links, walls and a wrench without previous training in such objects. We strongly believe our method can be used for class-independent object detection, with many real-world applications such as chain following and mine detection.

**Keywords:** Object detection · Detection proposals · Sonar image processing · Forward-looking sonar

## 1 Introduction

Autonomous Underwater Vehicles (AUVs) are increasingly being used for survey and exploration of underwater environments. For example, the oil and gas industry requires constant monitoring and surveying of seabed equipment, and marine researchers require similar capabilities in order to monitor ocean flora and fauna.

The perception capabilities of AUVs are not comparable to land and air vehicles. Most of the perception tasks, such as object detection and recognition, are done in offline steps instead of online processing inside the vehicle. This limits the applications fields where AUVs are useful, and strongly decreases the level of autonomy that this kind of vehicles can achieve.

Most of these limits on perception capabilities come directly from the underwater environment. Water absorbs and scatters light, which limits the use of

© Springer International Publishing AG 2016
F. Schwenker et al. (Eds.): ANNPR 2016, LNAI 9896, pp. 209–219, 2016.
DOI: 10.1007/978-3-319-46182-3_18

optical cameras, specially near coasts and shores due to water turbidity and suspended material. Typical perception sensors for AUV are different kinds of Sonar, which uses acoustic waves to sense and image the environment. Acoustic waves can travel great distances on water with small attenuation, depending on frequency, but interpreting an image produced by a sonar can be challenging.

One type of sonar sensor is Forward-Looking Sonar (FLS), where the sensor's field of view looks forward, similar to an optical camera. Other kinds of sonar sensors have downward looking fields of view in order to survey the seabed. This kind of sensor is appropriate for object detection and recognition in AUVs.

Object detection in sonar imagery is as challenging as other kinds of images. Methods from the Computer Vision community have been applied to this kind of images, but these kind of methods only produce class-specific object detectors. Most research has been performed on detecting marine mines [19], but constructing a class-agnostic object detector is more useful and will greatly benefit AUV perception capabilities.

Computer Vision literature contains many generic object detection algorithms, called detection proposals [1,6], but these techniques were developed for color images produced by optical cameras, and color-based techniques fail to generate correct proposals in sonar images. Convolutional and Deep Neural Networks are the state of the art for many computer vision tasks, such as object recognition [12], and they have also been used to generate detection proposals with great success [4,20,21].

The purpose of our work is to build an algorithm for detection proposal generation in FLS images, but our technique can still be used for other kinds of sonar images. Instead of engineering features that are commonly used for object detection, we propose to use a Convolutional Neural Network (CNN) to learn objectness directly from labeled data. This approach is much simpler and we believe that has better generalization performance than other object detection approaches.

## 2   Related Work

Detection Proposals [2] are class-agnostic object detectors. The basic idea is to extract all object bounding boxes from an image, and compute an objectness score [1] that can be used to rank and determine interesting objects, with the purpose of posterior classification.

Many methods to extract detection proposals in color images exist. Rathu et al. [18] uses cascade of objectness features to detect category-independent objects. Alexe et al. [1] use different cues to score objectness, such as saliency, color contrast and edge density.

Selective search by Uijlings et al. [24] uses a large number of engineered features and superpixel segmentation to generate proposals in color images, which achieves a 99 % recall on many datasets. Girshick et al. [5] combine Selective Search with a CNN image classifier to detect and recognize objects in a common pipeline.

Zitnick et al. [26] use edge information to score proposals from a sliding window in a color image. Kang et al. [10] use a data driver approach where regions are matched over a large annotated dataset and objectness is computed from segment properties.

Kuo et al. [13] shows how to learn objectness with a CNN with the purpose of reranking proposals generated by EdgeBoxes [26], with improved detection performance. A good extensive evaluation of many proposal algorithms is Hosang et al. [6].

More recent proposal approaches also use CNNs, such as Fast R-CNN [4] and Faster R-CNN [20]. Fast R-CNN uses bounding box regression trained over a convolutional feature map that can be shared and used for both detection and classification, but still using initial Selective Search proposals [24], while Faster R-CNN uses region proposal networks to predict proposals and objectness directly from the input image, while sharing layers with a classifier and bounding box regressor in a similar way that of Fast R-CNN.

Object detection in sonar images is mostly done with several kinds of engineered features over sliding windows and a machine learning classifier [3, 9, 19, 25], template matching [7, 16] is also very popular, as well as computer vision techniques like boosted cascade of weak classifiers [22]. In all cases this type of approach only produces class-specific detectors, where generalization outside of the training set is poor.

While proposal methods are been successful on computer vision tasks, color image features are not appropriate for sonar images, due to the different interpretation of the image content. Some methods such as EdgeBoxes [26] could be applied to sonar images, but it is well known that edges are unreliable in this kind of images due to noise and point of view dependence.

## 3   Forward-Looking Sonar Imaging

A Forward-Looking Sonar is an acoustic sensor that is similar to an optical camera [7], but with two major differences: Sound waves are emitted and the acoustic return is analyzed to discover scene structure, and the output image is similar to a top view of the scene instead of the typical front view of a optical camera. An FLS that uses high-frequency acoustic pulses can capture high resolution images of underwater scenes at distances ranging from 1 to 10 m at frame rates of up to 15 Hz. This kind of device is used for survey and object identification in underwater environments with AUVs [17].

An FLS has a fan-shaped field of view, with fixed viewing angles and configurable distances. The output of most sonar sensors is a one channel image that represents the amount of acoustic return from the scene, usually sampled along the sensor's field of view. A typical characteristics of acoustic images are shadow (dark) and highlight (light) regions, produced when objects block and reflect sound waves. The length of shadow regions depends on the height of the object. Ghost reflections are produced by spurious acoustic return from undesired sources, such as walls, the water surface, large material changes between

(a) FLS Image containing a Drink(b) Tin Can and Wall, note
Carton, Can and Bottles          the reflections in both ob-
                                 jects

**Fig. 1.** Sample FLS images from our dataset, captured with a ARIS Explorer 3000 sonar.

objects and interference inside small objects. The fan-shaped field of view introduces pixel shape distortions, since the size of a pixel in the image now depends on the distance from the origin. Farther distances map to bigger pixels that have increasing uncertainty, while closer distances have smaller uncertainties [17].

These features make FLS images hard to interpret. One important feature of sonar sensors is that in most cases distance and bearing to the object can be easily recovered directly from sonar data, but elevation information is lost. Figure 1 shows two FLS images, where each pixel represents approximately 3 mm, and some objects can be seen clearly (The tire and bottle in Fig. 1a), even with some fine details like the seams on the tire. Figure 1b shows typical sonar reflections.

## 4    Detection Proposals on FLS Imagery

Our proposed technique is similar in spirit to [1,13]. We propose to use a CNN to learn objectness scores of windows in a FLS image. We slide a $w \times w$ window over the image with a stride of $s$ pixels, but only consider windows that are inside the FLS's field of view. Our technique only requires a image dataset with labeled rectangles representing objects in the image. The objectness scores are estimated directly from the rectangles.

Each window is scored by our CNN and objectness is thresholded to select windows that contain objects. Different objectness threshold values $T_o$ will produce varying numbers of proposals. Control over the number of proposals is a desirable property of an proposal algorithm [6]. Windows with a low objectness score are discarded. Given a labeled dataset containing objects of interest, we construct a training set by running a sliding window over each image and cropping each window that has an intersection-over-union (IoU) score above a threshold. Given two rectangles $A$ and $B$, the IoU score is defined as:

$$\text{IoU}(A, B) = \frac{\text{area}(A \cap B)}{\text{area}(A \cup B)} \tag{1}$$

The IoU score is commonly used by the computer vision community to evaluate object detection algorithms [4,26].

For each cropped window, we estimate the ground truth objectness of that sub-image as a score based on the maximum IoU with ground truth. Our intuition for such scoring is that non-object windows have a very low IoU with ground truth, while object windows have a high IoU with ground truth, but multiple windows might contain the object. IoU decreases as the window moves farther from the ground truth, and we want the same behavior from our objectness scores. This technique also doubles as a data augmentation step, as many training windows will be generated for a single ground truth object. In practice we obtained up to 35 crops from one object.

The ground truth objectness for a window is computed as:

$$\text{objectness(iou)} = \begin{cases} 1.0 & \text{if iou} \geq 0.8 \\ \text{iou} & \text{if } 0.2 < \text{iou} < 0.8 \\ 0.0 & \text{if iou} \leq 0.2 \end{cases} \tag{2}$$

Equation 2 represents our desired behavior for the ground truth objectness score. Windows with a small IoU are very unlikely to contain an object, and this is reflected as zero objectness, while windows with a high IoU contain objects and get a objectness equals to one. Windows with IoU values in between are assigned objectness scores equals to the IoU.

Our CNN architecture is shown in Fig. 2. It is a 4 layer network that takes $96 \times 96$ pixel images as input, and has 1.4 million trainable parameters. The first layer convolves the input image with 32 $5 \times 5$ filters, then applies $2 \times 2$ Max-Pooling (MP), then the same process is repeated by another convolution layer with 32 $5 \times 5$ filters and $2 \times 2$ Max-Pooling. The classifier layers are one Fully Connected layer (FC) with 96 neurons, and another FC layer with one output neuron. All layers use the ReLU non-linearity except the last layer, which uses a sigmoid function to output objectness scores in the $[0, 1]$ range.

Our architecture was initially modeled to be similar to LeNet [14], by stacking Convolution, ReLU and Max-Pooling blocks for convolutional feature learning, and then produce objectness scores with a fully connected layer. This initial architecture worked best and generalizes very well over unseen data and out of sample images. Removing any layer decreases recall, and stacking more Conv-ReLu-MaxPool or adding fully connected layers leads to overfitting. We believe that this is an appropriate architecture for the problem and for the amount of data that we possess.

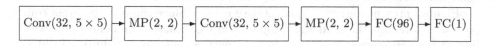

**Fig. 2.** Our neural network architecture

Batch Normalization [8] layers are inserted after each Max-Pooling layer, and after the first FC layer, in order to prevent overfitting and accelerate training. We also tested Dropout [23] but Batch Normalization gave better generalization performance. We trained our network with the standard mean square error (MSE) loss function and mini batch gradient descent, with a batch size of 32 images, and used the ADAM optimizer [11] with a initial learning rate $\alpha = 0.1$. We train for 10 epochs, and do early stopping when the validation loss stops improving (normally after 5 epochs). Our network is implemented in Python with Keras and Theano.

The dataset used to train this network is generated as follows. A $96 \times 96$ sliding window with stride of $s = 8$ pixels is run over each image in the dataset, and all windows that have IoU $\geq 0.5$ with ground truth are added to the training set, and objectness labels are determined with Eq. 2. This generates a variable number of crops for each image, up to 35 window crops. Some objects generated no crops, due to having IoU with ground truth less than 0.5, but were still included in the dataset to keep examples with intermediate objectness values. Negative samples were generated by selecting 20 random windows from the same sliding window with a maximum IoU $\leq 0.1$ with ground truth. Each negative crop was assigned objectness label of zero.

A common issue is that many intersecting proposals are present in the output, due to the use of a sliding window. These can be reduced by performing non-maxima suppression over intersecting proposals with a minimum IoU threshold.

## 5    Experimental Evaluation

In this section we evaluate our proposed approach. We captured 2500 FLS images with an ARIS Explorer 3000 sonar, containing different objects that are interesting for the underwater domain, such as garbage objects, a model valve, a small tire, a small chain, and a propeller. We performed a 70 %/30 % split of the image set, where the training set contains 1750 images, and the test set 750 images. The training set was 85 %/15 % split into train and validation sets, the latter set for purpose of early stopping. We introduce up-down and left-right flips of each training image for data augmentation.

To evaluate at test time, we run a sliding window over the image, with the same parameters used to generate the training set, and threshold the predicted objectness score from our CNN. Any window with an objectness score bigger than $T_o$ is output as a object proposal. $T_o$ is a parameter that can be tuned by the operator to decide the number of proposals to generate. A detection is considered correct if it has IoU $\geq 0.5$ with ground truth.

Figure 3 presents our main quantitative results. Since there are no other comparable algorithms that generate proposals in FLS images, we defined our baseline as random scoring of image windows, with a uniform distribution in $[0, 1]$. Our method is clearly superior than the baseline, except when the objectness threshold is high. We believe that our results show that our proposed technique is promising. At $T_o = 0.5$, we obtain 94 % recall, with an average of $62 \pm 33$

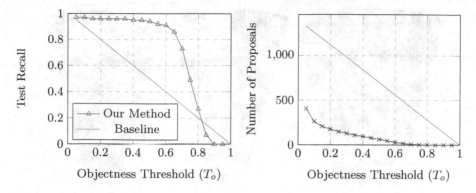

**Fig. 3.** Objectness threshold versus recall and number of proposals over our test set at 0.5 IoU with ground truth. Blue is recall, red is number of generated proposals, and gray is the baseline. Best viewed in color. (Color figure online)

generated proposals. The number of proposals is usually a weak indicator of the detector quality, since producing a large number of proposals will score a high recall but will be computationally expensive to evaluate such amount of proposals. A good detection proposal algorithm will generate a small amount of proposals that only cover objects. For comparison, the number of total windows to evaluated in a FLS image is around 1400.

Most missed detections correspond to inaccurate localization, due to the sliding window approach, specially for small objects. A large fraction of predicted scores are less than 0.75, which can be seen as the number of proposals drops to zero at $T_o = 0.8$, this show a bias in the learned network. We expected objectness scores to be more evenly distributed in the [0, 1] range.

Figure 4(a–c) shows sample detections over our test images. Proposals have a good cover of objects in the image, and surprisingly they also cover objects that are not labeled, which shows that the generalization performance of our approach is good. Multiple proposals cover each object, indicating a poor localization accuracy, but it can be improved by using non-maxima suppression.

Increasing $T_o$ will decrease the number of proposals, producing more accurate detections, but skipping untrained objects. Typically objects that are not seen during training receive lower objectness scores when compared to trained objects.

As qualitative results, we also generated a heatmap of scores produced by the network over the image. This is done by scoring each rectangle in the sliding window, and drawing a $s \times s$ rectangle on the proposal's center. This produces empty zones at edges of the sonar's field of view. Heatmaps are shown in Fig. 4(d–f).

The heatmaps show that high objectness scores are only produced over objects and not over background, which validates that our technique works appropriately.

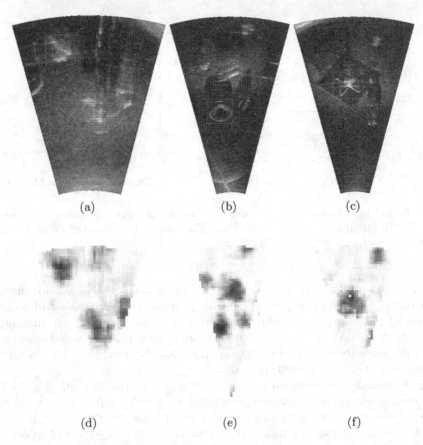

(a)          (b)          (c)

(d)          (e)          (f)

**Fig. 4.** Results on test images. Top row shows proposals generated with $T_o = 0.5$. Red rectangles are our proposals, while green is ground truth. Note how proposals cover all objects, even unlabeled ones. Bottom row shows heatmaps. White represents low objectness, and black is high objectness scores. Best viewed in color. (Color figure online)

We also evaluated our network's generalization abilities outside of the training set. We obtained images of objects that are not present in the training set, like a Chain (provided by the University of Girona), a Wall, and a Wrench. The proposals generated on such images are shown in Fig. 5(a–c). We expected that our technique would generate a low amount of proposals with very low position accuracy, but results show that the network can successfully generate proposals over object that it has not seen during training. We also provide heatmaps over these objects in Fig. 5(d–f).

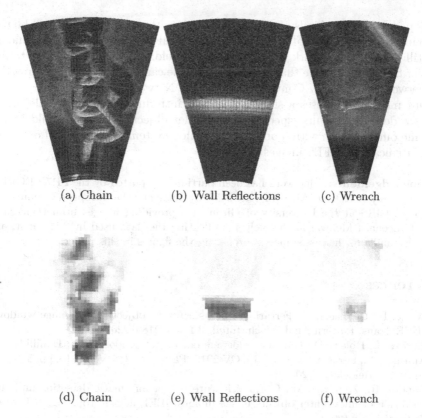

(a) Chain          (b) Wall Reflections          (c) Wrench

(d) Chain          (e) Wall Reflections          (f) Wrench

**Fig. 5.** Results on out-of-sample images. Top row shows proposals generated at $T_o = 0.5$. Our training set does not contain any object present in these images, but yet our network can generate very good proposals on them. Bottom row shows heatmaps. White represents low objectness, and black is high objectness scores. Best viewed in color. (Color figure online)

# 6  Conclusions and Future Work

In this work we have presented a CNN approach at objectness estimation and detection proposal generation for FLS images. Our approach is simple, only requiring images with labeled objects, and works surprisingly well, specially in objects that are not present in the training set, which is a very desirable property of any object detector.

Over our dataset we obtain a 94 % recall at objectness threshold of 0.5, and by lowering the threshold to 0.1 we obtain 97 % recall.

Neural network-based methods are responsible for large breakthroughs in computer vision for color images, and we believe that similar improvements are possible in other domains, such as FLS images that we used in this work. Detection proposal approaches imply that a generic object detector can be shared by many different perception tasks, while only a recognition stage has to be developed for specific kinds of objects. We believe that our approach will be useful to improve AUV perception capabilities, as well as increasing their autonomy.

Still there is much work to be done in this field. Our method is slow, taking 12 s per frame. We believe that computation time can be improved by converting our network into a Fully Convolutional Neural Network [15].

Our method only uses a single scale, and introducing multiple scales could improve detection results, specially for smaller objects. We also would like to combine our approach with a object recognition system for integrated detection and classification in FLS images.

**Acknowledgements.** This work has been partially supported by the FP7-PEOPLE-2013-ITN project ROBOCADEMY (Ref 608096) funded by the European Commission. We thank CIRS at the University of Girona for providing images from their chain dataset, Leonard McLean for his help in collecting the data used in this paper, and Polett Escanilla for her assistance in preparing the figures in this paper.

# References

1. Alexe, B., Deselaers, T., Ferrari, V.: Measuring the objectness of image windows. IEEE Trans. Pattern Anal. Mach. Intell. **34**(11), 2189–2202 (2012)
2. Endres, I., Hoiem, D.: Category independent object proposals. In: Daniilidis, K., Maragos, P., Paragios, N. (eds.) ECCV 2010, Part V. LNCS, vol. 6315, pp. 575–588. Springer, Heidelberg (2010)
3. Fandos, R., Zoubir, A.M.: Optimal feature set for automatic detection and classification of underwater objects in sas images. IEEE J. Sel. Top. Signal Process. **5**(3), 454–468 (2011)
4. Girshick, R.: Fast R-CNN. In: Proceedings of the IEEE International Conference on Computer Vision, pp. 1440–1448 (2015)
5. Girshick, R., Donahue, J., Darrell, T., Malik, J.: Rich feature hierarchies for accurate object detection and semantic segmentation. In: Proceedings of the IEEE Conference on Computer Vision and Pattern Recognition, pp. 580–587 (2014)
6. Hosang, J., Benenson, R., Dollár, P., Schiele, B.: What makes for effective detection proposals? (2015)
7. Hurtós, N., Palomeras, N., Nagappa, S., Salvi, J.: Automatic detection of underwater chain links using a forward-looking sonar. In: 2013 MTS/IEEE OCEANS, Bergen, pp. 1–7. IEEE (2013)
8. Ioffe, S., Szegedy, C.: Batch normalization: accelerating deep network training by reducing internal covariate shift. arXiv preprint arXiv:1502.03167 (2015)
9. Isaacs, J.: Sonar automatic target recognition for underwater UXO remediation. In: Proceedings of the IEEE Conference on Computer Vision and Pattern Recognition Workshops, pp. 134–140 (2015)
10. Kang, H., Hebert, M., Efros, A.A., Kanade, T.: Data-driven objectness. IEEE Trans. Pattern Anal. Mach. Intell. **37**(1), 189–195 (2015)
11. Kingma, D., Ba, J.: Adam: a method for stochastic optimization. arXiv preprint arXiv:1412.6980 (2014)
12. Krizhevsky, A., Sutskever, I., Hinton, G.E.: Imagenet classification with deep convolutional neural networks. In: Advances in Neural Information Processing Systems, pp. 1097–1105 (2012)

13. Kuo, W., Hariharan, B., Malik, J.: Deepbox: learning objectness with convolutional networks. In: Proceedings of the IEEE International Conference on Computer Vision, pp. 2479–2487 (2015)
14. LeCun, Y., Bottou, L., Bengio, Y., Haffner, P.: Gradient-based learning applied to document recognition. Proc. IEEE **86**(11), 2278–2324 (1998)
15. Long, J., Shelhamer, E., Darrell, T.: Fully convolutional networks for semantic segmentation. In: Proceedings of the IEEE Conference on Computer Vision and Pattern Recognition, pp. 3431–3440 (2015)
16. Myers, V., Fawcett, J.: A template matching procedure for automatic target recognition in synthetic aperture sonar imagery. IEEE Signal Process. Lett. **17**(7), 683–686 (2010)
17. Negahdaripour, S., Firoozfam, P., Sabzmeydani, P.: On processing and registration of forward-scan acoustic video imagery. In: Proceedings of the 2nd Canadian Conference on Computer and Robot Vision, pp. 452–459. IEEE (2005)
18. Rahtu, E., Kannala, J., Blaschko, M.: Learning a category independent object detection cascade. In: 2011 IEEE International Conference on Computer Vision (ICCV), pp. 1052–1059. IEEE (2011)
19. Reed, S., Petillot, Y., Bell, J.: Automated approach to classification of mine-like objects in sidescan sonar using highlight and shadow information. In: IEE Proceedings - Radar, Sonar and Navigation, vol. 151, pp. 48–56. IET (2004)
20. Ren, S., He, K., Girshick, R., Sun, J.: Faster R-CNN: towards real-time object detection with region proposal networks. In: Advances in Neural Information Processing Systems, pp. 91–99 (2015)
21. Ren, S., He, K., Girshick, R., Zhang, X., Sun, J.: Object detection networks on convolutional feature maps. arXiv preprint arXiv:1504.06066 (2015)
22. Sawas, J., Petillot, Y., Pailhas, Y.: Cascade of boosted classifiers for rapid detection of underwater objects. In: Proceedings of the European Conference on Underwater Acoustics (2010)
23. Srivastava, N., Hinton, G., Krizhevsky, A., Sutskever, I., Salakhutdinov, R.: Dropout: a simple way to prevent neural networks from overfitting. J. Mach. Learn. Res. **15**(1), 1929–1958 (2014)
24. Uijlings, J.R., van de Sande, K.E., Gevers, T., Smeulders, A.W.: Selective search for object recognition. Int. J. Comput. Vis. **104**(2), 154–171 (2013)
25. Williams, D.P., Groen, J.: A fast physics-based, environmentally adaptive underwater object detection algorithm. In: 2011 IEEE OCEANS, Spain, pp. 1–7. IEEE (2011)
26. Zitnick, C.L., Dollár, P.: Edge boxes: locating object proposals from edges. In: Fleet, D., Pajdla, T., Schiele, B., Tuytelaars, T. (eds.) ECCV 2014, Part V. LNCS, vol. 8693, pp. 391–405. Springer, Heidelberg (2014)

# Background Categorization for Automatic Animal Detection in Aerial Videos Using Neural Networks

Yunfei Fang[1(✉)], Shengzhi Du[2], Rishaad Abdoola[1],
and Karim Djouani[1]

[1] Department of Electrical Engineering, French South African Institute
of Technology, Tshwane University of Technology, Staatsartillerie Road,
Pretoria 0001, South Africa
fangyunfei08@gmail.com

[2] Department of Mechanical Engineering, Mechatronics and Industrial Design,
Tshwane University of Technology, Staatsartillerie Road,
Pretoria 0001, South Africa
DuS@tut.ac.za

**Abstract.** This paper addresses the problem of animal detection in natural environment from aerial videos. Since the natural environment is usually composed of several fundamental elements such as trees, grass, streams, etc., it is proposed to distinguish the animal by categorizing the background into several classes. From the manually labeled samples, texture as well as brightness features are extracted to train a feedforward Neural Network. Then the classifier is applied to filter the test frame to locate potential animal regions. Four texture measures calculated from Grey Level Co-occurrence Matrix (GLCM) are used for texture feature description. Instead of obtaining these texture measures from grey level images, it is proposed to carry out calculation for every channel of the RGB image. The implemented results illustrate that this feature extraction method works well and the texture feature is a decisive factor in background categorizing.

**Keywords:** Image segmentation · Background categorization · Texture analysis · Animal detection · Neural network

## 1 Introduction

### 1.1 Animal Detection in a Natural Environment

Applications of Unmanned Aerial Vehicles (UAVs) are of great benefit to the field of nature conservation. Compared to conventional ways of wildlife surveys, it is far more economical to collect information from the region by flying a UAV (mounted with a proper camera) across it. It is especially efficient in some real-time tasks, such as monitoring and anti-poaching. The possibility of applying computer vision techniques to automatically analyse the aerial videos are increasingly being investigated [1, 2].

In general, tasks that wildlife conservation may concern include: (1) Counting the number of animals to monitor the distribution and abundance of animal species.

© Springer International Publishing AG 2016
F. Schwenker et al. (Eds.): ANNPR 2016, LNAI 9896, pp. 220–232, 2016.
DOI: 10.1007/978-3-319-46182-3_19

(2) Identification and investigation of a particular animal. (3) Tracking and monitoring of a herd and risk estimation. Animal detection in the video is to decide whether or not an animal of specific species is present in the scene and where it is located.

Most object detection algorithms are based on machine learning mechanisms. Different views of an object were learned by a set of classifiers using positive and negative examples. By dividing the image into standard-sized sub-windows (patch) and passing them through trained filters, it can be then determined the existence and location of the object. For example, the neural networks [3], support vector machines [4], Bayesian networks [5] and deep learning approaches [6]. Neural networks play very important role in pattern recognition and have a wide range of applications in pedestrian detection [7], speech recognition [8], handwriting recognition [9], fingerprint analysis [10] and so on.

The modern object detection and classification tasks usually cope with ground perspective images with the scale of the object occupying the greatest portion of the image. The problem of animal detection in natural environment from aerial images has not significantly been addressed.

Based on the nature of the problem, it is proposed not merely learn the pattern of the animal but learn the pattern of background alongside by categorizing it into different classes. A feedforward neural network is trained to form a classifier because its architecture is prone to multi-class classification. Texture features are extracted as input for the neural network. By sliding the neural network filter across the testing video frame, the regions where animals exist are successfully highlighted.

## 1.2 Background Categorization

There are two hypotheses behind this research. Firstly, there are finite visual elements that a general natural environment can be divided into. Secondly, the classes, both the animal and subclasses of the background can be distinguished by Neural Network applying proper feature descriptor. It is proposed to us texture features along with brightness to discriminate different classes in the scene.

Texture describes the content of the object surface. The surface of an object can be considered as a composition of elementary structures, and it is made recognizable due to the balanced presence of some specific elementary structures in it.

Texture feature description is a key step in deciding performance of the Neural Network classifier. There are several texture feature extractors [5] such as Grey-Level Co-occurrence Matrices (GLCM), Local Binary Pattern operator (LBP), Local Phase Quantization (LPQ), and Gabor filters. GLCM measures some statistical characteristics of an image. It has proven to be a very powerful basis for texture feature extraction and has been widely used in areas such as texture classification, remote sensing, medicine, biology and agriculture [11–14].

The GLCM and the four texture measures calculated from it are used as texture feature descriptor to train the Neural Network classifier. It is proven to be effective in background categorization as well as animal detection.

## 2    Theoretical Background

### 2.1    GLCM

The Grey Level Co-occurrence Matrix (GLCM) is often used for a series of "second order" texture calculation, which considers pixel neighbouring relationships. It measures how often different combinations of pixel brightness values or grey levels occur in an image. GLCM is based on the relationship between two pixels. The spatial relationship between a reference pixel and its neighbour pixel defines the offset.

To reduce the size of the GLCM and increase the occupancy level of the matrix, the quantization level is set to 16.

To demonstrate the GLCM data structure, the image (300 × 300 pixels) shown in Fig. 1 is used as an example. The corresponding GLCM data is depicted in Fig. 1, where the grey level is set to 8 for convenient display. The highlighted cell at row 3, column 2 means the combination 3, 2 occurs 2956 times in the image, where the grey level of the reference pixel is 3 and its immediate neighbour on the left (defined by the offset) is 2.

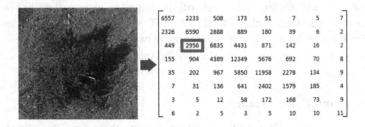

**Fig. 1.**  An example image and its corresponding GLCM

### 2.2    Texture Measures from the GLCM

Texture calculations are weighted averages of the normalized GLCM cell contents. Certain measures are based on the contrast information while others are based on the orderliness and the descriptive statistics of the GLCM texture measures. The following describes only the texture measures from GLCM which are used in this paper

$$Cn = \sum_i \sum_j |i - j|^2 P(i,j) \tag{1}$$

where $Cn$ is the measure of contrast, $P(i, j)$ is the content of the normalized GLCM at row $i$ and column $j$, illustrating the occurrence probability of gray level $i$ and $j$.

$$Co = \sum_i \sum_j \frac{(i - \mu_i)(j - \mu_j)P(i,j)}{\sigma_i \sigma_j} \tag{2}$$

where $Co$ is the measure of correlation, $\mu_i$, $\mu_j$, $\sigma_i$, $\sigma_j$ are the means and standard deviations respectively. Correlation is a measure of how correlated a pixel is to its neighbour over the whole image. A value of 0 implies that the pattern is uncorrelated, 1 implies perfect correlation and −1 implies that the spatial set exhibits a dissimilar, deterministic structure.

$$E = \sum_i \sum_j P(i,j)^2 \qquad (3)$$

where $E$ is the measure of energy. Energy will be equal to 1 for a constant image.

$$H = \sum_i \sum_j \frac{P(i,j)}{1 + |i - j|} \qquad (4)$$

where $H$ is the measure of homogeneity.the closeness of the distribution of elements in the GLCM to the GLCM diagonal. In the homogeneity measure, the weight values are the inverses of the contrast weight values. As we move further away from the diagonal, the weights decrease quadratically.

## 3  Proposed Method

Figure 2 shows the flow chart of the proposed method. With the aerial video captured from a natural environment, it is at the first place to investigate the visual elements existing in the scene so that the animal and elements of the background can be assigned a class name. Training data of the Neural Network classifier is obtained by taking some frames from the video. These video frames will be partitioned into sub-windows of standard size, which are also called "patches". A set of patches for each class will be achieved through manual labelling. The classes will include the animal and subclasses of the background. Features extracted from the patch are used as input for the Neural Network classifier.

In the testing procedure, for each frame of the video clip, a window of proper size will slide end-to-end across the whole image. The window region from the image will be taken and GLCM and texture measures will be calculated. The Neural Network classifier will assign each patch a class name. Patches classified as animal will be

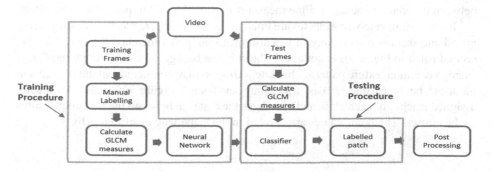

**Fig. 2.** Flow chart of the proposed method

highlighted to form the detection. Further processing such as morphological operations can be applied to reduce errors and improve the results.

The testing video is captured from a typical natural environment in South Africa. The original frame size is $1080 \times 1920$ pixels. Animals existing in the video are mostly blesboks. The method is implemented on this video for a detailed explanation.

### 3.1 Selection of Patch Size

The resolution and scale of the patch make the texture of objects' surfaces different, especially for an animal, which has a clear silhouette. Empirically, the training and testing patch size chosen here is 40-by-40 pixels. One reason is that in the testing video, texture in this size is discernible and can be described properly. Another reason is that, under such setup, animals shall contain hundreds of pixels, and an animal in this size will consist of several patches, it is more stable than forming a detection using a single patch. For comparison, results of 100-by-100-pixel patch size are illustrated in the next section.

It is left to the future work to train a classifier for animals that are viewed from a distance by supplying enough relevant training examples, or using adaptive patch sizes according to the estimated distance from the camera to the animals.

### 3.2 Background Categorization

The required number of classes needed for the Neural Network classifier depends on the classification performance. It will be sufficient to train a two-class (animal and background) classifier as long as the performance is good enough. Categorizing the background into different classes will help form a background model, so that the ani-mal is distinguished if it does not match any subclasses of the background. Figure 3(a) shows some example textures of different elements captured from different scenes. Potential classes are placed in bounding boxes of different colours. Notice that the 4th and 5th patches exhibit similar texture patterns, though the environment looks different. In Fig. 3(b), 3-D points representing texture features of each patch are plotted. The point value $P = [x, y, z]^T$ is derived by calculating three texture measures (Contrast, Correlation, Homogeneity) of the GLCM of each patch. It can be seen that the distance between different elements, and the measures of the 4th and 5th patch are quite close.

In our testing video, five classes are finally selected which are blesbok, bush, ground, gravel and shadow respectively. It is worth mentioning that the class "shadow" (as the second patch in Fig. 3(a)) is not a real object in the background, but an "element" that shares a common texture pattern. This categorization may not include all the elements in the scene, but is sensible and easy for training and testing. Figure 3(c) takes 10 manually assigned patches of each class and visualizes the distance between the texture measures of the classes. The point values are derived in the same way as in Fig. 3(b).

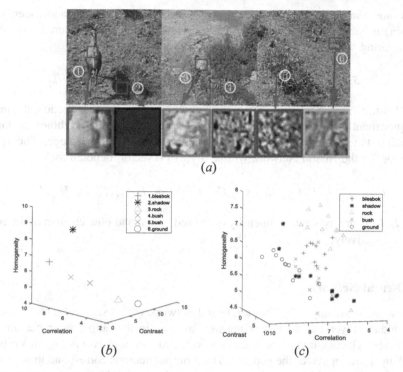

*(a)*

*(b)*          *(c)*

**Fig. 3.** (*a*) Texture of different elements and potential categorization (*b*) Texture measures of the six patches (*c*) Example texture measures from every class

## 3.3  Feature Extraction

Feature extraction is essential to the classification performance. As described before, the GLCM is a 16-by-16 matrix derived from the specific patch. The texture measures calculated from the GLCM can be used to describe the texture feature:

$$F1 = [Cn, Co, E, H]^{T} \tag{5}$$

where $Cn$, $Co$, $E$, $H$ are Contrast, Correlation, Energy and Homogeneity respectively.

As the GLCM is calculated on grey-scale level, it might neglect some useful information in the colour domain. In this case, it is proposed to calculate the GLCM and texture measures for every single channel of the RGB image. The descriptor vector is:

$$F2 = [Cn_r, Cn_g, Cn_b, Co_r, Co_g, Co_b, E_r, E_g, E_b, H_r, H_g, H_b]^{T} \tag{6}$$

where the subscripts r, g, b represents corresponding measures calculated from the 3 channels respectively. This feature extractor for 3 colour channels will be proven to have a higher distinguishing capacity compared to the traditional GLCM.

As shown in Fig. 2(c), the black asterisks representing the class shadow seem to not be distinguishable enough from other classes. Herein, brightness feature is introduced by calculating the average intensity of the grey level image:

$$F3 = \left[I, Cn_r, Cn_g, Cn_b, Co_r, Co_g, Co_b, E_r, E_g, E_b, H_r, H_g, H_b\right]^T \tag{7}$$

where I is the average intensity of the grey level image. In the next section, the feature descriptor mentioned will be compared, as well as describing the brightness feature by calculating the average intensity of 3 channels of the colour image. The feature descriptor for the Neural Network will be a 15-by-1vector, denoted as:

$$F4 = \left[I_r, I_g, I_b, Cn_r, Cn_g, Cn_b, Co_r, Co_g, Co_b, E_r, E_g, E_b, H_r, H_g, H_b\right]^T \tag{8}$$

where $I_r, I_g, I_b$ are the average intensity of the red, green and blue channel of the colour image respectively.

## 3.4    Neural Network

In this work, a tow-layer feedforward neural network with a sigmoid transfer function in the hidden layer and a softmax transfer function in the output layer classified the vectors well. The diagram of the neural network structure is shown in Fig. 4. Vectors of size 15-by-1 are inputs for the network. The 5 output neurons correspond to the 5 target classes. Experiments showed that 10 neurons in the hidden layer generates promising results. Most of the training algorithms do not have significant influence on estimation performance. And the training time and number of iterations do not matter much with such small data. The results illustrated in the next section are trained using the learning algorithm: Scaled Conjugate Gradient (SCG), as described in [15].

In the test, 25 frames are selected from the video sequence, after manually label-lingm the counts of the input data for each class are 322, 684, 1244, 524, 244 (blesbok, bush, ground, gravel, shadow). 70 % of the data was divided for training, 15 % for validation, and the last 15 % for testing.

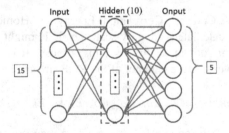

**Fig. 4.** Neural network structure diagram

# 4    Results and Discussion

For illustration purposes, Fig. 5(a), (b) and (c) are chosen from the testing video to demonstrate the results. The animals are from the same herd in subsequent frames of the testing video, but postures and sizes of the blesboks in Fig. 5(b) are different from (a), and the saturation in (c) is different from (a) and (b). These 3 scenarios are chosen to depict the effects of animal postures and the size on the proposed method as well as changes in lighting conditions. Inputs for the Neural Networks related in Fig. 5 are the same, which are texture measures calculated from R, G, B channels of the patch ($F2$ in (6)).

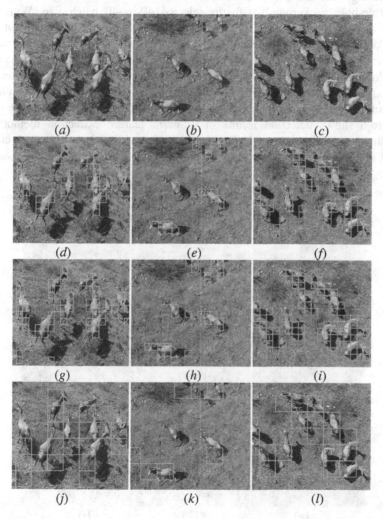

**Fig. 5.** Detection results when only take the texture measures from 3 colour channels as descriptor ($F2$ in (6)). (*a*), (*b*), (*c*) original images (*d*), (*e*), (*f*) detection results in 5 classes (*g*), (*h*), (*i*) detection results in 2 classes (*j*), (*k*), (*l*) detection results of 100-by-100 patch size and in 3 classes

Figure 5(d), (e) and (f) show the detection results in 5 classes (blesbok, bush, ground, gravel, shadow) and the 40-by-40 patch size as described in the previous section. The results look promising with nearly no part of the animal excluded from detection. Most parts of the animals' body are highlighted and due to the selection of the patch size, most of the animals are covered by at least 2 bounding boxes. This will make it easy to filter out false detections by requiring a minimum number of hits within region. Different postures of the animals have little effect on the detection performance. Variation in saturation also doesn't influence the detection of the animals much but the portion of background classified as animals does increase, as shown in (f).

Figure 5(g), (h) and (i) are the detection results of categorizing the output to 2 classes (blesbok, background) with the selected 40-by-40 patch size. With all the elements besides blesbok treated as "background", the results are obviously not as neat as in 5 classes, with more of the background being classified as part of the animal class.

Figure 5(j), (k) and (l) are the detection results in 3 classes (blesbok, background and shadow) using a 100-by-100 patch size. Using this patch size, a single detection of an animal will not contain more than 1 bounding box making it difficult to erase the false positive detection and the animal is easily lost, making the result noisy.

Figure 6 shows detection results of different feature extraction methods. The original images are the same as in Fig. 4, and output of the Neural Network classifier

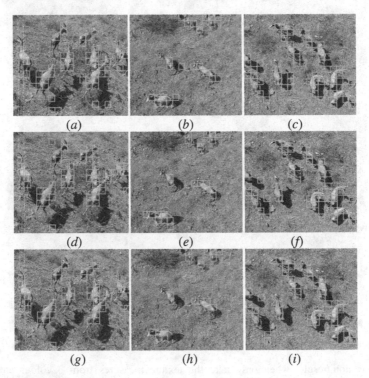

**Fig. 6.** Detection results all in 5 classes with different feature extraction methods. $(a)$, $(b)$, $(c)$ 4 texture measures calculated from GLCM $(F1in(5))$ $(d)$, $(e)$, $(f)$ 12 texture measures with average intensity of grey level image $(F3in(7))$ $(g)$, $(h)$, $(i)$ 12 texture measures with average intensity of each channel $(F4in(8))$.

consists of 5 classes. Figure 6(a), (b) and (c) take the 4 texture measures of the GLCM as input descriptor, as $F1$ in (5). Input descriptor vector for Fig. 6(d), (e) and (f) is $F3$ in (7), and $F4$ in (8) for Fig. 6(g), (h) and (i). Results generated by using Neural Network input vectors $F1, F2, F3, F4$ will be referred to as method I, II, III and IV respectively.

Introduction of brightness feature makes the class "shadow" more distinguishable from other classes. Difference is obvious in scenes without animals. Figure 7(a), (b) and (c) shows false positive detections using method I, III, IV separately. In Fig. 7(a), shadows of the bush are totally not distinguished from the class "blesbok". Results are promoted by introducing the average pixel intensity as input descriptor as in Fig. 7(b). Figure 7(c) achieves the best performance, where average intensity values of R, G, B

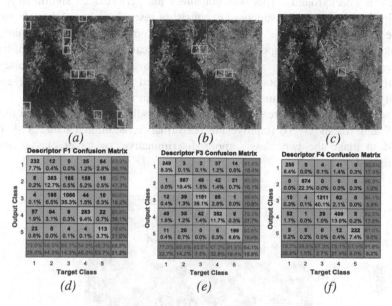

**Fig. 7.** (a), (b), (c) False positive detections of discriptor $F1, F3, F4$ seperately (d), (e), (f) Classification confusion matrix of discriptor $F1, F3, F4$ seperately.

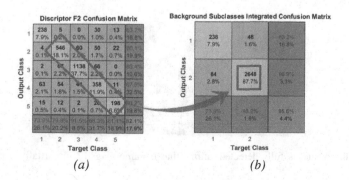

**Fig. 8.** Neural network classification confusion matrix (a) confusion matrix of descriptor $F2$ (b) confusion matrix of 4 background subclasses taken as background.

channels are calculated separately. Method IV also achieves the best classification accuracy as demonstrated by the confusion matrix in Fig. 7(d), (e) and (f).

The performance of method I using traditional GLCM for intensity image is far from satisfactory as depicted in the confusion matrix in Fig. 7(a). But performance of applying GLCM and corresponding texture measures in every channel of the colour image as proposed is promising. Comparison can be made between Figs. 7(a) and 8(a), which shows the classification confusion matrix of method II, where $F2$ is taken as the descriptor.

It is worthy to mention that the other 4 classes besides "blesbok" shall be taken as one single class "background" as shown in Fig. 8(b). Because the misclassification between the background elements doesn't affect detection of the animals. In Fig. 8(b), the false negative rate of "16.8 %" means that 16.8 percent of the "blesbok" patches are classified as "background". This will not affect the performance significantly if the constraint that each animal is supposed to consist of several patches of the selected size. The true positive rate "83.2 %" is fine enough to catch major parts of the animal's body and make a high rate of true positive in the animal scale. Unequal number of training examples from the 2 classes makes the false discovery rate "26.1 %" quite large. Generally, most of the false positive errors are isolated patches as shown in Fig. 9 that are easily reduced by restricting the number of hits in the region.

As seen in Fig. 8(a), there seems to be more misclassifications between the class "gravel" and "blesbok". To improve the performance, additional features should be included to make the two classes more discriminative, or target classes shall be reinvestigated.

**Fig. 9.** False positive detections

**Fig. 10.** A successfully detected animal that is camouflaged and partially occluded

Figure 10 shows an interesting fact during the testing procedure. A blesbok in a cluttered background is detected though it is barely noticeable intuitively. This shows that the proposed feature extraction and background categorization method makes the animal distinguishable enough to be detected.

# 5 Conclusion and Further Work

A specific natural environment is believed to be composed of finite visual elements. The presented method shows that it is possible to categorize these elements into several classes, and the proposed feature descriptor combined with a Neural Network form a robust classifier to distinguish the animals from subclasses of background. Future work will involve testing in various environments with different background features. The animal species shall be extended and whether the method is suitable for classification between the animals shall be investigated. Once an animal is detected it will be tracked to eliminate the need for further detections in the subsequent frames. Real-time capabilities will be investigated.

**Acknowledgements.** The authors wish to thank Glen Afric Country Lodge for their welcoming hospitality and the opportunity offered to collect data.

# References

1. Sirmacek, B., Wegmann, M., Cross, A., Hopcraft, J., Reinartz, P., Dech, S.: Automatic population counts for improved wildlife management using aerial photography. In: iEMSs (2012)
2. van Gemert, J.C., Verschoor, C.R., Mettes, P., Epema, K., Koh, L.P., Wich, S.: Nature conservation drones for automatic localization and counting of animals. In: Agapito, L., Bronstein, M.M., Rother, C. (eds.) ECCV 2014 Workshops. LNCS, vol. 8925, pp. 255–270. Springer, Heidelberg (2015)
3. Henry, A., Shumee, B., Takeo, K.: Neural network-based face detection. IEEE Trans. Pattern Anal. Mach. Intell. **20**(1), 23–38 (1998)
4. Felzenszwalb, P.F., Girshick, R.B., McAllester, D., Ramanan, D.: Object detection with discriminatively trained part-based models. IEEE Trans. Pattern Anal. Mach. Intell. **32**(9), 1627–1645 (2010)
5. Park, S., Deriche, R.: A hierarchical Bayesian network for event recognition of human actions and interactions. Multimed. Syst. **10**(2), 164–179 (2004)
6. Zhao, L., Thorpe, C.E.: Stereo-and neural network-based pedestrian detection. IEEE Trans. Intell. Transp. Syst. **1**(3), 148–154 (2000)
7. Lang, K.J., Waibel, A.H., Hinton, G.E.: A time-delay neural network architecture for isolated word recognition. Neural Netw. **3**(1), 23–43 (1990)
8. Lee, S.B.: Neural-network classifiers for recognizing totally unconstrained handwritten numerals. IEEE Trans. Neural Netw. **8**(1), 43–53 (1997)
9. Wilson, C.L., Candela, G.T.: Neural network fingerprint classification. Artif. Neural Netw. **1**, 2 (1993)
10. Zhang, J., Tan, T.: Brief review of invariant texture analysis methods. Pattern Recogn. **35**(3), 735–747 (2002)

11. Guo, Y., Zhao, G., Pietikinen, M.: Discriminative features for texture description. Pattern Recogn. **45**(10), 3834–3843 (2012)
12. Guo, B., Damper, R.I., Gunn, S.R.: A fast separability-based-feature-selection method for high-dimensional remotely sensed image classification. Pattern Recogn. **41**, 1653–1882 (2008)
13. Guang-ming, X.: An identification method of malignant and benign liver tumors from ultrasonography based on GLCM texture features and fuzzy SVM. Expert Syst. Appl. **37**, 6737–6741 (2010)
14. Huang, K.: Application of artificial neural network for detecting Phalaenopsis seedling diseases using color and texture features. Comput. Electron. Agric. **57**, 3–11 (2007)
15. Martin, F.M.: A scaled conjugate gradient algorithm for fast supervised learning. Neural Netw. **6**(4), 525–533 (1993)

# Predictive Segmentation Using Multichannel Neural Networks in Arabic OCR System

Mohamed A. Radwan$^{(\boxtimes)}$, Mahmoud I. Khalil, and Hazem M. Abbas

Computers and Systems Engineering Department, Faculty of Engineering,
Ain Shams University, Cairo, Egypt
mohamedat.Radwan@gmail.com

**Abstract.** This article offers an open vocabulary Arabic text recognition system using two neural networks, one for segmentation and another one for characters recognition. The problem of words segmentation in Arabic language, like many cursive languages, presents a challenge to the OCR systems. This paper presents a multichannel neural network to solve offline segmentation of machine-printed Arabic documents. The segmented characters are then used as input to a convolutional neural network for Arabic characters recognition. The accuracy of the segmentation model using one font is 98.9 %, while four-font model showed 95.5 % accuracy. The accuracy of characters recognition on Arabic Transparent font of size 18 pt from APTI data set is 94.8 %.

**Keywords:** Arabic segmentation · OCR · Convolutional neural networks

## 1 Introduction

In the classic topic of Arabic characters recognition, we are concerned about digitizing Arabic documents into electronic format. Since Arabic is cursive, so the range of research in the topic can be classified according to how the system recognizes words or sub-words. In [1–4], word level features are recognized to classify them into a word in a vocabulary set. On the other hand [5,6], recognize characters features by using a preprocessing step to segment the input word, then the segmented characters are recognized by a character recognition model. While [7,8], use a sliding window to recognize characters features.

This paper offers an open-vocabulary Arabic text recognition system using two neural networks, one for the segmentation and another one for characters recognition. Automatic segmentation of Arabic has always been a tough problem to solve [9]. Unlike Latin languages which are cursive mostly in handwritten text, Arabic and Farsi are cursive by nature, so typesetting is cursive in both machine generated and handwritten text. Segmentation of words to their constituting characters is a crucial step to the succeeding recognition phase. An Arabic character can have up to four different shapes according to its placement in the word: isolated, start, middle, and end (Fig. 1a). Some characters may only differ

© Springer International Publishing AG 2016
F. Schwenker et al. (Eds.): ANNPR 2016, LNAI 9896, pp. 233–245, 2016.
DOI: 10.1007/978-3-319-46182-3_20

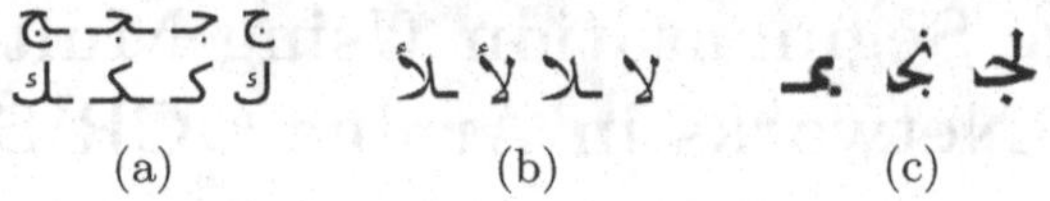

234    M.A. Radwan et al.

جخح
جكسكج    لالالالا    لجنج

(a)            (b)            (c)

**Fig. 1.** (a) Different shapes of two characters. (b–c) Horizontal and vertical ligatures.

in the number of diacritics. Characters also have different heights and widths. Defined combinations of certain characters can have special ligatures to connect them (Fig. 1b and c). Due to these characteristics, segmentation algorithms can fall short by over-segmentation of wide characters, or under-segmentation of interleaving characters.

Many algorithms devised for segmentation of cursive Arabic documents, made use of the structural pattern of lower pixels density between characters. Based on this pattern, a histogram of horizontal projection has been widely used for segmentation [10,11]. However, this method is prone to over-segmentation when a character is composed of several ligatures, or under-segmentation due to the overlapping of characters (Figs. 2a, b). Other structural segmentation algorithms [5,6], depend on thinning or contour tracing to extract strokes or angles and use them as features for extraction. This way they can solve the under-segmentation resulting from overlapping but suffers from over-segmentation. For characters recognition stage the problem becomes much easier, [12] uses a decision tree, [13] uses extracted moments and other shape features as an input for a Neural Network. In [7,8,14,15] Hidden Markov Models are used to learn characters features and do implicit segmentation.

Recently Neural Networks are shown to have state-of-the-art results in object and characters recognition tasks [16,17]. They learn hierarchical representations by stacking layers that learn features in each one from the output of the previous layer. In this paper, a multichannel neural network [18] is used to predict the likelihood of a window, sliding on a sub-word image, being on candidate cut place for characters segmentation. Afterwards, another convolutional neural network is used to recognize the segmented Arabic characters. Advances in training deep neural networks are exploited to reduce over-fitting. Convolutional layers [19] are used for learning features from the image. Regularization techniques like training data augmentation and dropout [20], are used to enhance network generalization.

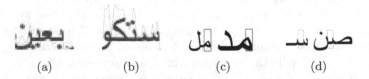

(a)            (b)            (c)            (d)

**Fig. 2.** (a) Threshold over-segmentation in left character. (b) Threshold over-segmentation in the right character, and under-segmentation in the middle. (c) Character context and scale. (d) Small window over-segmentation effect.

The contribution of this paper is a segmentation 3-windows neural network that is used in our OCR pipeline. The segmented output is then used as input to a characters recognition model. The proposed segmentation model can learn to explicitly segment Arabic words, of one font or preferably multiple fonts, into separated characters. The segmentation problem is formulated as a non-linear regression problem, assigning a sliding window values between $[-0.5, 0.5]$ indicating confidence in segmentation area (Fig. 4a). The model is blind to characters classes, as it only recognizes the features of segmentation windows. The character recognition model is formulated as a convolutional neural network that classifies a segmented character into one of the Arabic characters. The segmentation model established a 98.9 % accuracy and the character recognition model had 94.8 % accuracy against the APTI dataset subset.

The paper starts by introducing the segmentation model and its components in Sect. 2. The character recognition model is described in Sect. 3. The details of experiments carried out to test this model are listed in Sect. 4.

## 2   The Segmentation Model

The typical OCR system can easily segment a document into lines then words. The word or sub-word segmentation task, is to recognize boundaries in-between characters of a word, given its image. Using a scanning window on the word, the model predicts a likelihood of the current window to be a segmentation area. The variances due to the difference in characters dimensions add lots of aberrations to how characters appear inside a window (Fig. 2c). Data augmentation techniques (Sect. 4.1) are used to synthesize more data by applying some deformations on the training set [21]. The functions chosen are scaling down or up, and translation; so to simulate what might happen in test sets. To solve over-segmentation, a wider window is needed to recognize wide characters correctly (Fig. 2d). Nonetheless, increasing the window too much will add information other than the location of the segmentation, thus losing the cut localization, or under-segmentation. Taking into consideration these problems two rules are established:

1. Have as much context inside a window without losing the cut localization.
2. Maintain a design matrix that allows data augmentation without losing label consistency.

A neural network with a single wide window as input will fail the first point but allow the second. A Recurrent neural network for sequence labelling of every horizontal point in the window, aside from being hard to train, will fail the second point. A model that uses three windows as input to a multichannel neural network is developed (Fig. 3). So beside using a small window for segmentation, another two channels are added as a previous window and a next window. This will increase the network input context. Data augmentation is carried out by applying deformations to each window separately, without affecting the label consistency thus satisfying the second point as well.

236    M.A. Radwan et al.

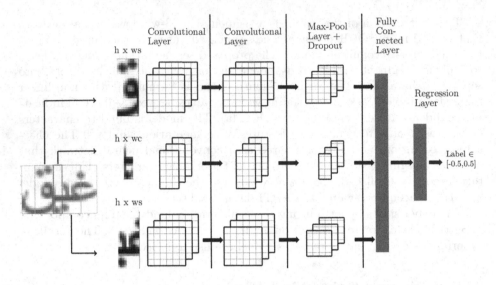

**Fig. 3.** Multichannel neural network for Arabic segmentation.

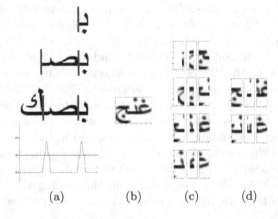

(a)          (b)          (c)          (d)

**Fig. 4.** (a) Finding the ground truth of segmentation, and the function used for assigning labels. (b) The data for this word is extracted by a sliding window. (c) Examples labelled as −0.5. (d) Examples labelled greater than 0.

## 2.1 The Convolutional Channels

Each channel learns low level features using convolutional layers [19]. A layer contains a set of features mapping input to output by convolving a filter $w_i^{(l)}$ of size $A \times B$. $Y_{ijk}^{(l)}$, the component $j, k$ of feature map $i$ in $l$-th convolutional layer is given by:

$$H_{ijk}^{(l)} = B_i^{(l)} + \sum_{i'=1}^{m} \sum_{a=0}^{A} \sum_{b=0}^{B} w_{iab}^{(l)} * Y_{i'(j+a)(k+b)}^{(l-1)} \tag{1}$$

$$Y_{ijk}^{(l)} = max(0, H_{ijk}^{(l)}) \qquad (2)$$

where $max(0, x)$ is a rectified linear activation function [22], $B_i^{(l)}$ is a bias term. The number of feature maps in the previous layer is $m$, $Y_{i'}^{(l-1)}$ is the output of the feature map $i'$ from the previous layer. A max pooling layer is added after the convolutional layers that sub-samples the max value of every $2 \times 2$ window. Then, a dropout is used to reduce over-fitting of data that is not complex enough. It drops some neurons and their connections during training and found to increase the regularization of convolutional neural nets [20].

## 2.2  The Fully Connected Layer

The model's fully connected layer learns correlations between patterns that appear in each channel separately. The left channel should learn the right parts of the characters, while the right channel learns the left parts of characters, and the middle learns the area in-between. A dense fully connected layer then learns the associations between each channel's specific features to find the correlation between the three of them.

$$Y_i^{(h)} = max(0, W_M Y_M + W_L Y_L + W_R Y_R + B_i^{(h)}) \qquad (3)$$

where $Y_x, W_x$ are the output of a channel and the weight associated with it for each $x \in \{L,M,R\}$ the left, middle, and right channels respectively. Left and right channels weights can be thought of as a bias that this layer learns for middle channel features. At the end there is a regression layer that learns the likelihood of the window to be a segmentation place. This architecture is similar to multi-modal neural networks [18].

Two models are constructed from training on two different training sets. First model was trained on one font, while the second model was trained on four fonts. Models parameters are listed in Table 1. Stochastic gradient descent with $5 \cdot 10^{-4}$ learning rate is used to minimize their mean squared error loss function.

Table 1. Segmentation model parameters.

|  | One-font model | Four-font model |
|---|---|---|
| Training fonts | 1 | 4 |
| Left and right channel filters | 48, 48 | 64, 64 |
| Middle channel filters | 24 | 32 |
| Convolutional window size | 3 | 3 |
| pool window size | 2 | 2 |
| Dropout | 0.25 | 0.25 |
| Fully connected layer size | 180 | 256 |

## 3   The Character Recognition Model

Our second model accepts a segment of a word containing a character as an input. The model training data is augmented as discussed below in Sect. 4.1 for better generalization and tolerance to shifts and errors resulting from segmentation model. It's based on convolutional neural network (Fig. 5), with two convolutional layers and two max pooling layers after each one. The convolutional layers are modelled as in Eq. (1). First convolutional layer has 64 filters, and second convolutional layer has 32 filters and both have $2 \times 2$ pooling layers. A fully connected layer (Eq. 3) with 64 outputs is used after the convolutional layers, it's then followed by 25 % dropout. The last layer is a logistic regression layer for classifying the input from the last layer into an Arabic character. Stochastic gradient descent, with $1 \cdot 10^{-4}$ learning rate, was also used as a learning algorithm.

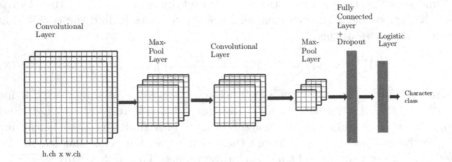

**Fig. 5.** Arabic character recognition neural network.

## 4   Experiments

In these experiments, we assume that the fonts to be recognized are known beforehand and that the input font size will be normalized to the font size we trained on. We conduct experiments on two 3-windows segmentation models, once trained on one font and another on 4 fonts. We then compare 3-windows model to 1-window model on common test set. Finally, A characters model is trained and then tested along with segmentation model on APTI dataset.

### 4.1   Data Augmentation

To improve generalization and reduce over-fitting we increase the training data by applying signal transformations to images. These transformations are selected so as to preserve input label, and add desired invariances to the model. For two dimensional objects recognition such as in recognizing handwritten digits from

MNIST, [21] proved that augmenting the training dataset resulted in significance accuracy improvement. For our problem, we added slight translation invariance on horizontal and vertical dimensions which should add tolerance to segmentation variability. Rotation and shear invariance with small angles were also used, which is useful for italic variations. In addition we used zoom in and out transformation to account for text size changes.

## 4.2  Segmentation Model

For segmentation model experiments[1], we used a set of defined fonts for training and testing; and a constant font size of 18pt for generating training data. For this size the mean characters width was found to be 12 pixels hence it was used afterwards as windows width, and the height for all windows was set to 26 pixels. To test the segmentation method, two instances of our proposed model were constructed, the first is trained on one font, while the second on four fonts. The right and left windows were of width 12 pixels, whereas the middle window width was 4 pixels. Another model with a single window was also constructed for comparison where the single window was of width 12 pixels.

For testing, we used an Arabic dictionary dataset[2]. Using this dataset, two test sets are constructed for evaluation; the first contains words written in one font and the other in four fonts.

**Data Generation.** The ground truth of a word segmentation (Fig. 4a) can be found in two steps:

1. Characters are drawn sequentially and width is measured each time to find the ideal segmentation (Fig. 4a).
2. For each segmentation place, we assign the label of the segmentation pixel and pixels around it by a Gaussian function, that has a mean at the center pixel from first step.

Then, examples are generated by sliding a window, of width 4 pixels and height 26 pixels, on the word. One example is added for each window step, consisting of: the sliding window as the middle window, plus the windows to its left and right. Each example will have label $-0.5$ (Fig. 4c) if it is inside a character's boundaries. Otherwise if the middle window starts in a segmentation place, then its label is given by the Gaussian function associated with this segmentation area found in step 2 above (Fig. 4d). This algorithm is repeated for 2100 words, each word having $Wn$ random characters where $Wn \in \{2, 3, 4, 5\}$. One third of these words are held out as a validation data set that is used for parameters selection.

The data selected for the training set is used for augmentation procedure. We use data augmentation technique described in Sect. 4.1, on the training data subset. For each training example, three of these deformations are selected randomly to apply one on each window. The resulting new three windows are added

---

as a new example with the same label. The size of data with labels greater than zero (segmentation areas) are increased explicitly to balance them with the data of labels less than zero, to ensure classes balance during training.

The first model uses Arial font in the generation. The second model uses the following four fonts: Arial, Tahoma, Thuluth, and Damas, which are very different typographically. The test sets are generated similarly yet using about 250,000 real Arabic words from a dictionary. These words range in length from two to six characters. Two test sets are generated using this dictionary, first using the Arial font, and the other test set is generated using same previous four fonts.

### 4.3   The Characters Model

Using similar generation technique as in segmentation model data, training and validation sets are generated for characters model. For each Arabic character, we generate examples for it in different contexts that applies to it: alone, to the right, to the left, and in the middle. The set is split to 66 % for training set and the remaining for validation set. The training set is used to generate more examples as described in Sect. 4.1.

### 4.4   Evaluation on APTI Data Set

Arabic Printed Text Image (APTI) data set is a large scale benchmark for recognition systems in Arabic. A subset of Arabic Transparent Font of size 18 is used to text a pipeline of Segmentation system and Characters recognition system (Fig. 6).

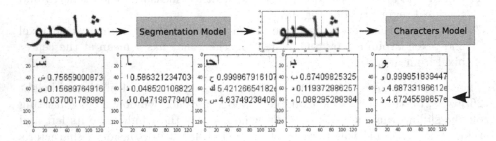

**Fig. 6.** Pipeline for APTI data set

### 4.5   Experimental Results

**Segmentation Model.** The loss function values of each trained model are listed in Table 2 for training, validation, and two test sets. Besides reporting the loss of the test sets, an accuracy is also reported. The test set is converted into classification by checking if the model would assign the segmentation ground truth (Fig. 4a) with labels greater than 0.2. This accuracy is reported on both

**Table 2.** Resulting values for loss functions and (Acc $L > 0$) accuracy of a test data with labels greater than 0.

|  | Recall | Validation | 1-Font test | | 4-Font test | |
|---|---|---|---|---|---|---|
|  | Loss | Loss | Loss | Acc $L > 0$ | Loss | Acc $L > 0$ |
| 1-Font 3-Window model | 0.0247 | 0.0315 | 0.0335 | 98.9% | 0.110 | 65.8% |
| 4-Font 3-Window model | 0.0301 | 0.0309 | 0.0431 | 98.7% | 0.0465 | 95.5% |

test sets in the table. The 1-Font model is able to segment 4-font test set with accuracy of 65.5%, so it is able to get around 40% of the other fonts' segments right.

The 3-Window model accuracy and experimental results from other segmentation models [9], are listed in Table 3. While the datasets used in all other experiment were not open and listed for reference, we compare our model directly to a 1-window mode on the same data, it's trained on one font data, by testing both on same 1-font test data. The 1-window model has 90.2% accuracy which shows the huge significance of a 3-window model and its versatility against confusing windows that cause under and over segmentation.

Figure 7a shows the correct segmentation of a sample document containing previously over-segmented and under-segmented characters being correctly segmented by the proposed model. The data examples that maximally activates neurons in the left channel are shown in (Fig. 7b), and the right channel in (Fig. 7c). The windows that maximally activate the left channel are mostly right

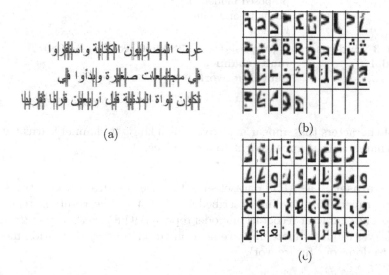

(a)

(b)

(c)

**Fig. 7.** (a) Segmentation of a document using the model. (b) Segments of words that have maximal neural response in the left channel, and the right channel in (c).

Table 3. Comparison results of different methods

|  | Method | Data | Accuracy |
|---|---|---|---|
| Zidouri, Abdelmalek [23] | Structural method | 200 images | 90 % |
| Broumandnia et al. [24] | Wavelet transform | 1000 words different sizes and fonts | 97.83 % |
| Nawaz et al. [13] | Vertical and horizontal projection | Many document images each containing about 200 characters | 76 % |
| Bushofa, BMF and Spann, M [25] | Contour information | 1,065 characters from each font are tested | 97.01 % |
| Hamid, Alaa and Haraty, Ramzi [26] | Structural features for Feed-forward Multilayer neural networks | 10,000 exemplars | 69.72 % |
| Touj et al. [15] | HMM with standard Hough transform as method 1 and with generalized Hough transform as method 2 | 6,400 characters | Method 1: 91 %, and method 2: 97 % |
| 1-Font one window model | Neural Network similar to the proposed model but with only one window | 250,000 words | 90.2 % |
| **1-Font 3 windows model** | **Our proposed multi-channel neural network** | **250,000 words** | **98.9 %** |

parts of characters that appear next to a cut. The right channel learns left parts of the character that appear right to a cut place.

**APTI Dataset.** The APTI dataset subset of size 18pt from Arabic transparent font was run in the pipeline described in Sect. 4.4. The result is reported in Table 4. The character recognition model reported 94.8 % accuracy (5.2 % error), which leaves room for more improvement on the character recognition model we intend to do in our future work.

**Table 4.** Character recognition error rate on size 18pt subset of APTI.

| System | Error (%) |
|---|---|
| Proposed pipeline | 5.2 |
| IPSAR [27] | 4.3 |
| UPV-REC1 [27] | 3.1 |
| Siemens [28] | 0.0 |
| THOCR [28] | 0.9 |

## 5  Conclusion

A multichannel neural network is used to segment Arabic documents. It incorporates a sliding window as a middle channel and another next and previous windows for more context. This model acts very well against over-segmentation and under-segmentation problems. A model trained to segment one Arabic font has an accuracy of 98.9 %. A model trained to segment four fonts has a 95.5 % accuracy. This model showed better accuracy than using a one-window model of the Neural Network. Then the segmentation model output is used to classify characters using a character recognition model. This proposed pipeline was tested against an APTI dataset subset showing 94.8 % accuracy. Future work would be around improving the character recognition process in the pipeline.

The pipeline is useful for automatically building an open vocabulary Arabic OCR system, as the generation and training process can be autonomous and has very few variables or features to be tuned.

## References

1. Al-Khateeb, J.H., Khelifi, F., Jiang, J., Ipson, S.S.: A new approach for off-line handwritten arabic word recognition using KNN classifier. In: 2009 IEEE International Conference on Signal and Image Processing Applications (ICSIPA), pp. 191–194, November 2009
2. Al-Badr, B., Haralick, R.M.: Segmentation-free word recognition with application to Arabic. In: Proceedings of the Third International Conference on Document Analysis and Recognition, vol. 1, pp. 355–359. IEEE (1995)
3. Khorsheed, M.S., Clocksin, W.F.: Multi-font Arabic word recognition using spectral features. In: 15th International Conference on Pattern Recognition, Proceedings, vol. 4, pp. 543–546. IEEE (2000)
4. Jelodar, M.S., Fadaeieslam, M.J., Mozayani, N., Fazeli, M.: A persian OCR system using morphological operators. In: WEC (2), pp. 137–140 (2005)
5. Märgner, V.: Sarat-a system for the recognition of Arabic printed text. In: Proceedings of the 11th IAPR International Conference on Pattern Recognition. Conference B: Pattern Recognition Methodology and Systems, vol. II, pp. 561–564. IEEE (1992)
6. Azmi, R., Kabir, E.: A new segmentation technique for omnifont Farsi text. Pattern Recognit. Lett. **22**(2), 97–104 (2001)

7. Khoury, I., Giménez, A., Juan, A., Andrés-Ferrer, J.: Window repositioning for printed Arabic recognition. Pattern Recognit. Lett. **51**, 86–93 (2015)
8. Ahmad, I., Mahmoud, S.A., Fink, G.A.: Open-vocabulary recognition of machine-printed Arabic text using hidden Markov models. Pattern Recognit. **51**, 97–111 (2016)
9. Alginahi, Y.M.: A survey on Arabic character segmentation. Int. J. Doc. Anal. Recognit. (IJDAR) **16**(2), 105–126 (2013)
10. Amin, A.: Off-line Arabic character recognition: the state of the art. Pattern Recognit. **31**(5), 517–530 (1998)
11. Zheng, L., Hassin, A.H., Tang, X.: A new algorithm for machine printed Arabic character segmentation. Pattern Recognit. Lett. **25**(15), 1723–1729 (2004)
12. Bushofa, B.M.F., Spann, M.: Segmentation and recognition of printed arabic characters using structural classification. Image Vis. Comput. **15**, 167–179 (1997)
13. Nawaz, S.N., Sarfraz, M., Zidouri, A., Al-Khatib, W.G.: An approach to offline arabic character recognition using neural networks. In: Proceedings of the 10th IEEE International Conference on Electronics, Circuits and Systems, ICECS, vol. 3, pp. 1328–1331. IEEE (2003)
14. Gouda, A.M., Rashwan, M.A.: Segmentation of connected Arabic characters using hidden Markov models. In: IEEE International Conference on Computational Intelligence for Measurement Systems and Applications, CIMSA, pp. 115–119. IEEE (2004)
15. Touj, S., Amara, N.B., Amiri, H.: Two approaches for arabic script recognition-based segmentation using the Hough transform. In: ICDAR, pp. 654–658. IEEE (2007)
16. Cireşan, D.C., Meier, U., Masci, J., Gambardella, L.M., Schmidhuber, J.: High-performance neural networks for visual object classification. arXiv preprint arXiv:1102.0183 (2011)
17. Lee, C.-Y., Xie, S., Gallagher, P., Zhang, Z., Zhuowen, T.: Deeply-supervised nets. arXiv preprint arXiv:1409.5185 (2014)
18. Ngiam, J., Khosla, A., Kim, M., Nam, J., Lee, H., Ng A.Y.: Multimodal deep learning. In: Proceedings of the 28th International Conference on Machine Learning (ICML 2011), pp. 689–696 (2011)
19. LeCun, Y., Bottou, L., Bengio, Y., Haffner, P.: Gradient-based learning applied to document recognition. Proc. IEEE **86**(11), 2278–2324 (1998)
20. Srivastava, N.: Improving neural networks with dropout. Ph.D. thesis, University of Toronto (2013)
21. Simard, P.Y., Steinkraus, D., Platt, J.C.: Best practices for convolutional neural networks applied to visual document analysis. In: Null, p. 958. IEEE (2003)
22. Glorot, X., Bordes, A., Bengio, Y.: Deep sparse rectifier neural networks. In: International Conference on Artificial Intelligence and Statistics, pp. 315–323 (2011)
23. Zidouri, A.: On multiple typeface Arabic script recognition. Res. J. Appl. Sci. Eng. Technol. **2**(5), 428–435 (2010)
24. Broumandnia, A., Shanbehzadeh, J., Nourani, M.: Segmentation of printed Farsi/Arabic words. In: IEEE/ACS International Conference on Computer Systems and Applications, AICCSA 2007, pp. 761–766. IEEE (2007)
25. Bushofa, B.M.F., Spann, M.: Segmentation of Arabic characters using their contour information. In: Proceedings of the DSP 97 13th International Conference on Digital Signal Processing, vol. 2, pp. 683–686. IEEE (1997)
26. Hamid, A., Haraty, R.: A neuro-heuristic approach for segmenting handwritten arabic text. In: ACS/IEEE International Conference on Computer Systems and Applications, pp. 110–113. IEEE (2001)

27. Al-Muhtaseb, H.A., Mahmoud, S.A., Qahwaji, R.S.: Recognition of off-line printed Arabic text using hidden Markov models. Signal Process. **88**(12), 2902–2912 (2008)
28. Slimane, F., Kanoun, S., El Abed, H., Alimi, A.M., Ingold, R., Hennebert, J.: ICDAR competition on multi-font and multi-size digitally represented Arabic text. In: 12th International Conference on Document Analysis and Recognition (ICDAR), pp. 1433–1437. IEEE (2013)

# Quad-Tree Based Image Segmentation and Feature Extraction to Recognize Online Handwritten Bangla Characters

Shibaprasad Sen[1(✉)], Mridul Mitra[1], Shubham Chowdhury[1],
Ram Sarkar[2], and Kaushik Roy[3]

[1] Future Institute of Engineering and Management, Kolkata, India
shibubiet@gmail.com, mitra.ml50495@gmail.com,
chowdhury.shubham@yahoo.co.in
[2] Jadavpur University, Kolkata, India
raamsarkar@gmail.com
[3] West Bengal State University, Barasat, India
kaushik.mrg@gmail.com

**Abstract.** In this paper, three different feature extraction strategies along with their all possible combinations have been discussed in detail for the recognition of online handwritten Bangla basic characters. Applying a quad-tree based image segmentation approach the target character has been dissected for the extraction of features. Out of these three techniques, one is computing area feature (using composite Simpson's rule) while other two are extracted local (mass distribution and chord length) features. Authors have also investigated optimal depth of the quad-tree (while segmenting an image), at which classifier reveals its best performance. The current experiment has been tested on 10,000 character dataset. Sequential Minimal Optimization (SMO) produces highest recognition accuracy of 98.5 % when all three feature vectors are combined.

**Keywords:** Online handwriting recognition · Bangla script · Quad-tree based image segmentation · Composite Simpson's rule · Mass distribution · Chord length

## 1 Introduction

Not only due to the increasing dependency on easily available handheld devices in the form of Smart phones, tablets, iPad, A4 Take Note, etc. which are available at reasonable cost, but also these devices now govern the human society because of their numerous applicability to make life easier. One of the interesting properties of such devices is that people can provide information freely on those devices and written information can be saved in the form of online information bearing the pixels information along trajectory path with pen up/down status. Adopting such online devices, people not only minimize the chances of mistyping that may arise when writing with a keyboard but also saves extra time that would have been required for typing the same information. In Online Handwriting the information are stored as real time coordinate data points. In contrast to that in offline Handwriting recognition, the information are

© Springer International Publishing AG 2016
F. Schwenker et al. (Eds.): ANNPR 2016, LNAI 9896, pp. 246–256, 2016.
DOI: 10.1007/978-3-319-46182-3_21

saved as images. So, in the later one the data are prone to quality degradation leading to high noise level in the information whereas in the former one, the information can never be prone to quality issues and hence efficient in contrast to offline. These benefits, in turn, make Online Handwriting Recognition (OHR) an upcoming research domain. Though in the literature, substantial amount of research publications are available for Devanagari script [1–8], but while talking about Bangla script, this statement is not defensible. Limited number of research works in the literature silently describes the truth. Hence, researchers in the Bangla OHR domain need to pay more attention. Bhattacharya et al. in [9] have shown the advantage of using direction code features to recognize the symbols from Bangla alphabet set. Authors in [10] have highlighted a novel approach to recognize handwritten Bangla characters in a 2D plane without being concerned about writing direction. From implementation point, authors have assumed that structural shape of a character can be represented by different skeletal convexity of its component strokes. In [11], authors have followed a different strategy where firstly component strokes are extracted at character level. Then, stroke level sequential and dynamic information have been computed to extract the feature values for stroke recognition purpose. Target character is then constructed from recognized strokes by matching the stroke sequences from a stored database. In [12], authors have described the benefits of the customized version of Distance Based Feature extraction technique towards recognition of handwritten Bangla characters. At pre-processing stage, characters are segmented into N number of divisions and then distance values have been computed from each segment point to rest of the points. Authors in [13] have proposed a stroke based recognition scheme where each stroke is described by a string of shape features. Then, Dynamic Time Warping (DTW) technique has been adopted in order to recognize an unknown stoke by comparing it with a previously prepared stroke database. Sen et al. in [14] have figured out the individual as well as combined impact of global and local information of an image. Authors in [15] have shown the success rate obtained by combining some online (point based and structural features) and offline (quad tree based longest run and convex hull) features for the recognition of online handwritten Bangla characters. Parui et al., in [16], have successfully managed to group 54 stroke classes from Bangla character alphabet set which is based on graphemes level shape similarity. Stroke level HMM has been constructed to recognize constituent strokes and then 50 look up tables has been prepared to identify the target character from recognized strokes. R. Ghosh, in [17], has divided all the constituent strokes of the sample character into nine rectangular zones and then has focused on building of an effective feature vector by considering standard deviation of x and y coordinates, writing direction, curvature, slope, curliness at stroke level. Produced feature vector are then fed to the Support Vector Machine (SVM) classifier to recognize the same.

In this paper, authors have concentrated on some pre-processing steps along with estimating strong discriminating feature set for the recognition of handwritten Bangla characters.

## 2  Data Collection and Pre-processing Step

For the current work, 100 different persons in West Bengal, India belonging to different professional and educational background, age, gender etc. have contributed 200 handwritten samples for each Bangla character. Considering 50 character symbols of Bangla script (see Fig. 1), size of the present database becomes 10,000. No strict instruction was given during data collection process; only contributors were asked to write the constituent strokes of the characters as part of basic stroke database [11]. In this experiment, at the beginning of pre-processing step, duplicate points are removed from character sample to minimize redundancy and then pixel points are scaled to fit into a window of size 512 × 512 in order to cope up size variability [12]. Scaled pixel points are then rearranged to obtain a new sequence of points which are unit distance apart, by applying 4-connected Bresenham's algorithm. Character image with irregular scaled coordinates and unit distant coordinates are shown in Fig. 2.

| অ | আ | ই | ঈ | উ | ঊ | ঋ | এ | ঐ | ও |
|---|---|---|---|---|---|---|---|---|---|
| ঔ | ক | খ | গ | ঘ | ঙ | চ | ছ | জ | ঝ |
| ঞ | ট | ঠ | ড | ঢ | ণ | ত | থ | দ | ধ |
| ন | প | ফ | ব | ভ | ম | য | র | ল | শ |
| ষ | স | হ | ড় | ঢ় | য় | ৎ | ং | ঃ | ঁ |

**Fig. 1.** Basic symbols in Bangla alphabet set

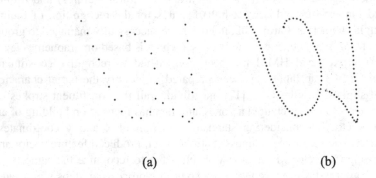

(a)                              (b)

**Fig. 2.**  (a) Scaled character showing non-uniform pixel intervals due to speed variation (b)same character with unit distant pixels after applying 4-connected Bresenham's algorithm

# 3    Feature Extraction

In this paper, quad-tree based image segmentation approach is followed to design three different feature extraction techniques for the recognition of online handwritten Bangla characters. At the first step, the sample character is divided into four (2 × 2) rectangular blocks. Then one area based feature using, *composite Simpson's* rule and two other local features namely *mass distribution* and *chord length* have been computed from therein. In each epoch, the level of quad-tree is increased by one, which can be achieved by dividing each rectangular block into four sub-blocks to get a structure of 4 × 4 blocks at depth two. The said features are again estimated from each block to witness the outcome with a closer view than before. In the current experiment, the observations are recorded by segmenting the image up to 64 blocks (i.e. quad-tree of depth four). Figure 3 reflects quad-tree based segmentation of the same character sample varying the depth from one to four. In this way 4, 16, 64 and 256 feature values have been produced at quad-tree of depth one, two, three and four respectively. The feature extraction methodologies are described in following subsections:

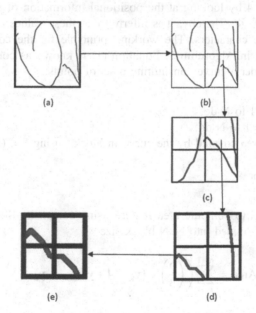

**Fig. 3.** Illustration of image segmentation using quad-tree based approach (a) original image. (b-e) segmented images produced at different depths of the tree

## 3.1    Area Feature

As the structural patterns of the alphabets in Bangla are supposedly unalike from each other, hence, it can be assumed that for different types of characters, pixel patterns constituting the samples must be dissimilar at different blocks. There may be some blocks where no pixels are found for any particular character sample. Therefore, when block-wise area under the curve is calculated then these calculations become distinct

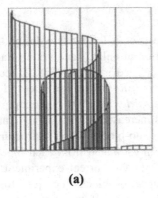
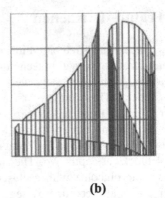

(a)                                        (b)

**Fig. 4.** Calculation of area under curve in the respective blocks for two different character samples using composite Simpson's rule in a quad-tree based image segmentation at depth two

for different character patterns. The truthfulness of this statement can be easily observed from Fig. 4 by looking at the positional information of selected blocks for character samples ২, ক. Thereby, it is inferred that these values could be useful for classification of the characters. The working principle for the computation of area feature is presented in Algorithm 1. Equation (1) is known as composite Simpson's rule to find area under a curve constituting a set of points.

**Step I**: BEGIN
**Step II**: for i = 1 to N do,
**Step III**: for j = 1 to N do,
**Step IV**: Find area covered by the curve in block$_{i,j}$ using Eq. (1)
**Step V:** End for
**Step VI:** End for
**Step VII:** End

**Algorithm 1:** Steps to compute area feature using composite Simpson's rule when sample character is divided into NxN block size

$$Area = \sum_{k=1}^{n-1}\left(\left(\frac{h}{3}\right) * (y_k + 4 * y_{middle} + y_{k+1})\right) \tag{1}$$

Where, each rectangular block contains varying number of pixels n, starting from $y_1$ to $y_n$. ($y_k$, $y_{k+1}$) denotes the measurement of y coordinate values between consecutive pixel points in that block. Values of h and $y_{middle}$ have been calculated as follows:

$$h = \frac{mod(x_{k+1} - x_k)}{2}$$

$$y_{middle} = \frac{(y_k + y_{k+1})}{2}$$

After execution of Algorithm 1, a total of N*N number of area values will be generated as an outcome when the character image is divided into NxN rectangular blocks. These measurements are taken as feature values in this experiment. This has been assumed that the blocks which do not have any pixel return zero as area value.

## 3.2    Local Feature

- Mass Distribution

Depending on structural pattern of the character, certain rectangular blocks are densely populated by the data pixels. Therefore, block-wise mass distribution information may carry an important role to distinguish different character patterns efficiently. Here mass distribution describes the pixel counts inside a block, produced by quad-tree image segmentation approach. Figure 5 shows mass distribution of the character samples হ and ক when the images are segmented into 16 blocks. Here, blue points are the pixels in the respective blocks. From these figures this can be easily understood that for different character patterns, a particular block has varied data pixels, which in turn produces discriminative feature towards online Bangla handwritten character recognition.

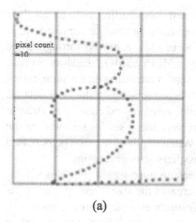 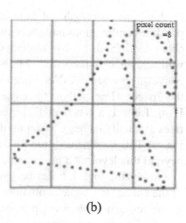

(a)                                                    (b)

**Fig. 5.** Mass distribution of two character samples হ and ক in a quad-tree based image segmentation at depth two (for each image, pixel counts are shown for a particular block)

- Chord Length

As compared to mass distribution, in this approach, the length of the contributed chord in each block has been considered. Dividing the character sample into a number of small chords/segments and storing block-wise chord length information as feature values play a vital role in this pattern classification problem. This is because these lengths vary significantly for different character patterns. Algorithm 2 describes the steps to compute block-wise chord length feature.

**Step-I**: BEGIN
**Step-II**: for i = 1 to N do,
**Step-III**: for j = 1 to N do,

$$\textbf{Step-IV: } C\ hord\_Length_{i,j} = \sum_{n=1}^{k-1} \sqrt{(x_n - x_{n+1})^2 + (y_n - y_{n+1})^2}$$

**Step-V:** End for
**Step-VI:** End for
**Step-VII:** End

**Algorithm 2:** Steps to compute block-wise *chord length* feature when sample character is divided into NxN blocks.

Let us assume that m number of varying pixels starting from $(x_1, y_1)$ to $(x_m, y_m)$ is there in each rectangular block as mentioned in Fig. 5. To find the chord length of $block_{i,j}$, summation of Euclidian distances between the consecutive pair of pixels have been measured in the said block as indicated in Step-IV of Algorithm 2. Figure 5 clearly demonstrates the fact that length of the chord remarkably changes for different characters in a particular block which enables the classifier to differentiate handwritten online Bangla characters successfully.

## 4   Result and Discussion

In this paper, strengths of individual feature set, mentioned earlier, as well as their possible combinations have been observed while recognizing handwritten online Bangla characters. Table 1(a-b) highlights the recognition accuracies, achieved through some well-known classifiers like Multi-Layer Perceptron (MLP), SMO, Random Forest, NaiveBayes and BayesNet when different quad-tree depths are followed for segmenting the images. Here, authors have used 5-fold cross validation scheme on the total dataset. From Table 1, it is noticed that irrespective of applied feature set and classifiers, success rates gradually increase as depth of the quad-tree structure increases for segmenting the image. This statement is true for depth up to three. When depth of the tree is increased beyond this level then performances of the said classifiers start decreasing.

Analyzing Fig. 3, it can be stated that with increasing quad-tree depth, closer view of the character sample is obtained which in turn decreases the size of component parts in the respective blocks. Up to depth level three, features estimated on obtained segments of the character sample, relatively smaller in size, become informative enough to classify them. When character is further divided into more number of blocks, by increasing depth of the tree, block-wise components are becoming so small in size. As a result, features estimated from these segments become less informative and thus fail to identify the character samples properly.

Gray cells, in Table 1, represent the depth of quad-tree where classifiers show their best performances for different feature combinations. Bold styled values specify the name of feature extraction procedure that reflects top recognition at a particular depth. From Table 1, this is observed that, in the present experiment, SMO outperforms all other classifiers to yield the recognition accuracy of 98.5 % (marked as bold styled and colored in red) when all three feature sets are combined and depth of the quad-tree is

**Table 1.** (a-b). Success rates (in %) of different classifiers for the recognition of online Bangla characters when individual and various combinations of estimated feature sets are applied for different levels of quad-tree based image segmentation (represented by block size)

| Feature sets applied | Classifiers applied at different quad-tree depth | | | | | | | | | |
|---|---|---|---|---|---|---|---|---|---|---|
| | 2 x 2 | | | | | 4 x 4 | | | | |
| | MLP | SMO | Random Forest | NaiveBayes | BayesNet | MLP | SMO | Random Forest | Naïve Bayes | BayesNet |
| Composite Simpson's Feature (1) | 58.6 | 47.2 | 62.7 | 50.5 | 50.5 | 87.9 | 89.7 | 95.5 | 80.5 | 88.4 |
| Mass Distribution (2) | 55.9 | 47.7 | 58.6 | 47.7 | 47.1 | 89.2 | 93.4 | 95.8 | 82.9 | 87.9 |
| Chord Length (3) | 60.5 | 45.8 | 68.3 | 46.4 | 52.9 | 90.7 | 92.5 | 96.4 | 80.9 | 89.9 |
| 1+2 | 84.3 | 77.3 | 86.7 | 68.5 | 67.9 | 96.1 | 97.5 | 97.1 | 86.3 | 91.7 |
| 1+3 | 82.1 | 66.9 | 83.3 | 57.7 | 62.6 | 93.6 | 94.6 | 96.6 | 82.4 | 91.0 |
| 2+3 | 86.1 | 75.3 | 87.1 | 62.4 | 66.0 | 97.3 | 97.6 | 97.4 | 85.8 | 92.2 |
| 1+2+3 | **92.6** | 85.6 | 90.9 | 69.2 | 71.3 | 97.6 | **98.1** | 97.6 | 85.9 | 92.4 |

(a)

| Feature sets Applied | Classifiers applied at different quad-tree depth | | | | | | | | | |
|---|---|---|---|---|---|---|---|---|---|---|
| | 8 x 8 | | | | | 16 x 16 | | | | |
| | MLP | SMO | Random Forest | Naïve Bayes | BayesNet | MLP | SMO | Random Forest | Naïve Bayes | BayesNet |
| Composite Simpson's Feature (1) | 88.5 | 96.3 | 97.0 | 84.2 | 92.2 | 87.2 | 96.2 | 83.7 | 81.6 | 92.1 |
| Mass Distribution (2) | 90.1 | 97.8 | 97.1 | 85.3 | 91.0 | 88.4 | 97.5 | 85.6 | 82.5 | 91.0 |
| Chord Length (3) | 91.4 | 97.2 | 97.3 | 85.5 | 93.1 | 89.2 | 96.9 | 89.2 | 83.9 | 93.1 |
| 1+2 | 98.1 | 98.2 | 97.2 | 86.4 | 92.6 | 96.0 | 97.8 | 93.4 | 85.9 | 92.6 |
| 1+3 | 96.8 | 97.5 | 97.4 | 85.6 | 93.3 | 94.3 | 97.1 | 93.1 | 84.3 | 93.3 |
| 2+3 | 97.9 | 97.9 | 97.5 | 86.6 | 93.2 | 94.5 | 97.8 | 92.8 | 85.1 | 93.2 |
| 1+2+3 | 98.2 | **98.5** | 97.8 | 86.8 | 93.5 | 96.4 | **97.9** | 93.5 | 86.1 | 93.5 |

(b)

three. Figure 6 graphically describes the outcomes of the different experimentations performed under the current work. Blue, red, green, and violet lines reflect the nature of the classifiers when features are extracted from the images segmented by quad-tree based approach at depth one, two, three and four respectively. It is clearly seen from Fig. 6 that green line is always at top position and the outcome for SMO classifier is best at this depth. Table 2 reveals the number of feature counts for all possible combination that are fed to the classifiers at varied depths of the tree.

Though the present system works well for the recognition of online handwritten Bangla characters, still certain misclassifications have been noticed. Table 3 shows the

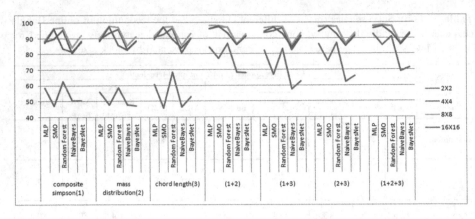

**Fig. 6.** Graphical behavior of the said classifiers for all combinations of feature estimation procedures considering different depths of the quad-tree (Color figure online)

**Table 2.** Number of feature counts for all possible combination at various quad-tree depths

| Features used | Quad-tree depth | | | |
|---|---|---|---|---|
| | One (2 × 2) | Two (4 × 4) | Three (8 × 8) | Four (16 × 16) |
| Composite Simpson's feature (1) | 4 | 16 | 64 | 256 |
| Mass distribution (2) | 4 | 16 | 64 | 256 |
| Chord length (3) | 4 | 16 | 64 | 256 |
| (1 + 2) | 8 | 32 | 128 | 512 |
| (1 + 3) | 8 | 32 | 128 | 512 |
| (2 + 3) | 8 | 32 | 128 | 512 |
| (1 + 2 + 3) | 12 | 48 | 192 | 768 |

**Table 3.** Most confusing character pairs

| Original character (200 samples) | Misclassified as (samples out of 200) |
|---|---|
| ল | ন(6) |
| ড | ভ(3) |
| গ | প (3) |
| চ | ঢ (3) |
| ঔ | উ (2) |
| ঘ | থ (2) |

**Table 4.** Compative study of some recently published works along with the proposed work

| Method | Features used | Accuracy (in%) |
|---|---|---|
| Bhattacharya et al. [9] | Directional Code Feature | 83.61 |
| Parui et al. [16] | Shape Based Feature (Shape and Size of stroke) | 87.7 |
| Roy et al. [11] | Structural + Point Based Feature | 88.23 |
| Sen et al. [15] | Combination of some online and offline Feature (point Based feature + quad tree based feature) | 83.92 |
| R. Ghosh [17] | Structural and Directional Features | 87.48 |
| Sen et al. [12] | Customized version of Distance based Feature | 98.2 |
| **Proposed technique** | Combination of Area, Mass Distribution and Chord Length Features | 98.5 |

character pairs which were difficult to handle by the proposed system. After a thoughtful analysis, authors have reached to the decision that due to strong structural resemblance between the characters belonging to a pair, almost similar feature values are estimated, thereby, classifiers have made mistakes during the recognition step.

To prove the effectiveness of the present system, Table 4 compares it with some past works.

## 5    Conclusion

In the present work, authors are intended to observe the effects of three feature extraction strategies along with their probable combinations for the recognition of online handwritten Bangla characters, when applied at different local regions of any character image. Authors have also tried to find the optimal depth of the quad-tree while segmenting a character image to estimate feature information. Although some misclassifications between character pairs have been observed, still reasonably large number of samples of Bangla alphabet set is identified properly by applying this approach. Hence, this technique could be included in OHR domain to solve pattern classification problems. In future, this feature extraction procedure will also be tested for stroke based character or word recognition purpose too.

## References

1. Connell, S.D., Sinha, R.M.K., Jain, A.K.: Recognition of unconstrained online Devanagari characters. In 15th International Conference on Pattern Recognition, pp. 368–371 (2000)
2. Joshi, N., Sita, G., Ramakrishnan, A.G., Deepu, V.: Machine recognition of online handwritten Devanagari characters. In: International Conference on Document Analysis and Recognition, pp. 1156–1160 (2005)
3. Swethalakshmi, H., Jayaraman, A., Chakravarthy, V.S., Sekhar, C.C.: On-line handwritten character recognition for Devanagari and Telugu scripts using support vector machines. In: International Workshop on Frontiers in Handwriting Recognition, pp. 367–372 (2006)

4. Swethalakshmi, H., Sekhar, C.C., Chakravarthy, V.S.: Spatiostructural features for recognition of online handwritten characters in Devanagari and tamil scripts. In: International Conference on Artificial Neural Networks, vol. 2, pp. 230–239 (2007)
5. Kumar, A., Bhattacharya, S.: Online Devanagari isolated character recognition for the iPhone using hidden markov models. In: International Conference on Students' Technology Symposium, pp. 300–304 (2010)
6. Tripathi, A., Paul, S.S., Pandey, V.K.: Standardization of stroke order for online isolated Devanagari character recognition for iPhone. In: IEEE International Conference on Technology Enhanced Education, pp. 1–5 (2012). DOI:10.1109/ICTEE.2012.6208657
7. Kubatur, S., Sid-Ahmed, M., Ahmadi, M.: A neural network approach to online Devanagari handwritten character recognition. In: International Conference on High Performance Computing and Simulation (2012). DOI:10.1109/HPCSim.2012.6266913
8. Lajish, V.L., Kopparapu, S.K.: Online handwritten Devanagari stroke recognition using extended directional features. In: IEEE 8th International Conference on Signal Processing and Communication System (2014). DOI:10.1109/ICSPCS.2014.7021063
9. Bhattacharya, U., Gupta, B.K., Parui, S.K.: Direction code based features for recognition of online handwritten characters of Bangla. In: International Conference on Document Analysis and Recognition, pp. 58–62 (2007)
10. Bag, S., Bhowmick, P., Harit, G.: Recognition of bengali handwritten characters using skeletal convexity and dynamic programming. In: International Conference on Emerging Application of Information Technology, pp. 265–268 (2011)
11. Roy, K.: Stroke-database design for online handwriting recognition in Bangla. In: International Journal of Modern Engineering Research, pp. 2534–2540 (2012)
12. Sen, S.P., Sarkar, R., Roy, K.: A simple and effective technique for online handwritten Bangla character recognition. In: 4th International Conference on Frontiers in Intelligent Computing: Theory and Application (2015)
13. Bandyopadhyay, A., Chakraborty, B.: Development of online handwriting recognition system: a case study with handwritten Bangla character. In: World Congress on Nature and Biologically Inspired Computing, pp. 514–519 (2009)
14. Sen, S.P., Bhattacharyya, A., Das, A., Sarkar, R., Roy, K.: Design of novel feature vector for recognition of online handwritten Bangla basic characters. In: 1st International Conference on First International Conference on Intelligent Computing & Communication (2016)
15. Sen, S.P., Paul, S.S., Sarkar, R., Roy, K., Das, N.: Analysis of different classifiers for on-line Bangla character recognition by combining both online and offline information. In 2nd International Doctoral Symposium on Applied Computation and Security Systems (2015)
16. Parui, S.K., Guin, K., Bhattacharya, U., Chaudhuri, B.B.: Online handwritten Bangla character recognition using HMM. In: International Conference on Pattern Recognition, pp. 1–4 (2008)
17. Ghosh, R.: A novel feature extraction approach for online Bengali and devenagari character recognition. In: International Conference on Signal Processing and Integrated Networks, pp. 483–488 (2015)

# A Hybrid Recurrent Neural Network/Dynamic Probabilistic Graphical Model Predictor of the Disulfide Bonding State of Cysteines from the Primary Structure of Proteins

Marco Bongini, Vincenzo Laveglia, and Edmondo Trentin[✉]

Dipartimento di Ingegneria dell'Informazione s Scienze Matematiche,
Università degli Studi di Siena, Siena, Italy
{bongini,laveglia,trentin}@dii.unisi.it

**Abstract.** Cysteines in a protein have a tendency to form mutual disulfide bonds. This affects the secondary and tertiary structure of the protein. Therefore, automatic prediction of the bonding state of cysteines from the primary structure of proteins has long been a relevant task in bioinformatics. The paper investigates the feasibility of a predictor based on a hybrid approach that combines the dynamic encoding capabilities of a recurrent autoencoder with the short-term/long-term dependencies modeling capabilities of a dynamic probabilistic graphical model (a dynamic extension of the hybrid random field). Results obtained using 1797 proteins from the May 2010 version of the Protein Data Bank show an average accuracy of 85 % by relying only on the sub-sequences of the residue chains with no additional attributes (like global descriptors, or evolutionary information provided by multiple alignment).

**Keywords:** Hybrid random field · Probabilistic graphical model · Recurrent neural network · Recurrent autoencoder · Disulfide bond · Cysteine bonding state

## 1 Introduction

The automatic prediction of the disulfide-bonding state of cysteines from sequences of amino-acids in proteins is a relevant task in bioinformatics [8,17]. It represents a significant step towards realizing functional and structural predictions from the primary structure of (generally, non-homologous) proteins [9]. Several statistical and machine learning approaches (e.g. artificial neural networks (ANN) [8], hidden Markov models (HMM) [1], support vector machines (SVM) [19], conditional random fields (CRF) [16]) have been applied to the task throughout the last couple of decades. A review is offered in [17]. In most cases, the accuracy of the different approaches could be improved by combining the bare sequence-level information (i.e., a representation of the sequence of amino-acids, or residues) with global descriptors of the chemical and physical

© Springer International Publishing AG 2016
F. Schwenker et al. (Eds.): ANNPR 2016, LNAI 9896, pp. 257–268, 2016.
DOI: 10.1007/978-3-319-46182-3_22

properties of the overall protein, or with the evolutionary information stemming from the multiple alignment of homologous individuals [8]. In their raw representation, the sequences at hand may be thought of as character strings built from a finite and discrete alphabet of individual symbols (which, in the typical case of eukaryotic organisms adds up to 20 symbols) corresponding to the different, standard abbreviations of the specific amino-acids. Some feature extraction process is required in order to map these symbols onto real-valued feature vectors whenever continuous-valued models (like ANNs or SVMs) are used. The length of the sequences varies significantly on a protein-by-protein basis, ranging from a few dozens residues to a few thousands. These sequences may contain a variable number of cysteines (from none to several dozens) which, in turn, may form a disulfide bond with another cysteine within the protein. Disulfide bonds affect the folding and the conformational stability of the protein, with the obvious implications on its secondary and tertiary structures (the quaternary structure is affected, in turn, as long as one or more bonds are formed with cysteines that are present in the environment, e.g. within neighboring proteins). It is an implicit consequence of the central dogma of molecular biology [7] that the sequences (i.e., the primary structure of the proteins) encapsulate the information that an automatic learner could exploit in order to predict accurately the disulfide bonding state of the cysteines. Still, the task remains by and large an open problem to date.

The paper proposes and evaluates a hybrid automatic disulfide bond predictor that stems from the combination of a recurrent neural network (RNN) and a dynamic probabilistic graphical model for sequences. The latter, namely the dynamic hybrid random field (D-HRF) that we first proposed in [4,18], is an extension to sequences of the hybrid random field (HRF) [10]. The RNN is used as a recurrent autoencoder that extracts dynamically a proper discrete representation of the residues that is suitable to be modeled via D-HRF (depending on the disulfide bonding state, and according to a maximum pseudo-likelihood criterion). The definition of D-HRF and the fundamentals of its algorithms for parameter and structure learning are reviewed in Sect. 2, while the hybrid RNN/D-HRF system is presented in Sect. 3. All in all, the machine is expected to exploit both local (residue-neighborhood level) and global (sequence level) information, by means of (1) a proper location-dependent encoding of the primary structure into properly defined random variables via RNN, and of (2) a short-term/long-term modeling of the probabilistic (in)dependencies among these variables realized by the D-HRF. The comparative results of experiments carried out on the dataset Protein Data Bank (PDB) [3] are reported on in Sect. 4. Conclusions are drawn in Sect. 5.

## 2   Review of Dynamic Hybrid Random Fields

The most popular instances of probabilistic graphical models are represented by Bayesian networks (BNs) [13], and by Markov random fields (MRFs) [12]. There are (in)dependence structures over any given set of random variables which can be modeled via BNs but not via MRFs, and vice-versa. Besides, established learning algorithms for BNs and MRFs present high computational complexity as the number of variables increases. Incidentally, no learning algorithm for MRFs has qualified as a standard reference so far. In [11], a probabilistic graphical model known as the hybrid random field (HRF) was introduced to overcome such drawbacks of traditional paradigms. The HRF was proven to subsume the modeling capabilities of both BNs and MRFs. A broad-sense definition of HRFs as a specific set of BNs is given in [10]. The reader is referred to the latter for all basic concepts of standard HRFs which are relevant to this paper. It is proven that HRFs posses the *modularity* property [10], that is they can factorize the overall joint probability of the modeled random variables as a product of local probabilistic quantities defined at the level of the individual BNs in the HRF. As a consequence, a strict-sense and simpler definition of HRF can be given, which roughly goes as follows: an HRF is a collection of Bayesian networks which possesses the modularity property (see, e.g., [11]). In this paper, for simplicity and coherence with the proposed algorithms, the strict-sense definition of HRF is assumed (without loss of generality).

A dynamic extension of HRFs to sequence modeling was proposed in [4,18]. The model is referred to as the dynamic hybrid random field (D-HRF), and it is formally defined as follows.

*Definition*: A dynamic HRF $\mathcal{DH}$ is a tuple $\mathcal{DH} = (\mathbf{X}, S, \pi, \mathcal{F}, \mathbf{a}, \mathcal{H})$ where

1. $\mathbf{X}$ is a set of (observable) random variables $X_1, \ldots, X_n$. Outcomes of the random variables depend on time $t = 1, \ldots, T$, that is we will write $X_i(t)$ whenever we need to make the dependency explicit.
2. $S$ is a set of $Q$ latent random variables, $S = \{S_1, \ldots, S_Q\}$. It is assumed that sequences of such latent variables are responsible for the generation of sequences of outcomes of the observable variables, and that the variables in $S$ can be thought of as the states of a discrete-time Markov chain (*latent Markov assumption*). We write $q_t$ to denote the state of the Markov chain at time $t$ for $t = 0, \ldots, T$.
3. $\pi$ is a probability distribution of the initial latent variables, i.e. $\pi = \{Pr(S_i \mid t = 0), S_i \in S\}$, where $t$ is the discrete time index. For instance, if the Markov chain over $S$ may equally-likely start with any latent variable, then $\pi$ is uniform over $S$. Contrariwise, if a certain $S_j$ can never occur at time $t = 0$, then $\pi(S_j) = 0$, etc.
4. $\mathcal{F} \subseteq S$ is the set of final states, i.e. the latent variables which can legitimately generate sets of outcomes of the observable variables at time $T$ (namely, at the end of sequences).
5. $\mathbf{a}$ is a probability distribution that characterizes the (allowed) transitions between latent variables, that is $\mathbf{a}_{ij} = \{Pr(S_j$ at time $t \mid S_i$ at time $t-1), S_i \in S, S_j \in S\}$ where the transition probabilities $\mathbf{a}_{ij}$ are assumed to be independent of time $t$. Note that the definition is meaningful due to the latent Markov assumption.
6. $\mathcal{H}$ is a set of HRFs over $\mathbf{X}$, $\mathcal{H} = \{\mathcal{H}_1, \ldots, \mathcal{H}_Q\}$, where $\mathcal{H}_q$ is uniquely associated with $q$-th latent variable $S_q$ such that the joint emission probability

$b(\mathbf{X}) = P(X_1, \ldots, X_n \mid S_q)$ is modeled via HRF $\mathcal{H}_q$ over $\mathbf{X}$, independently of time $t$, and we assume that the probability distribution of $\mathbf{X}(t)$ is independent of the probability of $\mathbf{X}(t')$ (for all $t' \neq t$) given the latent variable (*emission Markov assumption*). In this definition, bearing in mind the definition of HRF, it turns out that $\mathcal{H}_q$ is a set of Bayesian networks $BN_{q,1}, \ldots, BN_{q,n}$ (with directed acyclic graphs $\mathcal{G}_{q,1}, \ldots, \mathcal{G}_{q,n}$) such that:

(a) each $BN_{q,i}$ contains $X_i$ plus a subset $\mathcal{R}_q(X_i)$ of $\mathbf{X} \setminus \{X_i\}$, namely the set of relatives of $X_i$ in $BN_{q,i}$;

(b) for each $X_i$, $P(X_i | \mathbf{X} \setminus \{X_i, q\}) = P(X_i | \mathcal{MB}_{q,i}(X_i))$, where $\mathcal{MB}_{q,i}(X_i)$ is the set containing the parents, the children, and the parents of the children of $X_i$ in $\mathcal{G}_{q,i}$ (namely, the Markov blanket of $X_i$ in $BN_{q,i}$).

The Markov assumption holding in HRFs is referred to as the *observable Markov assumption* in the present framework.

Note that the overall D-HRF can be thought of as a probabilistic graphical model over the set of random variables $S \cup \mathbf{X}$. Nonetheless, this definition allows for separate sets of BNs (i.e., different HRFs) for each latent variable, meaning that it does not extend regular HRFs to sequences by defining them as sets of dynamic Bayesian networks in a straightforward manner.

Learning in D-HRFs occurs at two levels, i.e. parameter learning and structure learning. Let $\mathcal{DH}$ be a D-HRF, and let $O = O_1, O_2, \ldots, O_T$ be a training sequence of outcomes of the observable variables, i.e. $O_t = (x_1, \ldots, x_n)$ for $t = 1, \ldots, T$. The parameter learning algorithm exploits a recursive scheme [18], which is similar (to some extent) to the popular forward-backward procedure for HMMs [14]. Assuming that a (preliminary, at least) structure of the D-HRF is given, the algorithm aims at maximizing the pseudo-likelihood $P^*(O|\mathcal{DH})$, as in regular probabilistic graphical models (see [10] for a justification of why the pseudo-likelihood is used instead of the bare likelihood criterion). Each step of the forward-backward procedure involves learning the conditional probability tables (CPTs) [10] of the BNs modeling the local conditional distributions for each HRF $\mathcal{H}_1, \ldots, \mathcal{H}_Q$ associated with the latent variables. That is, for each latent variable $q = 1, \ldots, Q$ and for each observable variable $X_i$, the task is to learn the parameters of the conditional distribution $P(X_i | q, mb_{q,i}(X_i))$, for each state $mb_{q,i}$ of the variables in $\mathcal{MB}_{q,i}$, where $\mathcal{MB}_{q,i}$ is the Markov blanket [10] (in $\mathcal{H}_q$) for $i$-th observable variable and $q$-th latent variable. In short, the overall procedure relies on combining the CPTs learning algorithm of the individual HRFs [10] with the iterative forward-backward dynamic processing of the input sequence. The details of the algorithm are handed out in [18].

As for structure learning in a generic (say, the $q$-th) HRF within the D-HRF, it is formulated as the problem of learning, for each variable $X_i$, what other variables appear as nodes in the Bayesian network $BN_{q,i}$, and what edges are contained in the directed acyclic graph (DAG) $\mathcal{G}_{q,i}$. In other words, this means learning the structure of each Markov blanket $\mathcal{MB}_{q,i}(X_i)$ within $q$-th HRF. While parameter learning assumes that the Markov blanket of each variable has previously been fixed, the aim of structure learning is to identify each Markov blanket and to determine its graphical structure. A structure learning algorithm for D-HRFs, called dynamic Markov blanket merging (DMBM) was presented in [4]. The aim of DMBM is to find an assignment of Markov blankets $\mathcal{MB}_{q,1}(X_1), \ldots, \mathcal{MB}_{q,n}(X_n)$ to the nodes $X_1, \ldots, X_n$ (within $q$-th HRF) that maximizes the model pseudo-likelihood given the dataset. The basic idea behind DMBM is to start from a certain assignment of neighbors to the variables of the model, learn the local BNs of the model, and then to iteratively refine the assignment so as to come up with Markov blankets that increase the model pseudo-likelihood with respect to the previous assignment. This iterative procedure stops when no further refinement of the Markov blankets assignment increases the value of the pseudo-likelihood. In other words, DMBM is nothing but a local search algorithm exploring a space of possible Markov blanket assignments to the observable variables for each state of the D-HRF. The formalization of the algorithm is found in [4].

# 3   The Hybrid RNN/D-HRF

The idea behind the proposed approach is related to some extent to Bengio's hybrid ANN/HMM system for automatic speech recognition [2], where an ANN is trained to extract features for an HMM from an input sequence of acoustic observations. In the present study, a recurrent neural autoencoder is trained to develop suitable internal representations of the sequences of amino-acids. Like in regular autoencoders, this representation is a lower-dimensional encoding of any observed sub-sequence of residues up to the current location (assuming the primary structure is fed to the RNN sequentially, residue after residue, starting from the first amino-acid in the protein[1]). The encoding is expected to capture most of the relevant information necessary to reconstruct the amino-acid sequence at hand, filtering out noise and redundancies. The rationale behind the approach is fourfold: (1) reducing the local sparsity of the representation of the residues. In fact, a plausible vector-space representation of the $t$-th amino-acid in the protein would require a 20-dimensional vector having null components except for that corresponding to the residue itself. The autoencoder shall map this representation onto a corresponding, lower-dimensional, and denser one; (2) as a consequence of its recurrent connections, the RNN is expected to maintain a moving memory of the internal representations entailed by the presence of individual amino-acids met at an earlier time along the protein. Therefore, each residue can extend its influence over a mid-term period, not limited to its specific location; (3) reducing the noise and mis-alignments due to potential laboratory errors in protein sequencing and, above all, in the multiple alignment procedure used for mapping whole families of homologous proteins onto a single prototypical representative; (4) feeding the predictor (in this case, the D-HRF) with a representation which, besides the local information at time $t$, accounts for short- and mid-term dependencies among the input observations (long-term modeling, and the actual prediction of the bonding state are accomplished by the D-HRF, eventually).

---

[1] Bearing in mind this sequential dynamics of the model, we will occasionally refer to the RNN as operating "over time", such that it is fed with $t$-th residue at "time $t$".

Formally, let $\mathcal{T} = \{\mathcal{P}_1, \ldots \mathcal{P}_n\}$ be a dataset of $n$ proteins (or, prototypical outcomes of a multiple-alignment procedure over subsets of homologous proteins). For $i = 1, \ldots, n$ we have $\mathcal{P}_i = a_{i,1}, \ldots, a_{i,n(i)}$, that is a sequence of length $n(i)$ of amino-acids where each $a_{i,j}$ $(j = 1, \ldots, n(i))$ is represented via the usual (say) 3-letter abbreviation. In order to feed a RNN with the sequences, we resort to an equivalent real-valued representation in terms of a "one-hot" coding function $c : \{\text{"Ala"}, \text{"Arg"}, \ldots, \text{"Val"}\} \rightarrow \mathbb{R}^{20}$ defined as $c(\text{"Ala"}) = (1.0, 0.0, \ldots, 0.0)$, $c(\text{"Arg"}) = (0.0, 1.0, \ldots, 0.0)$, etc. The input to the RNN at time $j$ over sequence $i$ is denoted accordingly as $\mathbf{x}_{i,j} = c(a_{i,j})$. For notational convenience, from now on we write $\mathbf{x}_1^T$ to denote a generic RNN input sequence of length $T$ obtained this way.

The RNN architecture is a plain multilayer perceptron (MLP) backbone with an additional set of context units (which, at time $t + 1$, back-up the inputs at time $t$), such that each input unit has a corresponding context unit which is laterally connected to the former, and each context unit has a self-feeding recurrent connection from and to itself. In so doing, the context units form a decaying memory of the filtered history of the previous inputs, keeping a trace of the amino-acids observed in the past. Feed-forward connection stem from the context units and feed the hidden layer of the RNN (in a fully connected manner, i.e. between all possible context-to-hidden units pairs). The RNN has 20 input and 20 output units. When fed with the generic input vector $\mathbf{x}_t$ at time $t$, the RNN returns an output $\mathbf{y}_t$ which depends on $\mathbf{x}_t$ and on the history of previous inputs. Instead of prescribing $\mathbf{y}_t \simeq \mathbf{x}_t$, like in regular autoencoders, we rather enforce the mid-term memory and the time-smoothing mechanisms expected of the resulting machine by filtering the sequence of target outputs $\hat{\mathbf{y}}_1, \ldots, \hat{\mathbf{y}}_T$ for backpropagation though-time (BPTT) as follows. First (at time $t = 1$), for $k = 1, \ldots, 20$ we let $\hat{y}_{1,k} = x_{1,k}$. Then, for $t = 2, \ldots, T$ we let $\hat{y}_{t,k} = \min(x_{t,k} + \rho \hat{y}_{t-1,k}, 1.0)$, where $\rho \in (0, 1)$ is a memory decay factor. In practice, the appearance of a certain residue at time $t$ casts a (vanishing) shadow onto the next time steps, whose length is determined by $\rho$. The initialization at time $t = 1$ simply sets the initial target equal to the initial input (as in a regular autoencoder). Any time an amino-acid is found along the input sequence, the corresponding component of the target output is set to 1 (making sure that the target values never exceed 1, regardless of the past history: that is the rationale behind the presence of the $\min(.)$ function in the equation). Starting from the next time step $t + 1$, the $k$-th target progressively lowers its value, with a persistence that depends on $\rho$, vanishing eventually.

Once training is accomplished, the RNN realizes a time-dependent mapping $m : \mathbb{R}^{20} \times \mathbb{N} \rightarrow \mathbb{R}^d$ that projects the generic 20-dimensional input vector $\mathbf{x}_{i,t}$ at time $t$ onto a $d$-dimensional vector $\mathbf{z}_{i,t} = m(\mathbf{x}_{i,t}, t)$ (assuming $d << 20$) accounting for such time-dynamics. Both $d$ and $\mathbf{z}_{i,t}$ depend on the dimensionality and nature of the hidden layer in the RNN (the "bottleneck" typical of autoencoders) used for realizing the encoding. Since $\mathbf{z}_{i,t}$ is continuous-valued, it cannot be feed into the D-HRF directly. A discretization is accomplished first. Assuming the usual logistic sigmoids $\sigma(.)$ are used in the encoding hidden layer of the RNN, having the interval $(0,1)$ as their counterdomain, we quantized the output $z_{k,t} = \sigma_k(.)$ of the generic $k$-th hidden unit at time $t$ onto $q$ intervals $I_1, \ldots, I_q$ such that $I_\ell = (a_\ell, b_\ell)$ for $\ell = 1, \ldots, q$, $a_1 = 0.0$, $b_q = 1.0$, and $a_{\ell+1} = b_\ell$ if $\ell < q$. A discrete alphabet of $q$ symbols $\{S_1, \ldots, S_q\}$ is associated uniquely with the intervals and used to represent the discretized value $disc(z_{k,t})$ of $z_{k,t}$ as $disc(z_{k,t}) = S_s$ iff $z_{k,t} \in I_s$. Note that the construction of the intervals $I_1, \ldots, I_q$ shall account for the fact that the logistic sigmoid entails a non-uniform distribution of its values, and that even small differences in close-to-zero outputs of the RNN activation functions may make a difference.

At this point, each input protein is encoded as a sequence of $d$-tuples of symbols of the alphabet $\{S_1, \ldots, S_q\}$, which may be thought of as the time-specific outcomes of a set of $d$ observable random variables $X_1, \ldots, X_d$ and, in turn, modeled via a D-HRF. Bearing in mind the notation introduced in the previous section, the training sequence $O = O_1, O_2, ..., O_T$ for the D-HRF is then defined by letting $O_t = (disc(z_{1,t}), \ldots, disc(z_{d,t}))$ for $t = 1, \ldots, T$.

The prediction of the disulfide bonding state of the cysteines can be stated formally in terms of a 2-class classification problem (namely, bonding/non-bonding). The latter is faced within the framework of Bayes decision theory, by assigning each cysteine to the class predicted according to the maximum class-posterior probability. The posterior probabilities are estimated relying on Bayes' theorem, where the prior probabilities (estimated from the relative frequencies of the two classes in the training set) are combined with the class-conditional probabilities estimated via D-HRF (two class-specific D-HRFs are trained independently of each other from the training sub-sequences of the corresponding class).

## 4   Results

Experiments were carried out using proteins from the crystallographic database Protein Data Bank (PDB) [3], possibly the most popular benchmark for the prediction of disulfide bonds between cysteines [17]. As in [15], a subset of the May 2010 release of PDB (available publicly at http://www.biocomp.unibo.it/savojard/PDBCYS.ssbonds.txt) was used. It consists of 1797 sequences (Eukaryotic proteins) having an average length of 273 residues and containing a variable number of cysteines, ranging from 2 to 50 (roughly 6 cysteines per protein on average), for an overall number of 10813 cysteines (3194 bonding, 7619 non-bonding). A 10-fold cross-validation evaluation strategy was applied. For each individual fold, 80 % of the cysteines were used for training, 10 % for validation (i.e., model selection), and the remaining 10 % for test.

Application of the proposed RNN/D-HRF hybrid occurred according to the following steps and architectures/hyper-parameters selection. First, windows of 31 contiguous residues were extracted from the overall sequences for all cysteines, each window being centered at the location of the corresponding cysteine (the length of the windows was selected by cross-validation during a preliminary trial). If the number of residues proceeding (or, following) a certain cysteine was not enough for covering the whole window length, an adequate number of additional null items were used to pad the initial (or final, respectively) portion of the window up to the required length. The residues in th windows were encoded via RNN into sequences of 3 discrete random variables to be modeled via D-HRF, letting $\rho - 0.9$ and using $q = 12$ discrete symbols represented by the capital letters from "A" to "L" (the possible outcomes of the random variables), as explained in the previous section. The quantization of the real-valued outputs of the corresponding 3-unit hidden layer of the RNN was defined as $disc(z) = A$ if $z < 10^{-3}$, $disc(z) = B$ if $z \in (10^{-3}, 10^{-2})$, $disc(z) = C$ if $z \in (10^{-2}, 10^{-1})$, and finally $disc(z) = D, \ldots, L$ at regular intervals between 0.1 and 1.0. The RNN had logistic sigmoids with adaptive bias in the hidden units, and linear output units. A single RNN was used (regardless of the bonding state of individual cysteines) and trained for 1000 epochs of BPTT with a (non-adaptive) learning rate $\eta = 0.001$ and no momentum, starting from a uniformly random initialization of the forward connection weights (and, of the bias of the sigmoids) over the $(-10^{-3}, 10^{-3})$ interval and an initial weight equal to 0.9 for the lateral and self-recurrent connections.

As for the D-HRF, two class-specific models were created, having 3 latent variables each (performance could not be improved when resorting to longer left-to-right Markov chains, while the complexity of the resulting machine increased). A Viterbi segmentation [14] of the input sequences was used for the initialization, where the MBs for the BNs in the state-specific HRFs (found via the chi-square test of correlation among the variables) turned out to be the complete cliques over the 3 encoding variables $X_1, X_2$, and $X_3$ at any given location within the window. This initialization was followed by an application of the parameter and structure learning algorithms. Further re-iterations could increase the pseudo-likelihood at the expense of the accuracy, due to the non-discriminative nature of the maximum likelihood criterion and its likely tendency to over-explain (that is, overfit) the individual training data.

The results (in terms of average over the many-fold $\pm$ standard deviation, when available) are reported in Table 1, which includes comparisons with respect to major established prediction techniques that were evaluated on the PDB. For each technique the table reports, in the order, the model (acronyms are explicated below), the specific input attributes (i.e., the information fed to the machine), the version of PDB used, the corresponding dataset size (expressed as $n_p(n_c)$ where $n_p$ is the number of proteins and $n_c$ is the number of cysteines), the bibliographic source, and the accuracy. The attributes are grouped into three major types, namely RC (the bare Residue Chain, like in the present study), PSSM (the real-valued Position Specific Score Matrix, obtained by multiple alignment via BLAST and conveying evolutionary information [8]), and AI (Additional Information, like global descriptors [9], subcellular information [15], etc.).

The first two rows report on the highest accuracies yielded by MLPs [8], obtained when using windows of (best) length 13. Based on the gap between the results obtained with (PSSM) or without (RC) evolutionary information, the Authors of [8] drew strong conclusions on the relevance of the latter to

**Table 1.** Predicting the disulfide bonding state of cysteines from the PDB (*Legend for attributes: RC = residue chain, PSSM = position specific score matrix, AI = additional information*).

| Model | Attributes | Dataset | Dataset size | Source | Avg. accuracy (%) ± std. dev. |
|---|---|---|---|---|---|
| MLP | RC | PDB Oct 1997 | 641 (2452) | [8] | 71.80 ± 0.9 |
| MLP | PSSM | PDB Oct 1997 | 641 (2452) | [8] | 81.0 ± 0.8 |
| HMM | RC | PDB May 2010 | 1797 (10813) | This paper | 78,24 ± 5,70 |
| SVM | RC | PDB 2002 | 969 (4136) | [19] | 81 |
| 2-stage SVM | PSSM + AI | PDB Sep 2001 | 716 (4859) | [9] | 82.96 |
| 2-stage SVM (w/o tuning) | PSSM + AI | PDB 2002 | 969 (4136) | [6] | 85.2 |
| CRF | RC | PDB Sep 2001 | 716 (4859) | [16] | 84 |
| **RNN/D-HRF** | RC | PDB May 2010 | 1797 (10813) | This paper | 85,02 ± 0,41 |
| RNN + SVM + FSA | PSSM + AI | PDB Jul 2005 | NA | [5] | 88 ± 1 |
| 2-stage SVM (w/tuning) | PSSM + AI | PDB 2002 | 969 (4136) | [6] | 90.7 |
| CRF + FSA | PSSM | PDB May 2010 | 1797 (10813) | [15] | 91 |

the end of predicting the bonding state. The gap helps us positioning correctly the degree of difficulty of the task we are facing (that is, prediction relying on protein-specific sequence-level information only).

The next row reports the result we obtained using a standard discrete HMM trained over windows of (best) length equal to 31. Two class-specific left-to-right HMMs were used, having 5 states each, trained via Baum-Welch on the 20 abbreviations ("Ala", ..., "Val") of the different amino-acids. The result may be considered as baseline for the D-HRF, as well, since a D-HRF with MBs limited to singletons (i.e., a single random variable $X_1$ for the HRFs at each time $t$) reduces to a discrete HMM.

The results yielded by SVMs are reported in the next three rows. The first is referred to the highest accuracy reported in [19] using only windows of the RC (best window length: 15). Results with SVM could be improved by combining SVM-based classifiers trained in two stages over different input features in [6,9]. As for the latter approach, the expression "w/o tuning" refers to the setup with no grammatical post-processing (i.e. "tuning", as presented in [6]) aimed at preventing higher-level inconsistent predictions (e.g., an uneven number of bonding cysteines within a given protein). The accuracy yielded by the same approach with tuning (penultimate row of the table) is amongst the highest to date. Further improvements to the SVM-based predictor were presented in [5], where the SVM benefits from a bi-directional RNN preprocessing, as well as from a post-processing via finite state automaton (FSA) which, similarly to the afore-mentioned "tuning" [6], acts a s language model hampering unlikely/impossible sets of predictions within any given protein.

The comparison with the conditional random fields (CRFs) [16] is interesting, since CRFs are, like D-HRFs, probabilistic graphical models. Besides, CRFs are intrinsically discriminative, since they model conditional probability densities instead of joint distributions of the random variables. Nevertheless, the RNN/D-HRF approach compares favorably with the results handed out in [16]. Still, the highest accuracy reported on (91 %, last row of the table) is yielded by another variant of CRFs called Grammatical-Restrained Hidden Conditional Random Field [15], basically a CRF followed by an FSA-based post-processing (again, to ensure satisfaction of the aforementioned grammatical constraint). The difference between the accuracies yielded by the two variants of CRFs ([16] versus [15]) underlines the gap in terms of performance due to (1) using PSSM instead of bare RC, as well as to (2) post-processing via the language model.

The result obtained via the proposed RNN/D-HRF ($85.02 \pm 0.41$ % accuracy, with average 82 % sensitivity and 86 % specificity) is substantially aligned with the performance offered by most major established techniques, regardless of the input attributes adopted (and, in spite of the fact that we tested the algorithm on the largest and most complex version of the PDB available to date). In particular, to the best of our knowledge the accuracy yielded by the RNN/D-HRF is the highest to date when relying on the RC only.

# 5    Conclusions

The paper presented the first application of D-HRFs to real-life data. It was shown that the model is effective, and that its hybridization with the recurrent autoencoder is a promising direction for learning and prediction-making over the primary structure of proteins. Albeit preliminary, the results obtained are (at least) aligned with those offered by major, state-of-the-art approaches, especially in the light of he fact that the predictions are based on the sequence of residues only (while most established methods rely on additional attributes, like global descriptors of chemical/physical properties of the proteins, and/or evolutionary information). Future research activities are going to focus on evaluating the approach using PSSM and/or additional attributes, as well as on the investigation of the effects of post-processing the output of the D-HRF according to a language model.

# References

1. Baldi, P., Brunak, S.: Bioinformatics: The Machine Learning Approach. MIT Press, Cambridge (1998)
2. Bengio, Y.: Neural Networks for Speech and Sequence Recognition. International Thomson Computer Press, London (1996)
3. Berman, H.M., Westbrook, J., Feng, Z., Gilliland, G., Bhat, T.N., Weissig, H., Shindyalov, I.N., Bourne, P.E.: The protein data bank. Nucleic Acids Res. **28**(1), 235–242 (2000)
4. Trentin, E., Bongini, M.: Towards a novel probabilistic graphical model of sequential data: fundamental notions and a solution to the problem of parameter learning. In: Mana, N., Schwenker, F., Trentin, E. (eds.) ANNPR 2012. LNCS, vol. 7477, pp. 72–81. Springer, Heidelberg (2012)
5. Ceroni, A., Passerini, A., Vullo, A., Frasconi, P.: DISULFIND: a disulfide bonding state and cysteine connectivity prediction server. Nucleic Acids Res. **34**(Web-Server-Issue), 177–181 (2006)
6. Chung, W.-C., Yang, C.-B., Hor, C.-Y.: An effective tuning method for cysteine state classification. In: Proceedings of the National Computer Symposium, Workshop on Algorithms and Bioinformatics, Taipei, Taiwan, 27–28 November 2009
7. Crick, F.: Central dogma of molecular biology. Nature **227**(5258), 561–563 (1970)
8. Fariselli, P., Riccobelli, P., Casadio, R.: Role of evolutionary information in predicting the disulfide-bonding state of cysteine in proteins. Proteins **36**(3), 340–346 (1999)
9. Frasconi, P., Passerini, A., Vullo, A.: A two-stage svm architecture for predicting the disulfide bonding state of cysteines. In: Proceedings of the IEEE Workshop on Neural Networks for Signal Processing, pp. 25–34 (2002)
10. Freno, A., Trentin, E.: Hybrid Random Fields: A Scalable Approach to Structure and Parameter Learning in Probabilistic Graphical Models. ISRL, vol. 15. Springer, Heidelberg (2011)
11. Freno, A., Trentin, E., Gori, M.: A hybrid random field model for scalable statistical learning. Neural Netw. **22**, 603–613 (2009)
12. Ross Kindermann, J., Snell, L.: Markov Random Fields and Their Applications. American Mathematical Society, Providence (1980)

13. Pearl, J.: Bayesian networks: a model of self-activated memory for evidential reasoning. In: Proceedings of the 7th Conference of the Cognitive Science Society, pp. 329–334. University of California, Irvine, August 1985
14. Lawrence, R.: Rabiner: a tutorial on hidden markov models and selected applications in speech recognition. Proc. IEEE **77**(2), 257–286 (1989)
15. Savojardo, C., Fariselli, P., Alhamdoosh, M., Martelli, P.L., Pierleoni, A., Casadio, R.: Improving the prediction of disulfide bonds in eukaryotes with machine learning methods and protein subcellular localization. Bioinformatics **27**(16), 2224–2230 (2011)
16. Shoombuatong, W., Traisathit, P., Prasitwattanaseree, S., Tayapiwatana, C., Cutler, R.W., Chaijaruwanich, J.: Prediction of the disulphide bonding state of cysteines in proteins using conditional random fields. IJDMB **5**(4), 449–464 (2011)
17. Singh, R.: A review of algorithmic techniques for disulfide-bond determination. Brief. Funct. Genomic Proteomic **7**(2), 157–172 (2008)
18. Trentin, E., Bongini, M.: Towards a novel probabilistic graphical model of sequential data: fundamental notions and a solution to the problem of parameter learning. In: Mana, N., Schwenker, F., Trentin, E. (eds.) ANNPR 2012. LNCS, vol. 7477, pp. 72–81. Springer, Heidelberg (2012)
19. Chen, Y.C., Lin, Y.S., Lin, C.J., Hwang, J.K.: Prediction of the bonding states of cysteines using the support vector machines based on multiple feature vectors and cysteine state sequences. Proteins **55**(4), 1036–1042 (2004)

# Using Radial Basis Function Neural Networks for Continuous and Discrete Pain Estimation from Bio-physiological Signals

Mohammadreza Amirian$^{(\boxtimes)}$, Markus Kächele, and Friedhelm Schwenker

Institute of Neural Information Processing, Ulm University, Ulm, Germany
{mohammadreza.amirian,markus.kaechele,friedhelm.schwenker}@uni-ulm.de

**Abstract.** In this work we present extensions for Radial Basis Function networks to improve their ability for discrete and continuous pain intensity estimation. Besides proposing a mid-level fusion scheme, the use of standardization and unconventional loss functions are covered. We show that RBF networks can be improved in this way and present extensive experimental validation to support our findings on a multi-modal dataset.

## 1 Introduction

Physiological and pathophysiological pains are survival mechanisms generated by the brain in order to stimulate protective behavior. Accordingly, pain can be considered as a measure of medical health and elementary pain based treatments have been shown to be beneficial in 20 % to 70 % of the cases [15]. However, not all clinical patients are able to identify their level of pain such as neonates and somnolent patients. It is expressed that 30 % to 70 % of the patients were suffering from pain after undergoing surgery [22]. Therefore, automatic pain estimation as an element of electronic health surveillance has recently received increasing attention.

Initial pain quantification methods mostly used facial expressions for pain estimation. Different feature sets such as Principal Components and Gabor features had been combined with Support Vector Machines (SVMs) and Relevance Vector Machine (RVM) classifiers in a binary classification scenario (pain versus no pain) in [2,5,12,14]. Three class classification and continuous quantification of pain had also been done in [9,13], respectively.

Multi-modal pain classification [21] in binary and multi-class scenarios illustrates that promising cues exist in bio-physiological signals as well as video to recognize pain. Improved results accompanied by continuous pain estimation verified the merit of the bio-physiological signals in [7,8]. The recent study in [7] confirmed that pain is a very personal sensation and some subjects' physiological behavior has been shown to be more similar to specific subjects than to others.

The single modality approaches relying only on the facial expressions are highly sensitive to face detection. In the clinical setup a permanent face detection is rarely feasible. Thus, uninterrupted pain estimation is not always possible

© Springer International Publishing AG 2016
F. Schwenker et al. (Eds.): ANNPR 2016, LNAI 9896, pp. 269–284, 2016.
DOI: 10.1007/978-3-319-46182-3_23

using video features. Furthermore, in the case of pain estimation using dynamic classification approaches (such as Echo State Networks [6]), miss detections not only reduce the estimation accuracy of the current sample but also introduce uncertainty to the memory of the subsequent samples. It should be noted that commercially available fitness watches can measure some of the physiological signals such as blood volume pressure and galvanic skin response. Pain evaluators using bio-physiology can simply be attached to such devices in order to improve health monitoring.

The Radial Basis Function (RBF) neural networks [18] are used in this paper for continuous and discrete pain estimation through the physiological signals. The multi-class problem is investigated alongside the binary classification of the pain versus no pain. Back-propagation, Huber loss function, personalization and various fusion schemes are introduced and combined with the RBF networks to improve the accuracy of the pain estimation. Additionally, the application of the RBF nueral network is extended form classification to regression and the pain is estimated continuously.

The remainder of this paper is organized as follows: The BioVid experiment and database on which the experiments here are based is explained in Sect. 2, the physiological signal segmentation and feature extraction is presented in Sect. 3. We briefly explain Radial Basis Networks (RBF) in Sect. 4, before the experimental results are provided in Sect. 5 and the paper is concluded in Sect. 6.

## 2   The BioVid Experiment and Pain Database

The BioVid heat experiment [21] was conducted in 2013 at the University of Ulm. The main idea of the experiment was to stimulate the subjects with different heat levels to record bio-physiological signals namely, electrocardiography (ECG), electromyography (EMG) and skin conductance level (SCL) signals as well as video. One of the experiment objectives is to predict the pain level by multimodal processing of the physiological signals and the video. Furthermore, the stimulus signal is also recorded during the experiment.

Heat was applied to 96 subjects of this experiment using a thermode which was attached to their right hand (Fig. 1a). Subsequently, different levels of heat were considered as various levels of pain. In order to reduce the different tolerances of subjects, a calibration had been done before the experiment. Initially the temperature of 32 °C is assumed as the pain free temperature or level 0 of pain. Then the temperature is increased gradually and the subject is asked to react once he/she feels the pain. The initial feeling threshold of heat is considered as the first level of pain. The temperature is raised again afterwards until the highest endurable level is declared by the subject. The highest temperature is assumed as pain level 4. Two other levels of stimulation were selected corresponding to the temperatures linearly spaced between level 1 and 4.

After the calibration procedure the subjects were stimulated with different temperatures evaluated as various pain levels. The selection of heat levels was random but the subjects were stimulated 20 times with each heat level in the first

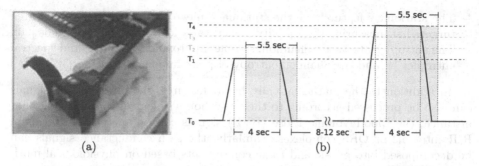

**Fig. 1.** (a) The thermode attached to the right wrist (the image is taken from [21]). (b) The heat stimulation baseline temperature ($T_0$) versus the pain threshold ($T_1$) and maximum endurance level ($T_4$). The signal segmentation for feature extraction is illustrated by the green window. Image taken from [7].

part of the experiment. The duration of the stimulation was 4 s and there was a 8–12 s stimulation free time between each two consecutive stimulus (Fig. 1b). The same experiment had been done in another section after 40 min of heat relaxation. The facial electromyography (EMG) signals were recorded in the second part in addition to the previous modalities of the first part of the experiment. Complementary details of the BioVid experiment are provided in [21].

## 3   Bio-physiological Feature Extraction

After the experiment, by removing all the defected signals, a dataset with 86 participants was left for analysis. Moreover, misplacement of the biopotential sensors as well as prolongation of the experiment degrades the pain estimation performance in the second part. Therefore, we have just analyzed the physiological signals of the first part of pain stimulation.

Considering the delay in the response of the sympathetic nervous system, the biopotential signals were segmented by a 1 s latency relative to the beginning of the stimulus (Fig. 1b) [7]. According to the physiological basis and motivations a window of length 5.5 s was used for the signal segmentation. It should be mentioned that the signals for the level zero of pain (baseline) were extracted within duration of 5.5 s right before the first stimulus pulse after the pain level one. This assumption was considered to minimize the effect of previous stimulations on the baseline.

A wide range of features were extracted from the segmented signals in order to be used in the classification and regression tasks. These features can be categorized into four main mathematical groups as follows:

- The time domain features such as Willison amplitude, V-order, log detector and so on aimed at measuring the intensity [16].
- The frequency group features include mode, mean, central and median frequencies, bandwidth and low pass to very low pass ratio in order to evaluate the rate of the vibration of muscles [10].

- The stationary features are the stationary mean, variance, median and area targeted at measuring the stability of the statistical properties of signals [20].
- The entropy features are Shannon, sample, approximation, fuzzy and spectral entropies [4] and the Shannon entropy of peak frequency shifting [3].

In addition to the mathematically based features, the physiological signals can also be processed according to their psychophysiological characteristic. For instance, the Blood Volume Pulse (BVP) signals can be analyzed in terms of R-R intervals or QRS complexes. Similarly, the skin conductance signals can be decomposed into phasic and tonic components based on physiological motivations [1]. The final sets of features used in this paper per modality are as follows:

- **EMG**: The set of the EMG features includes the statistical, time, frequency and stationarity related features of the electromyography signals [8].
- **BVP1**: The approximation coefficients of a four level wavelet decomposition using Daubechies wavelets of the blood volume pulse signals [23].
- **BVP2**: This feature set includes the amplitude of different points, time differences and angles of the PQRST complex of the heart signals [19].
- **SCL1**: The same features as for the EMG channel and 7 additional statistical features including skewness and kurtosis of the SCL signals formed this set of the features.
- **SCL2**: The last set of the features are based on the phasic and tonic decomposition of the SCL signals. These features include the number of SCL responses, latency of the first response, average of the phasic driver and etc. The mentioned features are derived according to the Ledalab project [1].

## 4    Radial Basis Function (RBF) Neural Networks

The remainder of this paper focuses on discrete and continuous quantification of pain using Radial Basis Function (RBF) neural networks. Besides classification, we extend the application of RBF networks to regression problems.

As shown in Fig. 2 the RBF networks are a three layer neural network including input, hidden (Radial Basis Function) and output layers. The general learning procedure of this network can consist of three different steps:

- The first step is an unsupervised learning procedure to learn the Radial Basis Function centers and width which is mainly done using the k-means algorithm.
- The second step is a supervised learning procedure to learn the output weights to create an efficient mapping from activation values to the target vector of the classes. This step is supposed to be done using pseudo-inversion.
- The third step is the simultaneous optimization of all parameters including output weights, Radial Basis Function center and width using the backpropagation algorithm.

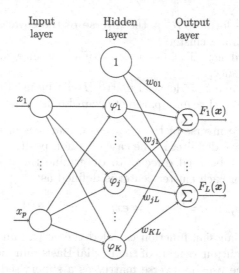

**Fig. 2.** The architecture of the Radial Basis Function (RBF) neural networks.

In previous literatures (for instance [18]) the 1-phase and 2-phase learning procedures were also considered. However, it is notable that 1-phase learning using backpropagation and proper initialization of the RBF parameters is possible. Moreover, the RBF network can be trained using k-means and the pseudo-inverse in a 2-phase learning procedure.

Before explaining the algorithms, we would like to express the notation that will be used later on in this report which is corresponding to Fig. 2 as follows:

- The $\mu^{th}$ input feature vector is denoted by $x^\mu \in \mathbb{R}^p$ with $\mu = 1, \ldots, M$.
- The $j^{th}$ cluster center of length $p$ is expressed by $c_j \in \mathbb{R}^p$ with overall $K$ clusters.
- The $j^{th}$ Radial Basis Function corresponding to an arbitrary modality (EMG, BVP or SCL) is denoted by $\varphi_j^{modality}$ and $\zeta_0 = 1$ is the bias term.
- The output of the RBF neural network and the target vector are denoted by $F_k(x)$ and $Y_k(x)$ respectively. Here, we consider $L$ output classes.

### 4.1 The Gaussian Kernel, Width and Distance

After the k-means algorithm which calculates the cluster centers and members, we need to define a function to evaluate the distance of a data point $x^\mu$ from a cluster center $c_j$. Let's define a positive definite matrix $R_j$ and accordingly assume the distance of a data feature vector and cluster center as:

$$\| x^\mu - c_j \|_{R_j} = \sqrt{(x - c_j)^T R_j (x - c_j)} \tag{1}$$

The different situations for the matrix $R_j$ which lead to the use of various distances, are as follows:

1. The most general form of $R_j$ is the inverse of the covariance matrix of the data samples within a cluster.
2. Using the main diagonal of the inverse of covariance matrix leads to the Mahalanobis distance.
3. Instead of $R_j$, a factor of identical matrix $I$ can be used for each cluster.
4. Alternatively, a global scaling parameter can be used for all clusters.

It is notable that the inverse of the covariance matrices of the feature vectors are not always positive definite. In the case of a non-positive definite covariance matrix Eq. 1 can not be used directly to determine the distance. Finally the Gaussian function for each cluster center is defined as:

$$\varphi_j(x^\mu) = h_j(x^\mu) = exp(- \parallel x^\mu - c_j \parallel_{R_j}) \tag{2}$$

The output of the Gaussian function is called the activation value and denoted by $h_j$ in Eq. 2. The output weights of the Radial Basis Function neural networks are trained using the pseudo-inverse matrix as a supervised learning based on the target labels.

## 4.2   Early, Mid-level and Late Fusions

According to the multi-modality of the BioVid database and the experiment, various fusion schemes can be applied. The first scheme is the early fusion where the new feature vectors are formed by concatenation of all the features of all modalities for every sample and the classifier is trained based on these new feature vectors.

The other commonly used method is late fusion where different classifiers are trained for each modality and the decisions of all classifiers are fused for the final decision. In the case of late fusion we used the mean of the confidence values of all classifiers as a criterion for the final decision in this paper. It should be considered that before late fusion, in order to avoid random extreme values, a soft-max function is applied to the output of each RBF classifiers.

In addition to the commonly used fusion methods, we propose mid-level fusion for RBF neural networks in this paper. In this case the unsupervised part of the learning procedure (clustering) has been done in every feature set independently. Afterwards, the activation values from all clusters of different modalities ($[\varphi_1^{EMG}, \ldots, \varphi_K^{EMG}, \varphi_1^{BVP}, \ldots, \varphi_K^{BVP}, \varphi_1^{SCL}, \ldots, \varphi_K^{SCL}]$) are concatenated and finally the output layer weights are trained based on the activations of all clusters of every feature set.

## 4.3   Using Huber Loss Function in RBF Neural Networks

The normal Radial Basis Function (RBF) neural networks perform inefficiently in the case of non-standardized data. After analyzing the activation values, it turns out that for some data points just one feature can dominate the whole activation value in a cluster. In other words, if one data sample is close to a cluster center in all feature dimensions expect one but very far just in that one dimension, the activation value of that cluster will be dominated by that single feature. To compensate for this effect and to obtain a robust classifier we used the Huber loss function instead of the euclidean distance in the argument of the Gaussian function. The Huber loss function is depicted in Fig. 3 for different free parameters ($\delta$) and can be mathematically expressed as follows:

$$L_\delta(a) = \begin{cases} \frac{1}{2}a^2, & \text{for } |a| \leq \delta \\ \delta\left(|a| - \frac{1}{2}\delta\right), & \text{otherwise} \end{cases} \tag{3}$$

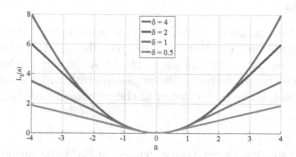

**Fig. 3.** The Huber loss function for different free parameters ($\delta$).

## 4.4 Backpropagation

Backpropagation can be used to improve the training procedure. The main idea of the approach is to minimize an error function which is the sum of squared deviations of the output of the classes ($\boldsymbol{F}_k$) from the target vectors ($\boldsymbol{y}_k$). This error function is defined as [18]:

$$E = \frac{1}{2} \sum_{\mu=1}^{M} \sum_{k=1}^{L} (y_k^\mu - F_k^\mu)^2 \tag{4}$$

The minimization task for Eq. 4 has been done iteratively using the gradient descent procedure. In other words, all the parameters of the network are shifted in the direction in which the error function has the quickest decrease. The value of this shift is computed as a multiplication of the learning rate ($\eta$) to the partial derivative of the error function with respect to the parameter that is optimized. The adaption rules for different parameters of the RBF network are as follows [18]:

$$\Delta w_{jk} = \eta \sum_{\mu=1}^{M} h_j(\boldsymbol{x}^\mu)(y_k^\mu - F_k^\mu) \tag{5}$$

$$\Delta c_{ij} = \eta \sum_{\mu=1}^{M} h_j(\boldsymbol{x}^\mu) \frac{x_i^\mu - c_{ij}}{\sigma_{ij}^2} \sum_{k=1}^{L} w_{jk}(y_k^\mu - F_k^\mu) \qquad (6)$$

$$\Delta \sigma_{ij} = \eta \sum_{\mu=1}^{M} h_j(\boldsymbol{x}^\mu) \frac{(x_i^\mu - c_{ij})^2}{\sigma_{ij}^3} \sum_{k=1}^{L} w_{jk}(y_k^\mu - F_k^\mu) \qquad (7)$$

where, $\sigma_{ij}$ is the width of the Gaussian kernel of the $j^{th}$ cluster on the $i^{th}$ coordinate of the features. In fact the $\sigma_{ij}$ are the elements on the main diagonal of the matrix $\boldsymbol{R}_j$ mentioned in Eq. 1.

## 4.5   Regression

The regression task is aimed at evaluating an input feature vector in terms of a continuous regression value instead of discrete classes. Accordingly, we might expect only one output neuron in the output layer of Fig. 2. Consequently we will have only one target vector in the regression task which should include all the classes. We define the target vector of the regression task ($\boldsymbol{y}_{reg}$) as:

$$y_{reg}^\mu = k, \quad \forall \; y_k^\mu = 1 \qquad (8)$$

Based on the regression target vector defined in Eq. 8 the output weights will be trained. Ultimately, the output of the regression task could be scaled and mapped to the desired values which are temperatures.

## 5   Simulations Results

The classification and regression tasks in this paper are defined as leave one subject out cross validation. In other words, in both cases the samples related to one participant are left out of the training set for testing and then the neural network or classifier is trained using the rest of the dataset. The left out subject is used as the test case for the trained classifier or regressor. All the present results in the following classification tables are the average value of such a cross validation classification for all 86 participants of the experiment.

### 5.1   Baseline Results

First, as a baseline to compare the RBF classifier results, we conduct the classification using RBF classifier with 50 clusters on the extracted features. The result of the classification for some pairs of the classes and the multi-class classifier are illustrated in Table 1. The baseline result shows that the classification accuracy barely surpasses 80 % in the binary problem of the pain level 0 versus 4 and the five class classification only shows 32.1 % of accuracy. It is notable that the skin conductance level features derived from phasic and tonic decomposition is the best feature set in all the classification scenarios and the result of early fusion hardy exceed the a-priori probabilities of the classification tasks.

**Table 1.** The classification results using RBF classifier with 50 clusters and non-standardized data. Bold values indicate the maximum of each row.

| RBF (50) | EMG | BVP1 | BVP2 | SCL1 | SCL2 | Late fusion | Mid fusion | Early fusion |
|---|---|---|---|---|---|---|---|---|
| 0 2 | 0.508 | 0.517 | 0.505 | 0.504 | **0.627** | 0.596 | 0.589 | 0.507 |
| 0 4 | 0.520 | 0.532 | 0.490 | 0.553 | **0.801** | 0.744 | 0.768 | 0.523 |
| 1 4 | 0.519 | 0.523 | 0.481 | 0.553 | **0.735** | 0.694 | 0.711 | 0.523 |
| 2 4 | 0.526 | 0.505 | 0.508 | 0.549 | **0.676** | 0.644 | 0.656 | 0.520 |
| Multi-class | 0.214 | 0.200 | 0.205 | 0.269 | **0.321** | 0.292 | 0.280 | 0.206 |

After exploring the baseline result one can simply realize that the skin conductance level features are much more discriminant for this classification tasks than the other channels. However, it is interesting to notice that the accuracy of the RBF network considerably deteriorates for the case of non-standard or badly scaled features. That is to say, applying the same kernel width to high dimensional feature vectors in which the variances of the features are considerably different will not lead to an accurate RBF classifier. For instance, we can observe such an effect in the SCL1 feature set which includes a wide range of features with different means and variances. This effect can be observed in the poor early fusion results of the RBF neural networks as well. However, in the case of the other SCL2 feature set which is not badly scaled the result in much more accurate. In the reminder of this paper we will propose standardization schemes to improve the classification accuracy.

## 5.2  Standardization and Personalization

The Radial Basis Function neural networks are sensitive to non-standardized data. Therefore, we will standardize the features in two ways before conducting the classification.

**Standardization.** Initially, we will standardize the data in a way that every final feature has a mean of zero and variance of one over the whole data set for all participants. This method is the normal standardization. The classification results are provided in Table 2. As can be seen in the Table 2, standardization improves the maximum of the classification accuracy in all tasks. It should be mentioned that standardization raises the classification accuracy in the case of early fusion noticeably.

**Personalization.** Another level of standardization called person dependent standardization or personalization can be conducted to improve the classification results. In this proposed method *every feature for each independent person* is standardized to zero mean and unit variance. The result of the classification using personalized features for RBF neural networks classifier is presented in Table 3.

**Table 2.** The classification results using RBF classifier with 50 clusters and standardized data.

| RBF (50) | EMG | BVP1 | BVP2 | SCL1 | SCL2 | Late fusion | Mid fusion | Early fusion |
|---|---|---|---|---|---|---|---|---|
| 0 2 | 0.525 | 0.508 | 0.488 | 0.608 | **0.637** | 0.619 | 0.589 | 0.626 |
| 0 4 | 0.569 | 0.503 | 0.524 | 0.771 | **0.817** | 0.781 | 0.799 | 0.799 |
| 1 4 | 0.567 | 0.525 | 0.532 | 0.717 | **0.751** | 0.731 | 0.729 | 0.729 |
| 2 4 | 0.556 | 0.517 | 0.520 | 0.667 | **0.685** | 0.674 | 0.665 | 0.667 |
| Multi-class | 0.237 | 0.205 | 0.209 | 0.315 | **0.326** | 0.314 | 0.289 | 0.322 |

**Table 3.** The classification results using RBF classifier with 50 clusters with personalization.

| RBF (50) | EMG | BVP1 | BVP2 | SCL1 | SCL2 | Late fusion | Mid fusion | Early fusion |
|---|---|---|---|---|---|---|---|---|
| 0 2 | 0.524 | 0.569 | 0.521 | 0.641 | 0.659 | **0.672** | 0.656 | 0.664 |
| 0 4 | 0.606 | 0.493 | 0.571 | 0.833 | **0.844** | 0.830 | **0.843** | 0.841 |
| 1 4 | 0.601 | 0.545 | 0.590 | 0.775 | **0.813** | 0.792 | 0.796 | 0.803 |
| 2 4 | 0.587 | 0.570 | 0.587 | 0.719 | 0.750 | **0.751** | 0.738 | 0.749 |
| Multi-class | 0.230 | 0.226 | 0.229 | 0.346 | 0.367 | 0.369 | 0.353 | **0.371** |

The classification result using standardized and personalized feature clarifies that both methods lead to improvement in the pain estimation accuracy. Whereas, the personalized features perform better than the standardized ones. The reason is the difference in baseline of the physiological signals for the different participants. For example, a participant with higher regular heart rate might have a heart rate corresponding to what the others experience in the first level of pain. The mid-level fusion structure proposed in this project shows a promising accuracy is some cases such as classification of the class 0 versus 4 based on personalized data. Furthermore, the late fusion in two binary classification scenarios and early fusion for multi-class problem outperforms the classification only based on the skin conductance level features.

### 5.3  Further Optimizations

Up to this point we have some acceptable results using RBF classifiers for standardized and personalized features. However, classification of non-standardized features is still an open problem.

**Mahalanobis Distance.** According to the literature a possible solution for dealing with non-standard or badly scaled features is using the Mahalanobis distance. The classification results using Mahalanobis distance is shown in Table 4. Furthermore, the classification results for non-standardized data using Mahalanobis distance is provided in Table 5.

**Table 4.** The classification results using RBF classifier with Mahalanobis distance with personalization.

| RBF (50) | EMG | BVP1 | BVP2 | SCL1 | SCL2 | Late fusion | Mid fusion | Early fusion |
|---|---|---|---|---|---|---|---|---|
| 0 2 | 0.533 | 0.539 | 0.520 | 0.613 | **0.655** | 0.651 | 0.640 | 0.592 |
| 0 4 | 0.592 | 0.527 | 0.565 | 0.792 | **0.849** | 0.820 | 0.833 | 0.749 |
| 1 4 | 0.592 | 0.585 | 0.588 | 0.744 | **0.794** | 0.783 | 0.785 | 0.724 |
| 2 4 | 0.577 | 0.574 | 0.565 | 0.692 | **0.747** | 0.729 | 0.721 | 0.672 |
| Multi-class | 0.239 | 0.229 | 0.227 | 0.336 | **0.373** | 0.369 | 0.356 | 0.327 |

**Table 5.** The classification results using RBF classifier with Mahalanobis distance and non-standardized data.

| RBF (50) | EMG | BVP1 | BVP2 | SCL1 | SCL2 | Late fusion | Mid fusion | Early fusion |
|---|---|---|---|---|---|---|---|---|
| 0 2 | 0.501 | 0.504 | 0.509 | 0.559 | **0.628** | 0.611 | 0.597 | 0.557 |
| 0 4 | 0.544 | 0.516 | 0.527 | 0.725 | **0.802** | 0.768 | 0.776 | 0.631 |
| 1 4 | 0.528 | 0.518 | 0.532 | 0.681 | **0.745** | 0.716 | 0.721 | 0.661 |
| 2 4 | 0.543 | 0.517 | 0.522 | 0.633 | **0.688** | 0.666 | 0.647 | 0.610 |
| Multi-class | 0.225 | 0.203 | 0.211 | 0.281 | **0.336** | 0.308 | 0.302 | 0.278 |

**Table 6.** The confusion matrix of the multi class classification using RBF network, Mahalanobis distance with personalization. The vertical classes are the outputs and the horizontal ones show the target classes.

|   | 0 | 1 | 2 | 3 | 4 | total |
|---|---|---|---|---|---|---|
| 0 | 0.50 | 0.32 | 0.21 | 0.12 | 0.06 | 0.24 |
| 1 | 0.22 | 0.26 | 0.22 | 0.14 | 0.06 | 0.18 |
| 2 | 0.15 | 0.20 | 0.22 | 0.21 | 0.09 | 0.17 |
| 3 | 0.08 | 0.14 | 0.21 | 0.28 | 0.18 | 0.18 |
| 4 | 0.05 | 0.08 | 0.14 | 0.25 | 0.61 | 0.23 |

Using Mahalanobis distance increases the classification accuracy in the multi-class and class 0 versus 4 tasks up to 37.3 % and 84.9 % respectively. Moreover, RBF neural networks with this scale-invariant distance outperform (Table 5) the same network with euclidean distance (Table 1) on non-standardized data. The maximum classification accuracy based on the non-standardized features of all tasks improves using Mahalanobis distance. The most notable improvement can be seen in the classification result of early fusion scheme and the first feature set for skin conductance level in Table 5 compared with Table 1.

It will also be informative to investigate at least one of the confusion matrices of the these classifiers. The confusion matrix of the multi-class classification task using Mahalanobis distance is provided in Table 6. It is obvious that the RBF classifier doesn't produce all the classes as an output with the same probability.

The reason is that the middle levels of pain including level 1 to 3 are highly overlapping. Additionally, Table 6 shows that there is a high probability of confusion between the pain threshold (level 1) and no pain. The best classified levels on pain in the multi-class problem are highest level of pain (level 4) and the no pain (class 0). Learning the RBF neural network based on fusion mapping can improve this situation [17].

**Huber Loss Function.** The free parameter of the Huber loss function can be changed to reach a good classification result. Figure 4 illustrates the classification result of non-standardized data using the RBF neural network for different free parameters ($\delta$). It can be observed from Fig. 4 that using the RBF neural networks with the free parameter of $\delta = 3$ can improve the maximum classification accuracy for the non-standardized. The classification result using Huber loss function in RBF network is represented in Table 7. It is visible that the classification accuracy of the pain level 0 versus 4 is improved up to 1.8 % compared to the accuracy achieved by using Mahalanobis distance for non-standardized features. However, the classification accuracy for the multi-class problem is slightly below what have been reached through using Mahalanobis distance.

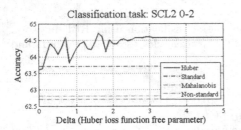 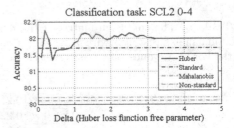

**Fig. 4.** Classification result for RBF neural network and Huber loss function.

**Table 7.** The classification results using RBF classifier with Huber loss function with free parameter of $\delta = 3$ and non-standardized data.

| RBF (50) | EMG | BVP1 | BVP2 | SCL1 | SCL2 | Late fusion | Early fusion |
|----------|-----|------|------|------|------|-------------|--------------|
| 0 2 | 0.498 | 0.519 | 0.496 | 0.498 | **0.646** | 0.633 | 0.497 |
| 0 4 | 0.522 | 0.540 | 0.508 | 0.520 | **0.820** | 0.760 | 0.495 |
| 1 4 | 0.496 | 0.532 | 0.497 | 0.505 | **0.749** | 0.713 | 0.503 |
| 2 4 | 0.517 | 0.501 | 0.506 | 0.520 | **0.697** | 0.671 | 0.502 |
| Multi-class | 0.237 | 0.214 | 0.222 | 0.319 | **0.335** | 0.214 | 0.205 |

**Backpropagation.** The performance of the Radial Basis Function neural networks can be improved using back-propagation. The considerable drawback of this method is the high computational complexity. Some classification results for RBF using back-propagation for standardized data are illustrated in Table 8. It is observable in Table 8 that personalization accompanied by Mahalanobis distance and backpropagation boosts the baseline result of classification of class 0 versus 4 from 80.1 % up to 85 %.

**Table 8.** The classification results using RBF classifier with back-propagation with personalization

| RBF (50) | EMG | BVP1 | BVP2 | SCL1 | SCL2 | Late fusion |
|----------|------|------|------|------|------|-------------|
| 0 4 | 0.602 | 0.496 | 0.576 | 0.833 | **0.850** | 0.828 |
| 1 4 | 0.607 | 0.559 | 0.592 | 0.775 | **0.807** | 0.787 |

It can be seen from Table 8 that the backpropagation optimization improves the classification accuracy of the RBF classifier compared with the base one (Table 4) in most of the cases. The base network here is a Radial Basis Function (RBF) neural network with Mahalanobis distance. It is beneficial to use various learning rates for different parameters (cluster center position, cluster width and output weights) optimized by the backpropagation as well.

## 5.4 Regression

Finally, we deal with the continuous estimation of the pain level using regression and RBF neural networks. For this purpose, we primarily used the initial setup of the RBF neural network with personalization. As we mentioned regarding signal segmentations, the training sample signals are segmented starting 1 s after the stimulus starts. Accordingly, a suitable shifting in labels is required.

There are two reasons for such a shift. First, as mentioned above the regressor is not trained based on the stimulus sequence but the segments with a 1 s lag relative to the start of the stimulus. In other words, the regression scenario here is not a sequence mapping from feature space to the stimulus which can compensate the lag of the predictions. Secondly, the heat stimulation does not affect all the physiological signal immediately and even with the same delay in time. Complementary research can be done to optimize the window length and delay of the signal segments compared to the stimulus for feature extraction.

The result of the regression for continuous pain level estimation versus the stimulus signal with and without proper shifting are illustrated and compared in Fig. 5.

Ultimately, different regression results are quantitatively compared in term of root mean squared error, normal and concordance correlation coefficients [11]. The result of this comparison is also provided in Table 9. The temporal missmatch between the prediction and label is obvious from the negative correlation

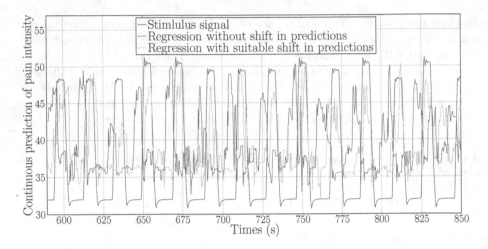

**Fig. 5.** Regression result versus the stimulus signal.

coefficient reported in the Table 9. Having said that, after a suitable tempo-ral shift the predictions shows a promising correlation of 0.48 with the labels. However, the concordance correlation coefficient can be improved by using more robust scaling methods.

**Table 9.** Quantitative comparison of different regression schemes.

|                               | RMSE   | CC      | CCC     |
|-------------------------------|--------|---------|---------|
| RBF with out shifting         | 8.6435 | −0.1027 | −0.0815 |
| RBF with suitable shifting    | 6.7175 | 0.4818  | 0.3460  |

# 6    Conclusion

The continuous and discrete pain level estimation is presented in this paper according to the BioVid heat database using Radial Basis Function neural net-works. The RBF classifier using Mahalanobis distance alongside personaliza-tion reaches a good compromise between accuracy and complexity. However, using backpropagation accompanied by a softmax at the output of the classi-fier improves the classification performance at the expense of imposing a high amount of computational complexity. The proposed mid-level fusion outperforms the early and late fusion for some tasks. The Huber loss function improves the classification results slightly for standard features. Ultimately, it is recommended to use a combination of the aforementioned methods to run the RBF networks up to highest performance. We showed that RBF networks have potential for classifying and predicting discrete and continuous pain intensity levels. Furthermore, the network provides informative output about the samples such as confidence values. This information can be used in active learning and also used to built another layer for late fusion using pseudo inversion to improve the fusion results in multi-modal scenarios.

**Acknowledgement.** This paper is based on work done within the Transregional Collaborative Research Centre SFB/TRR 62 *Companion-Technology for Cognitive Technical Systems* funded by the German Research Foundation (DFG). Markus Kächele is supported by a scholarship of the Landesgraduiertenförderung Baden-Württemberg at Ulm University. The work is furthermore supported by the *SenseEmotion* project funded by the German Ministry of Science, Research and Arts and was performed using the computational resource *bwUniCluster* funded by the Ministry of Science, Research and Arts and the Universities of the State of Baden-Württemberg, Germany, within the framework program *bwHPC*.

# References

1. Benedek, M., Kaernbach, C.: Decomposition of skin conductance data by means of nonnegative deconvolution. Psychophysiology **47**(4), 647–658 (2010)
2. Brahnam, S., Chuang, C.F., Shih, F.Y., Slack, M.R.: Machine recognition and representation of neonatal facial displays of acute pain. Artif. Intell. Med. **36**(3), 211–222 (2006)
3. Cao, C., Slobounov, S.: Application of a novel measure of EEG non-stationarity as 'Shannon-entropy of the peak frequency shifting' for detecting residual abnormalities in concussed individuals. Clin. Neurophysiol. **122**(7), 1314–1321 (2011)
4. Chen, W., Zhuang, J., Yu, W., Wang, Z.: Measuring complexity using FuzzyEn, ApEn, and SampEn. Med. Eng. Phys. **31**(1), 61–68 (2009)
5. Gholami, B., Haddad, W.M., Tannenbaum, A.R.: Agitation and pain assessment using digital imaging. In: Conference Proceedings: Annual International Conference of the IEEE Engineering in Medicine and Biology Society, IEEE Engineering in Medicine and Biology Society, Conference, vol. 2009. NIH Public Access (2009)
6. Jaeger, H.: The "echo state" approach to analysing and training recurrent neural networks. GMD report 148, GMD - German National Research Institute for Computer Science (2001)
7. Kächele, M., Thiam, P., Amirian, M., Schwenker, F., Palm, G.: Methods for person-centered continuous pain intensity assessment from bio-physiological channels. IEEE J. Sel. Top. Sign. Process. **13**(9), 1–11 (2016)
8. Kächele, M., Thiam, P., Amirian, M., Werner, P., Walter, S., Schwenker, F., Palm, G.: Multimodal data fusion for person-independent, continuous estimation of pain intensity. In: Iliadis, L., Jayne, C. (eds.) EANN 2015. CCIS, vol. 517, pp. 275–285. Springer, Heidelberg (2015)
9. Kaltwang, S., Rudovic, O., Pantic, M.: Continuous pain intensity estimation from facial expressions. In: Bebis, G., et al. (eds.) ISVC 2012, Part II. LNCS, vol. 7432, pp. 368–377. Springer, Heidelberg (2012)
10. Kim, J., André, E.: Emotion recognition based on physiological changes in music listening. IEEE Trans. Pattern Anal. Mach. Intell. **30**(12), 2067–2083 (2008)
11. Lin, L.I.: A concordance correlation coefficient to evaluate reproducibility. Biometrics **45**(1), 255–268 (1989)

12. Littlewort, G.C., Bartlett, M.S., Lee, K.: Automatic coding of facial expressions displayed during posed and genuine pain. Image Vis. Comput. **27**(12), 1797–1803 (2009)
13. Lucey, P., Cohn, J.F., Prkachin, K.M., Solomon, P.E., Chew, S., Matthews, I.: Painful monitoring: automatic pain monitoring using the UNBC-McMaster shoulder pain expression archive database. Image Vis. Comput. **30**(3), 197–205 (2012)
14. Lucey, P., Cohn, J.F., Prkachin, K.M., Solomon, P.E., Matthews, I.: Painful data: the UNBC-McMaster shoulder pain expression archive database. In: 2011 IEEE International Conference on Automatic Face and Gesture Recognition and Workshops (FG 2011), pp. 57–64. IEEE (2011)
15. Moore, R.A., Wiffen, P.J., Derry, S., Maguire, T., Roy, Y.M., Tyrrell, L.: Nonprescription (OTC) oral analgesics for acute pain-an overview of cochrane reviews. Status and Date, New published in (10) (2013)
16. Phinyomark, A., Limsakul, C., Phukpattaranont, P.: A novel feature extraction for robust EMG pattern recognition. arXiv preprint arXiv:0912.3973 (2009)
17. Schwenker, F., Dietrich, C., Thiel, C., Palm, G.: Learning of decision fusion mappings for pattern recognition. Int. J. Artif. Intell. Mach. Learn. (AIML) **6**, 17–21 (2006)
18. Schwenker, F., Kestler, H.A., Palm, G.: Three learning phases for radial-basis-function networks. Neural Netw. **14**(4), 439–458 (2001)
19. Tang, X., Shu, L.: Classification of electrocardiogram signals with RS and quantum neural networks. Int. J. Multimedia Ubiquit. Eng. **9**(2), 363–372 (2014)
20. Tkach, D., Huang, H., Kuiken, T.A.: Research study of stability of time-domain features for electromyographic pattern recognition. J. Neuroeng. Rehabil. **7**, 21 (2010)
21. Walter, S., Gruss, S., Limbrecht-Ecklundt, K., Traue, H.C., Werner, P., Al-Hamadi, A., Diniz, N., da Silva, G.M., Andrade, A.O.: Automatic pain quantification using autonomic parameters. Psychol. Neurosci. **7**(3), 363 (2014)
22. Wiebalck, A., Vandermeulen, E., Van Aken, H., Vandermeersch, E.: Ein konzept zur verbesserung der postoperativen schmerzbehandlung. Der Anaesthesist **44**(12), 831–842 (1995)
23. Zhao, Q., Zhang, L.: ECG feature extraction and classification using wavelet transform and support vector machines. In: International Conference on Neural Networks and Brain, 2005, ICNN&B 2005, vol. 2, pp. 1089–1092. IEEE (2005)

# Active Learning for Speech Event Detection in HCI

Patrick Thiam[(⊠)], Sascha Meudt, Friedhelm Schwenker, and Günther Palm

Institute of Neural Information Processing, University of Ulm,
James-Franck-Ring, 89081 Ulm, Germany
{patrick.thiam,sascha.meudt,friedhelm.schwenker,guenther.palm}@uni-ulm.de

**Abstract.** In this work, a pool-based active learning approach combining outlier detection methods with uncertainty sampling is proposed for speech event detection. Events in this case are regarded as atypical utterances (e.g. laughter, heavy breathing) occurring sporadically during a Human Computer Interaction (HCI) scenario. The proposed approach consists in using rank aggregation to select informative speech segments which have previously been ranked using different outlier detection techniques combined with an uncertainty sampling technique. The uncertainty sampling method is based on the distance to the boundary of a Support Vector Machine with Radial Basis Function kernel trained on the available annotated samples. Extensive experimental results prove the effectiveness of the proposed approach.

**Keywords:** Active learning · Supervised learning · Support Vector Machines · Support Vector Data Description · Gaussian Mixture Model · Rank aggregation

## 1 Introduction

The performance of Supervised Learning techniques relies strongly on the quality and the quantity of the available annotated data. Meanwhile, the annotation process of a huge dataset is known to be very cumbersome and expensive. This explains the scarceness of large labeled datasets, while, on the other hand, a large amount of unlabeled data is available. In the present work, we propose an active learning technique designed with the goal of attaining similar results as classifiers trained on fully annotated datasets by carefully selecting and annotating a strongly reduced subset.

Event detection can enhance the capabilities of an emotion recognition system. Once an event is detected, it can be subsequently given a label depicting the

P. Thiam—This paper is based on work done within the Transregional Collaborative Research Centre SFB/TRR 62 Companion-Technology for Cognitive Technical Systems funded by the German Research Foundation (DFG). The author is supported by the BMBF within the project *SenseEmotion*. We gratefully acknowledge the support of NVIDIA Corporation with the donation of the Tesla K40 GPU used for this research.

© Springer International Publishing AG 2016
F. Schwenker et al. (Eds.): ANNPR 2016, LNAI 9896, pp. 285–297, 2016.
DOI: 10.1007/978-3-319-46182-3_24

specific affective disposition of the user. For instance, laughter can be perceived as a sign of happiness or frustration. Heavy breathing can also be interpreted as a sign of despair or relief. Therefore, a cascaded classification consisting in first localizing the occurring events before proceeding with a specific classification would enhance the capabilities of the recognition system.

In this work, we focus on event detection in speech signals. Events are defined as atypical utterances in the speech signal such as laughter, heavy breathing or other expressions of dissatisfaction, frustration or despair. A pool-based active learning approach is proposed that combines uncertainty sampling with outlier detection techniques in order to select the most informative samples. By carefully selecting and annotating the samples from which a classification model is trained, the cumbersome and time expensive process of annotating the whole available dataset is avoided while no significant degradation of the performance of the generated system is observed. The proposed approach is tested on a subset of the Ulm University Multimodal Affective Corpus (uulmMAC) database, using uniquely the speech modality. We assess the effectiveness of the developed active learning approach by comparing its performance with a model trained on a fully annotated training set.

The remainder of this work is organised as follows. In the next section the utilised dataset is described, as well as the manual annotation process, followed by a description of the feature extraction process and the first assessment of the annotated set through supervised learning. The active learning approach is explained in the following section, followed by the presentation and the discussion of the experimental results. The work is finally concluded in the last section.

## 2    Dataset Description

The utilized dataset is a subset of nine participants of the Ulm University Multimodal Affective Corpus (uulmMAC database). The interaction scheme of uulm-MAC is based on the work of Schüssel et. al. who proposed a gamified experimental setup close to everyday life Human Computer Interaction [16] (e.g. looking for a specific menu entry or button followed by selecting or interacting with this element). The very generic paradigm allows a lot of variations in order to investigate different research questions. uulmMAC consists of 60 participants in 100 recording sessions of about 45 min each. The recordings constitute of synchronous multimodal data containing three video streams (frontal HD, frontal webcam, rear webcam), three audio lines (headset, ambiance and directional microphone), MS Kinect 2 (depth, infrared, video, audio, posture) and biophysiology (EMG, ECG, SCL, respiration and temperature). Figure 1 shows an overview of the experimental setup.

In the uulmMAC participants were asked to play a series of games to test their reaction to both overwhelming and boring difficulty levels. The task of each game sequence was to identify the singleton element, i.e. the one item that is unique in shape and color (number 36 and 2 in Fig. 2). The difficulty was set by adjusting the number of shapes shown per sequence and increasingly less money earned the longer they needed to answer. If the given answer was incorrect, the participant received no reward at all for that particular round. To modify the difficulty further, the time given per sequence was also adjusted. Each participant completed an introductory sequence and four game sequences of decreasing difficulty. The first sequence was designed to induce overload ($6 \times 6$ board with 6 s to provide an answer, see Fig. 2), the second was a mix of $5 \times 5$ and $4 \times 4$ with 10 s to provide an answer, the third was set to $3 \times 3$ with 100 s to provide an answer. Sequence 4 was designed to induce underload and the game setting was therefore set to an easy $3 \times 3$ mode with 100 s answering time. The last sequence induced frustration, e.g. by purposely logging in a wrong answer.

After each accomplished sequence the participants answered a series of questions. The aim of those questions was to determine valence, arousal and dominance [15] experienced in the particular sequence. Firstly, the participants answered in their own words how they felt during the game. Afterwards each

**Fig. 1. The experimental setup** with sensors MS Kinect 2 (1), front cameras (2), headset (3) and biosensors (4) (some physiologic sensors are placed under the shirt). **Left**: frontal camera view. **Right**: rear camera view.

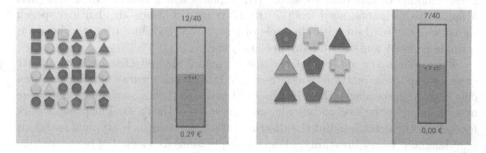

**Fig. 2. The game mode. Left**: a very complex game mode causing a high mental load. **Right**: a simple game mode causing the player to relax more.

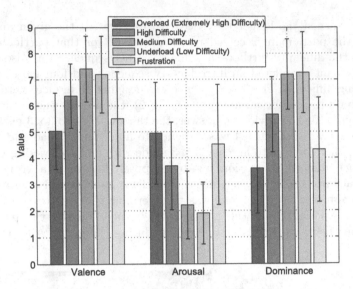

**Fig. 3. Mean VAD over all game sequences.** It can be seen, that the reported VAD varies significantly between the sequences, especially between sequences 1 and 4.

participant was presented with three scales and was asked to choose the value according to their experience (Self Assessment Scale (SAM)) [3].

Figure 3 shows how valence, arousal and dominance are evolving over the five sequences. The results suggest that the reported $(V, A, D)$ experiences differ significantly between the mental overload and underload sequences. It seems that valence is higher in mental underload, arousal is higher in mental overload and dominance is higher in the mental underload sequence. According to this, mental overload and underload can be expressed using the $(V, A, D)$ space as overload: $(V-, A+, D-)$ and underload: $(V+, A-, D+)$.

In a preliminary step of the current study, a set of 9 participants was selected based on their grade of expressiveness by observing their demeanour throughout the different phases of the experiment. The present work focuses on the headset audio channel of this reduced set of participants which consists of 4 male and 5 female participants, aged from 20 to 27 years old. Since an identical speech signal was recorded on both channels of the stereo recording from the headset, a single channel was used for further analysis.

The preliminary analysis of the speech signal showed that each candidate expressed sporadic reactions when they failed to give an accurate answer to a specific task, in particular during the phases of overload throughout which they were overwhelmed by the level of difficulty and the speed of the tasks. The present work focuses on the detection and the discrimination between such atypical reactions (e.g. laughter, heavy breathing, idiomatic expressions as a signal of boredom, dissatisfaction, frustration and despair) and the answers provided by the candidates which are principally spoken indexes of selected cells within a grid displayed during the tasks. In the present work, the atypical utterances described earlier will be referred to as events and the provided answers as normal utterances.

In order to be able to assess the performance of the developed methods, a second step was undertaken during which the extracted speech signal specific to each of the 9 candidates was manually annotated.

## 2.1  Manual Annotation

A manual annotation had to be carried out in order to provide ground truth labels that are essential for a further analysis and characterization of the disposition of each participant during the experiments, as well as for the proper assessment of the developed methods. The annotation step was preceded by an automatic segmentation step consisting in distinguishing between voice active segments and voice inactive (silence or noise) segments. This step was necessary since the study focuses on the aforementioned two classes of voice active segments (events and normal utterances). Since the recordings were performed in a relatively clean (noise-free) environment, a simple unsupervised voice activity detection [1] based on the energy of the speech signal was applied and yielded satisfactory results.

Subsequently, based on the results of the voice activity detection, each detected voiced segment was manually annotated as being either an event or a normal utterance. Furthermore, the boundaries of the voiced regions were manually adjusted in order to acquire a precise segmentation of the utterances

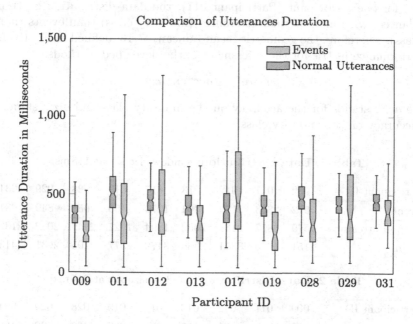

**Fig. 4. Manual Annotation Results**: participant dependent assessment of the duration of the annotated utterances (events and normal utterances). The figure depicts a great variation of utterance duration among the different participants.

and to substantially reduce the amount of noise in each segment. The whole annotation process was performed within the ATLAS annotation tool [14].

Following the annotation of the speech signal, a first assessment of the labeled speech samples was undertaken by comparing the duration of the normal utterances to those of events (see Fig. 4). The first observation is the specificity of the duration of the utterances to each participant. Each participant acts differently depending on his or her actual affective state. Secondly, for most of the participants the annotated events depict a greater variance in duration in comparison to normal utterances. This can be explained by the fact that the answers to the different tasks are basically pronounced numbers while events would vary from very short locutions to entire phrases spoken as a sign of frustration. Furthermore, the median duration of normal utterances is higher than the median duration of events for almost all participants. These observations describe the great diversity of the demeanour of the participants depending on their mental or affective disposition.

Moreover, the overall minimum duration of an event is situated somewhere between 30 and 50 ms, while the overall minimum duration of a normal utterance is situated somewhere between 150 and 200 ms. Based on these findings, 2 windows of respectively 115 and 215 ms were selected for the segmentation of the annotated speech samples in order to proceed with the feature extraction. The windows were shifted with a fixed offset of 65 ms. The resulting data distribution can be observed in both Tables 1 and 2. In both segmentation settings, except for one participant (Participant 011), the data distribution is strongly imbalanced, with more normal utterances (majority class) than events (minority class). Therefore, the geometric mean (gmean) [7,13] defined in Eq. 1 is used as performance metric for the assessment of the developed methods:

$$gmean = \sqrt{acc^+ \times acc^-} \tag{1}$$

where $acc^+$ stands for the accuracy on the minority class and $acc^-$ stands for the accuracy on the majority class.

**Table 1. Data Distribution**: window size set at 115 ms

| Participant ID | 009 | 011 | 012 | 013 | 017 | 019 | 028 | 029 | 031 |
|---|---|---|---|---|---|---|---|---|---|
| Events | 572 | 2863 | 1174 | 304 | 1276 | 450 | 656 | 849 | 281 |
| Normal utterances | 1579 | 2817 | 2137 | 2062 | 1894 | 1927 | 2449 | 2082 | 2132 |
| Total | 2151 | 5680 | 3311 | 2366 | 3170 | 2377 | 3105 | 2931 | 2413 |

**Table 2. Data Distribution**: window size set at 215 ms

| Participant ID | 009 | 011 | 012 | 013 | 017 | 019 | 028 | 029 | 031 |
|---|---|---|---|---|---|---|---|---|---|
| Events | 350 | 2206 | 890 | 195 | 1036 | 297 | 439 | 632 | 196 |
| Normal utterances | 1033 | 2162 | 1574 | 1464 | 1330 | 1363 | 1851 | 1503 | 1555 |
| Total | 1383 | 4368 | 2464 | 1659 | 2366 | 1660 | 2290 | 2135 | 1751 |

## 2.2   Feature Extraction

Following the segmentation of the annotated speech samples, a set of audio features was extracted from the different segments with fixed frames of 25 ms length sampled at a rate of 10 ms: 8 linear predictive coding coefficients (LPC) [10]; 5 perceptual linear prediction cepstral coefficients (PLP-CC) [8], each with delta and acceleration coefficients; 12 Mel frequency cepstral coefficients (MFCC) [9], each with delta and acceleration coefficients; fundamental frequency (F0); voicing probability; loudness contour; log-energy with its delta and acceleration [11]. Thus, each frame is represented by a 65 dimensional feature vector. Consequently, each labeled speech segment is represented by a $10 \times 65$ feature matrix for a window size of 115 ms, and $20 \times 65$ feature matrix for a window size of 215 ms. The features were extracted using the openSMILE feature extraction tool [6].

Subsequently, in order to compute the features at the segment level, 14 statistical functions (mean, median, standard deviation, maximum, minimum, range, skewness, kurtosis, first and second quartile, inter quartile, 1 %-percentile and 99 %-percentile, range of 1 % and 99 % percentile), were applied on each of the extracted frame level feature resulting in a 910 dimensional feature vector for each segment. These segment level feature vectors are used for the assessment of the developed methods.

## 2.3   Dataset Assessment: Supervised Learning Experiments

An important step of the study is the assessment of the generated dataset in order to ensure the relevance of the extracted feature vectors for the task at hand. Therefore, 3 classifiers were selected to perform a 5-fold participant dependent blocked cross validation [2] of the dataset for each segmentation setting: a Random Forest classifier with a fixed size of 300 trees; a Support Vector Machine (SVM) with a Gaussian Radial Basis Function (RBF) kernel; a multilayer perceptron (MLP) with a unique hidden layer of 100 neurons, each with a radial basis transfer function.

During the blocked cross validation the dataset is partitioned sequentially (not randomly) into several subsets. Therewith overlapping segments belonging to a specific utterance are all contained either in the training or in the testing set except for a few segments situated at the border of the partitions. Since we perform a 5-fold cross validation, such occurrences are minimal. Thus they are not discarded and kept in the sets. Following, each subset is used within each cross validation iteration as a test set while the remaining sets are used as training sets. Before the classification, each set is first preprocessed to get rid of noisy data. Subsequently, Principal Component Analysis (PCA) is performed to reduce the dimensionality of the dataset. The first 10 components are used at each iteration. Then the Synthetic Minority Over-Sampling Technique (SMOTE) [5] is applied on the training set to balance the data before the learning process takes place.

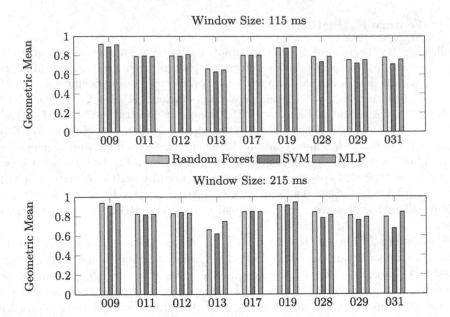

**Fig. 5. Supervised Learning Results**: person dependent 5-fold blocked cross validation. The upper figure depicts the results for a window size of 115 ms, while the lower figure depicts the results for a window size of 215 ms. Both figures depict similar performances amongst the participants despite the different window sizes with the Random Forest classifier performing in most cases slightly better than the MLP and the SVM.

The classification results can be seen in Fig. 5. The upper figure shows the results for the window size of 115 ms. The lower figure shows the results for the window size of 215 ms. The selected classifiers perform rather similarly in both segmentation settings with the best performances achieved by the Random Forest classifier in most cases, followed by the MLP and finally the SVM. These results validate the relevance of the generated feature set, as well as the applicability of different classifiers, to this specific speech event detection task. Based on these findings, an active learning method is proposed with the goal of reaching similar classification performances as depicted in Fig. 5, while reducing the costs of annotating the whole set.

## 3  Proposed Active Learning Approach

We propose a pool-based active learning approach that combines outlier detection with uncertainty sampling to select the most informative samples that are annotated by an oracle and subsequently used to train a classification model. The proposed approach is a further iteration of the methods presented by Thiam et al. in [18,19].

More specifically, the method consists in first selecting and ranking a subset of samples from the unlabeled set based on outlier detection techniques. Secondly, a SVM with a RBF kernel is trained on the available labeled set and used to select and rank another subset of samples from the same unlabeled set based on the distance from those samples to the decision boundary of the trained model. Subsequently, rank aggregation is performed on the resulting subsets and the $k$ highest ranked samples are selected for the annotation process. Two outlier detection techniques have been experimented with.

## 3.1 Outlier Detection with an Ensemble of Support Vector Data Description

This approach consists in generating an ensemble of Support Vector Data Description (SVDD) [17] models and performing the selection and ranking of the samples based on the voting count of the ensemble. As opposed to the method presented by Thiam et. al. [18], the models of the ensemble are not randomly generated. Based on the work of Chang et al. [4], a total of $m \times n$ SVDD models are generated by choosing $m$ values for the parameter $C$, equally spaced within the interval $\left[\frac{1}{N}, 1\right]$ where $N$ is the number of samples in the unlabeled set, and $n$ values for the parameter $\gamma$ of the RBF kernel, equally spaced within the interval $\left[\frac{1}{f}, 1\right]$ where $f$ is the dimensionality of the feature vectors. In this way, the grid of possible values for both parameters is covered and the diversity in the ensemble is ensured. Furthermore a threshold is used to prune the generated ensemble, based on the reclassification results of each generated model. Models with outlier classification rates higher than the specified threshold are discarded. Subsequently, the samples of the unlabeled set are ranked in descending order of the voting count.

## 3.2 Outlier Detection with a Gaussian Mixture Model

This approach consists in generating a Gaussian Mixture Model (GMM) from the unlabeled set and using the Mahalanobis distance from each sample to each Gaussian component center to determine the outliers. A specified threshold is also used in this case for the purpose of detecting outliers. Samples having a Mahalanobis distance higher than the specified threshold are considered as outliers. Once this has been done for each component, each sample in the subset of detected outliers is ranked in ascending order of the probability density function of the GMM.

## 3.3 Rank Aggregation and Sample Selection

The selection of the samples to be annotated is performed by applying Borda's geometric-mean rank aggregation method [12] on both subsets resulting from the outlier detection technique and the SVM model trained on the available labeled set and applied on the unlabeled set. Subsequently, the $k$ highest ranked samples are selected for the annotation.

## 3.4 Experimental Settings and Results

The proposed approach is tested in a 5-fold blocked cross validation setting during which after the sequential partition of the dataset is performed, each partition is used within each cross validation iteration as a test set and the remaining partitions are used as training set. During each iteration, the baseline is first computed by training a SVM with RBF kernel on the training set and performing the classification using the generated model on the test set. Next, the proposed approach is applied on the training set. After each active learning iteration, a SVM with RBF kernel is trained on the annotated samples and applied on the test set. A fixed number of 25 SVDD models ($m = 5, n = 5$) is generated for the committee and the number of components for the GMM is fixed at 5. Moreover, a maximum of 50 samples is selected during each active learning round.

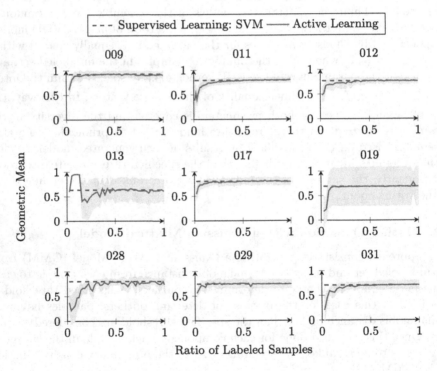

**Fig. 6. Active Learning with SVDD Outlier Detection Results**: person dependent 5-fold blocked cross validation with a window size of 115 ms. The dashed blue line corresponds to the averaged geometric mean of the fully supervised learning results computed with a SVM with RBF kernel. The continuous red line corresponds to the averaged supervised learning results computed with a SVM with RBF kernel trained on the available annotated samples after each active learning round. The lighter coloured corridors correspond to the standard deviation of the results after each active learning round. (Color figure online)

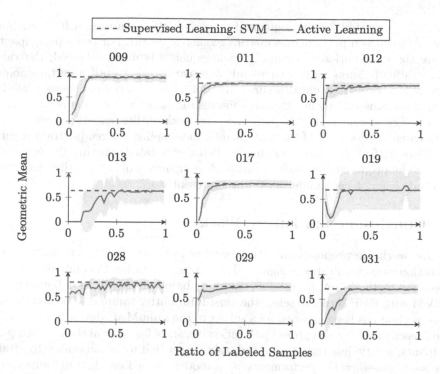

**Fig. 7. Active Learning with GMM Outlier Detection Results**: person dependent 5-fold blocked cross validation with a window size of 115 ms. The dashed blue line corresponds to the averaged geometric mean of the fully supervised learning results computed with a SVM with RBF kernel. The continuous red line corresponds to the averaged supervised learning results computed with a SVM with RBF kernel trained on the available annotated samples after each active learning round. The lighter coloured corridors correspond to the standard deviation of the results after each active learning round. (Color figure online)

Figures 6 and 7 show the results of the approach, based respectively on the ensemble of SVDD models and the GMM, applied on the data specific to each participant, with the window size of 115 ms (the same experiments have been conducted with the window size of 215 ms yielding similar results). The averaged geometric mean of the supervised learning classification of the blocked cross validation is plotted in blue and serves as the baseline in each case. The averaged geometric mean of the supervised learning classification performed using the annotated samples after each active learning round and applied on the test set is plotted in red. Furthermore, the standard deviation of the classification is plotted in lighter coloured corridors to depict the variance of the achieved results after each active learning round.

The figures specific to the participants 013 and 019 depict a great degree of variance during the whole active learning process in both outlier detection settings. This is due to the highly imbalanced distribution of events within each partition generated by the sequential cross validation, causing a high fluctuation of the chosen performance metric (geometric mean). For those participants where the distribution of events in each sequence is more balanced, the variance gradually sinks with each round of active learning until a certain point, then remains constant until the final iteration. This observation shows that the system remains stable once the most interesting samples are annotated.

The figures also depict the performance of both outlier detection methods. In both cases and for most of the participants, the baseline is already attained with less than half of the whole training set being annotated, proving the effectiveness of the proposed approach. A thorough comparison of both outlier detection methods has not been undertaken in the present work.

## 4    Conclusion and Future Work

In this work, we proposed an active learning approach for the annotation and detection of events in speech signals. The approach consists in combining outlier detection methods and uncertainty sampling based on the decision boundary of a SVM with RBF kernel to select the most informative samples to be annotated. The method has been assessed on a subset of the uulmMAC dataset using solely the speech modality and proved to be effective, since for most of the assessed participants, a little less than half of the training set had to be annotated to attain the same classification performance as a model trained on the fully annotated training set.

An application of the designed method in HCI could be the annotation and classification of a specific category of emotion (e.g. anger, sadness, happiness) in a one against all scenario, without having to go through the process of annotating an entire dataset. Another application of the method would be an exploratory and cascaded annotation of a dataset consisting in first detecting various atypical states in the dataset before providing specific labels to the detected events in a further step.

Further experiments are to be undertaken, involving the assessment of the applicability of the presented approach on other modalities (e.g. video and bio-physiological signals). Moreover, several outlier detection methods are also to be assessed, as well as different combination methods for the selection of the most informative samples.

## References

1. Alam, M.J., Kenny, P., Ouellet, P., Stafylakis, T., Dumouchel, P.: Supervised/unsupervised voice activity detector for text-dependent speaker recognition on RSR2015 corpus. In: Odyssey Speaker and Language Recognition Workshop (2014)
2. Bergmeir, C., Benítez, J.M.: On the use of cross-validation for time series predictor evaluation. Inf. Sci. **191**, 192–213 (2012)
3. Bradley, M.M., Lang, P.J.: Measuring emotion: the self-assessment manikin and the semantic differential. J. Behav. Ther. Exp. Psychiatry **25**(1), 49–59 (1994)

4. Chang, W.C., Lee, C.P., Lin, C.J.: A revisit to support vector data description (SVDD). Technical reports (2013)
5. Chawla, N.V., Bowyer, K.W., Hall, L.O., Kegelmeyer, W.P.: Smote: synthetic minority over-sampling technique. J. Artif. Intell. Res. **16**, 321–357 (2002)
6. Eyben, F., Weninger, F., Gross, F., Schuller, B.: Recent developments in opensmile, the Munich open-source multimedia feature extractor. In: ACM Multimedia (MM), pp. 835–838, October 2013
7. Gu, Q., Zhu, L., Cai, Z.: Evaluation measures of the classification performance of imbalanced data sets. In: Cai, Z., Li, Z., Kang, Z., Liu, Y. (eds.) ISICA 2009. CCIS, vol. 51, pp. 461–471. Springer, Heidelberg (2009)
8. Hermansky, H.: Perceptual Linear Predictive (PLP) analysis of speech. J. Acoust. Soc. Am. **87**(4), 1738–1752 (1990)
9. Jagan Mohan, B., Ramesh Babu, N.: Speech recognition using MFCC and DTW. In: 2014 International Conference on Advances in Electrical Engineering (ICAEE), pp. 1–4, January 2014
10. Krothapalli, S.R., Koolagudi, S.G.: Emotion recognition using vocal tract information. In: Krothapalli, S.R., Koolagudi, S.G. (eds.) Emotion Recognition using Speech Features. SpringerBriefs in Electrical and Computer Engineering, pp. 67–78. Springer, New York (2013)
11. Krothapalli, S.R., Koolagudi, S.G.: Speech emotion recognition: a review. In: Krothapalli, S.R., Koolagudi, S.G. (eds.) Emotion Recognition using Speech Features. SpringerBriefs in Electrical and Computer Engineering, pp. 15–34. Springer, New York (2013)
12. Lin, S.: Rank aggregation methods. Wiley Interdisc. Rev. Comput. Stat. **2**(5), 555–570 (2010)
13. Lòpez, V., Fernàndez, A., Garcìa, S., Palade, V., Herrera, F.: Strategies for learning in class imbalance problems. Pattern Recogn. **36**(3), 849–851 (2003)
14. Meudt, S., Bigalke, L., Schwenker, F.: Atlas - an annotation tool for HCI data utilizing machine learning methods. In: Proceedings of the 1st International Conference on Affective and Pleasurable Design (APD 2012) (Jointly with the 4th International Conference on Applied Human Factors and Ergonomics (AHFE 2012)), pp. 5347–5352 (2012)
15. Russel, J.A.: Core affect and the psychological construction of emotion. Pyschological Rev. **110**(1), 145–172 (2003)
16. Schüssel, F., Honold, F., Bubalo, N., Huckauf, A., Traue, H., Hazer-Rau, D.: In-depth analysis of multimodal interaction: an explorative paradigm. In: Kurosu, M. (ed.) HCI 2016. LNCS, vol. 9732, pp. 233–240. Springer, Heidelberg (2016)
17. Tax, D.M., Duin, R.P.: Support vector data description. Mach. Learn. **54**(1), 45–66 (2004)
18. Thiam, P., Kächele, M., Schwenker, F., Palm, G.: Ensembles of support vector data description for active learning based annotation of affective corpora. In: 2015 IEEE Symposium Series on Computational Intelligence, pp. 1801–1807, December 2015
19. Thiam, P., Meudt, S., Kächele, M., Palm, G., Schwenker, F.: Detection of emotional events utilizing support vector methods in an active learning HCI scenario. In: Proceedings of the 2014 Workshop on Emotion Representation and Modelling in Human-Computer-Interaction-Systems, ERM4HCI 2014, pp. 31–36. ACM, New York (2014)

# Emotion Recognition in Speech with Deep Learning Architectures

Mehmet Erdal, Markus Kächele[(⊠)], and Friedhelm Schwenker

Institute of Neural Information Processing, Ulm University,
James-Franck-Ring, 89081 Ulm, Germany
markus.kaechele@uni-ulm.de

**Abstract.** Deep neural networks (DNNs) became very popular for learning abstract high-level representations from raw data. This lead to improvements in several classification tasks including emotion recognition in speech. Besides the use as feature learner a DNN can also be used as classifier. In any case it is a challenge to determine the number of hidden layers and neurons in each layer for such networks. In this work the architecture of a DNN is determined by a restricted grid-search with the aim to recognize emotion in human speech. Because speech signals are essentially time series the data will be transformed in an appropriate format to use it as input for deep feed forward neural networks without losing much time dependent information. Furthermore the Elman-Net will be examined. The results shows that by maintaining time dependent information in the data better classification accuracies can be achieved with deep architectures.

## 1 Introduction

Paralinguistic information like the intonation are important parts in a conversation. We can consider these kinds of information as the semantics of a spoken utterance. For example, the word "yes" is basically an expression of agreement, but with a contemptuous intonation it can mean exactly the opposite namely rejection and this can be an evidence that the speaker is angry. Hence it is possible to perceive the emotional state of the speaker with paralinguistic information conveyed in the speech signal. Because emotions could be crucial for the interpretation of a spoken utterance, efforts are made to give computers the ability to recognize emotion in speech to improve the human-computer interaction (cf. [15]). Nowadays this is a growing field of research which is known as *affective computing*. Therefore the aim of speech emotion recognition is to identify the high-level affective state of an utterance from the low-level features. The task here is to recognize specific pattern as sequences in the speech signal and to categorize them into several classes of emotions.

There are several machine learning models that can be used for classification. In machine learning theory a model is an algorithm which learns from data to tackle a specific task without having to have been explicitly programmed. The learning process is often called training. One of those models are artificial

© Springer International Publishing AG 2016
F. Schwenker et al. (Eds.): ANNPR 2016, LNAI 9896, pp. 298–311, 2016.
DOI: 10.1007/978-3-319-46182-3_25

neural networks (ANN), which are slightly inspired by the functioning of the human brain. A deep neural network is an ANN with many layers of nonlinear processing units. The field of research that studies methods to train ANNs with deep architectures is called *deep learning*. Deep learning architectures (DLAs) have been shown to exceed preliminary state-of-the art results in several tasks including emotion recognition in speech [1–3].

## 2  Related Work

For a long time, DNNs were considered to be hard to train, because gradient-based learning algorithms with a random start initialization produced often poor solutions. Therefore an unsupervised greedy layer-wise learning algorithm for so called deep belief networks (DBNs) is proposed in [4]. A DBN is a composition of simple learning modules which have a layer of visible units to capture the input data and a layer of stochastic hidden units that learn to represent high-level abstractions of the input data. A so trained DBN serves as a start initialization by transforming it into a DNN. The DNN is then trained via backpropagation (bpp) to find a better solution than a randomly initialized network. The training algorithm for the DBN is therefore called pre-training and the phase with bpp is called fine-tuning.

In the work [5] a DBN is used to classify emotion in speech. Like in the present work the Berlin Database of Emotional Speech (EmoDB) with the Mel frequency cepstral coefficients (MFCCs) as speech features is used in the experiments. A multilayer perceptron (MLP) with one hidden layer served as model with a shallow network structure and as baseline. The best result of 60.32 % accuracy is obtained by the DBN in a speaker independent scheme. That was an improvement of 8.67 % over the baseline. In [6] a DBN is used to extract characteristic features for emotional expressions in the speech signal. An SVM is then used as classifier for the extracted features. In [1] a MLP with more than one hidden layer is used as DLA to recognize emotion in speech. Similar to the presented work the number of hidden layers and units are selected via cross validation. However the MLP is there used as feature-extractor and an extreme learning machine is then used to classify the data. In [17] a convolutional deep belief network (CDBN) was used for feature learning from audio data. The so generated features performed often better then MFCCs on several audio classification tasks. Using a more diverse set of features (i.e. including voice quality), the accuracy on the EmoDB dataset can be improved to over 88 % [23].

## 3  Deep Learning Architectures and Recurrent Neural Networks

In the present work DNNs are used as classifier to recognize emotion in speech. DNNs are not only standalone models, but also the basis of other deep learning models. For example, the deep RNN that is also used in this work, can be viewed as a special form of a DNN. In this section the basics of DNNs and RNNs are explained. Furthermore it will be shown how they can be used as classifiers.

## 3.1 Deep Neural Networks

Deep learning is essentially a method to approximate a parametric function via neural networks with many hidden layers. For this purpose a neural network represents the function $f(x; \theta)$ where $x$ is the input vector and $\theta$ a set of parameters. To be more precise $f$ is a composition of functions. Therefore the smallest unit of a neural network is a so called neuron. It maps the weighted sum $\sum_{i=1}^{k} x_i w_i$ to an activation value via the function $f_{act}(x^T w)$ where $x$ is the vector of inputs for the neuron and $w$ a vector of parameters denoted as weights. A layer of the network is a set of neurons that usually uses the same activation function. In this case a layer can be represented as function $f^{(i)}(x; W^{(i)}) = f_{act}(W^{(i)T} x)$ where $x$ is the input vector of the layer, $W \in \mathbb{R}^{k \times l}$ the matrix that contains the weights of $l$ neurons (each column of $W$ represents the weights of a neuron) and $i$ is the number of the layer. Putting all together such a neural network represents a composition of the layer functions with the parameters $\theta = \{W^{(1)}, W^{(2)}, \ldots, W^{(n)}\}$ (cf. [9]):

$$f(x; \theta) := f^{(n)}(\ldots f^{(2)}(f^{(1)}(f^{(0)}(x); W^{(1)}); W^{(2)}) \ldots; W^{(n)}) \tag{1}$$

As it can be seen, the output of a layer is the input of the next layer. This can be considered as forward propagation of the input through the network. Hence one speaks of a feed forward neural network. The first layer is called input layer and is only there to receive the input with the identity function $f^{(0)}(x) = x$. The last layer $f^{(n)}$ is called output layer. All other layers are called hidden layers. A feed forward neural network with more than one hidden layer is called deep neural network. A neural network can be modeled as acyclic directed graph as shown in Fig. 1.

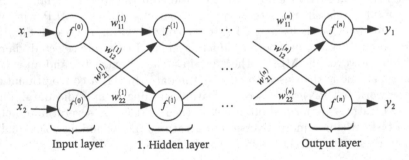

**Fig. 1.** DNN as directed acyclic graph. For simplification all layers have only two neurons.

It has been shown that it is very effective for DNNs to use the rectified linear function $relu(x^T w) = max(x^T w, 0)$ as activation for the hidden layers [7,8]. The activation function of the output layer depends on the task to tackle. To approximate a specific function $f^*$ the backpropagation-algorithm is used to learn the parameters $\theta$ that result in the ideal case in the best approximation $\mathring{f}(x; \theta) \approx f^*(x)$. This algorithm is based on stochastic gradient descent and computes recursively the gradients beginning with the last layer back to the first hidden layer. The gradients are then used to update the parameters. For this purpose a training set of examples $X_{train} = \{(x_i, y_i)\}_i^m$ is iteratively processed by the bbp-algorithm. A detailed description of the bbp-Algorithm can be found in [22].

## 3.2 Dropout Regularization

DNNs have a high number of adjustable parameters. Such powerful models tend to memorize the training set which is called overfitting. This phenomena is also well known for other classification models and hence there are several regularization techniques to overcome this problem. For DNNs the so called dropout regularization has been proven to be an effective tool [18]. The key idea is to deactivate a certain percentage of neurons. This can be performed as follows for all layers $i = 2, 3 \ldots, l$:

$$x^{(i+1)} = f^{(i)}(W^{(i)^T}(x^{(i)} * r^{(i)})) \text{ with } r^{(i)} \sim Bernoullli(\rho) \qquad (2)$$

$r^{(i)}$ is a vector of Bernoulli distributed random variables which have the value 1 with probability $\rho$. Such a sampled vector is element-wise multiplied with the result of the previous layer $x^{(i)} = f^{(i-1)}(\ldots)$. Therefore a 0 in this vector deactivates one respective neuron. This constitutes noise in the training data that makes it more difficult to memorize the training set.

## 3.3 Deep Recurrent Neural Networks

A drawback of the DNN is that it cannot take into account the inputs from the past of a time series, because it has no memory. Hence it is principally not suitable to process temporal data like time series. A better model for this purpose is the Elman-Network, a recurrent neural network invented by Elman [10]. It has only one hidden layer and parallel to the input layer a so called context layer that is fully connected to the hidden layer which serves as additional input. At timestep $t$ the context layer contains the output of the hidden layer at timestep $t - 1$. The RNN represents following recursive function:

$$h(t) = f_h(W^T x(t) + C^T h(t - 1)) \qquad (3)$$
$$y(t) = f_o(V^T h(t)) \qquad (4)$$

Here $x(t)$ is input vector, $h(t)$ the output vector of the hidden layer and $y(t)$ the output of the whole network at time $t$. $W, C, V$ are the weight matrices for input to hidden layer, context to hidden layer and hidden to output layer. $f_h$ and $f_o$ are the activation function for the hidden layer and output layer. Figure 2 illustrates the recursive flow of $h(t)$ between the context and hidden layer.

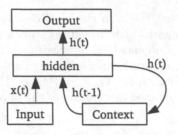

**Fig. 2.** Structure of a RNN.

After $n$ time steps a DNN with $n$ hidden layers is computed. In order to demonstrate this we concatenate $W$ and $C$ to one weight matrix $U$. To concatenate two vectors $x \in \mathbb{R}^n$, $h \in \mathbb{R}^m$ we write $[x; h]$ with the vector $(x_1, \ldots, x_n, h_{n+1}, \ldots, h_{n+m})$ as result. For a easier understanding we consider only 3 time steps. Now we can write the hidden activation function as follows:

$$h(3) = f_h^{(3)}(U^T[x(3); \underbrace{f_h^{(2)}(U^T[x(2); f_h^{(1)}(U^T[x(1); f_h^{(0)}])])}_{\text{hidden layers of the previous 2 time steps}}]) \tag{5}$$

$f_h^{(t)}$ is here the hidden activation function at time step $t$ and $f_h^{(0)}$ is an initial null vector. At time step $t = 3$ the additional input is $f_h^{(2)}$ which computes two hidden layers with the previous inputs $x(2)$ and $x(1)$. In this way the temporal data of the past is taken into account. With the hidden layer of the current time step and the output layer $f_o$ a DNN with overall 3 hidden layers is computed. To train such a network a slightly modified bpp-algorithm is used called backpropagation through time [11].

### 3.4 Deep Neural Networks as Classifier

In general a classifier is a function $y = f^*(x)$ that maps an n-dimensional input $x$ to a category $y$. To us a neural network for multinomial classification the softmax function is often used in the output layer. Each neuron represents one category $y \in \{c_1, \ldots, c_n\}$ and with the softmax function a probability for each category is predicted as follows:

$$P(y = c_i | x) = \frac{e^{x^T w_i}}{\sum_{j=1}^n e^{x^T w_j}} \tag{6}$$

Consequently a neural network classifier $p = f(x; \theta)$ maps an input $x$ to a vector of probabilities $p$. The element $p_i \in p$ with the highest probability stands for the predicted class $k_i$. To asses the performance of a classifier the classification accuracy with a test set $X_{test} = \{(x_i, y_i)\}_i^m$ with $X_{train} \cap X_{test} = \emptyset$ can be used as cost function:

$$accuracy := \frac{1}{m} \sum_{i=1}^m L(k_i, c_i) \quad \text{with} \quad L(k_i, c_i) = \begin{cases} 1 & \text{if } c_i = k_i \\ 0 & \text{else} \end{cases} \tag{7}$$

## 4 Feature Extraction and Preprocessing

The speech signal is provided as time series of amplitudes which is the result of an analog digital conversion. A time series is a sequence of scalars or vectors which depends on time. In case of the raw digital speech signal $\{x[t]\}$ the elements are scalars:

$$\{x[t]\} = \{x(t_0), x(t_1), \ldots, x(t_{i-1}), x(t_i), x(t_{i+1}), \ldots\} \tag{8}$$

The raw speech signal contains to much unnecessary information that is an obstacle to recognize certain emotion pattern. Hence a crucial step towards emotion recognition in speech is to extract appropriate features from the speech signal. Furthermore feature extraction reduces the high dimensionality of the raw speech signal. In several studies it has been shown that the Mel-Frequency Cepstral Coefficients (MFCCs) are very useful features to recognize emotion in speech [12]. One reason for that could be that MFCC mimics some parts of the human speech production and speech perception. In the presented work the MFCC-Algorithm is treated as black box which delivers very useful features. The MFCC-algorithm divides the speech signal into overlapping frames with a window of sufficient size $win_{mfcc}$ to compute the features. This window is shifted with a time step over the signal. The computed features for each frame are vectors. The result of the MFCC-algorithm is therefore a time series $\{v[t]\}$ of vectors with time step $\delta t$:

$$\{v[t]\} = \{v(0), v(\delta t), v(2\delta t), \ldots, v(n\delta t)\} \tag{9}$$

Here $n$ is the number of frames. The vector $v(i) = (v_1, \ldots, v_m)$ contains the $m$ extracted features of the $i$-th frame. This time series can be represented as feature matrix $V \in \mathbb{R}^{n \times m}$ whereby the $i$-th row is the $i$-th element of the time series. A typical value for $\delta t$ is 10 ms and since utterances have usually a duration of a few seconds, that results in feature matrices with several hundred rows. One simple approach to avoid this problem is to reduce the feature matrix to a vector $\overline{v}$ by computing the mean of all rows of $V$ (cf. [13]):

$$\overline{v} = \frac{1}{n}(\textstyle\sum_{i=0}^{n} v_{i1}, \ldots, \sum_{i=0}^{n} v_{im}) \tag{10}$$

But thereby the number of data points are the number of utterances and this is a sub-optimal solution, because one of the main problems of speech-based emotion recognition systems are the small number of available patterns [14]. Another problem is, that by building the mean of the time series much time dependent information are lost. Since emotions are generally expressed over time in an utterance, this could make it more difficult to recognize the emotion. Therefore we compute the mean over each $1 < l < n$ rows and get a new feature matrix $U \in \mathbb{R}^{\lfloor \frac{n}{l} \rfloor \times m}$ whose rows build a new time series $\{u[t]\}$ with time step $l$. Then each element $u(t_i)$ represents a section of $win_{mfcc} \cdot l$ ms of the origin speech signal $\{x[t]\}$.

# 5    Classification of Time Series

## 5.1    Windowing for Deep Feed Forward Neural Networks

Unlike RNNs feed forward neural networks are not constructed to process sequential data like time series. One method to take time distributed information into account is to window the time series before using it as input for the neural network. This is achieved by sliding a window $win_i$ of size $k$ over several steps along the times series $\{u[t]\}$. In this way a section of $k$ elements of $\{u[t]\}$ are concatenated to a new vector $x_{i-1} = [u(t_{(i-1)s}); \ldots; u(t_{k-1})]$ whereby $i$ is the number of the window and $s$ the step size.

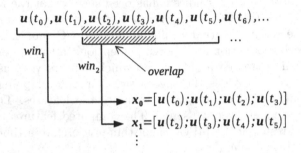

**Fig. 3.** Two steps of windowing the time series $\{u[t]\}$ with a window size of $k = 4$ and step size of $s = 2$.

As it can be seen in Fig. 3 if $s < k$ an overlap of $k - s$ elements in the newly generated data $x_i$ is created. By this overlap the neural network receives some information more than once.

## 5.2    Voting Based Classification

To classify the utterance $\{u[t]\}$ with the label $y_j$ a majority voting scheme can bee used to predict a class $c_l$ based on the associated windowed data set $\{(x_1, y_j), \ldots (x_m, y_j)\}$ as follows:

$$c_l = \arg \max_{l=1\ldots N} V_{c_l} \tag{11}$$

$V_{c_l}$ are the votes for class $c_l$ and determined for all available classes $l = 1 \ldots N$ with:

$$V_{c_l} = \sum_{i=0}^{m} G(x_i, c_l) \quad \text{with} \quad G(x_i, c_l) = \begin{cases} 1 & \text{if } f(x_i, \theta) \text{ predicted } c_l \\ 0 & \text{else} \end{cases} \tag{12}$$

# 6    Datasets

The used data is a significant component to asses an emotion recognition system. In this paper the widely used Berlin Database of Emotional Speech (EmoDB) serves as dataset [16]. The EmoDB consists of professional audio-recordings of seven acted emotions (anger, boredom, disgust, fear, happiness, sadness, neutral) spoken by 10 different actors in 10 different sentences. Overall 535 recorded and labeled utterances are provided. As mentioned above we used MFCCs as speech features. This was done by using window size of $win_{\text{mfcc}} = 25\,\text{ms}$ and a time step of $\delta t = 10\,\text{ms}$ to extract the first twelve cepstral coefficients with the energy as 13-th feature per window. Furthermore the first and second order derivatives of the cepstral coefficients are computed for each feature vector $v(t) = (v_1, \ldots, v_{13})$ for $i = 1, \ldots, 12$:

$$\Delta v_i = \frac{1}{2}(v_{i+1} - v_{i-1}) \tag{13}$$

$$\Delta\Delta v_i = \frac{1}{2}(\Delta v_{i+1} - \Delta v_{i-1}) \tag{14}$$

$v_i$ is here the $i$-th element in the feature vector. As additionally statistical feature the standard deviation $\sigma_j$ was computed for each column $j = 1, \ldots, 37$ of the feature matrix $M \in \mathbb{R}^{n \times 37}$ as follows:

$$\sigma_j = \sqrt{\frac{1}{n}\sum_{i=1}^{n}(v_{ij} - \overline{v}_{m,j})^2} \tag{15}$$

Here $\overline{v}_{m,j}$ is the $j$-th element of the $m$-th mean vector over every $l$ rows of the feature matrix with $m = 1, \ldots, \lfloor \frac{n}{l} \rfloor$. Hence for each utterance a feature matrix $U \in \mathbb{R}^{\lfloor \frac{n}{l} \rfloor \times 74}$ was computed whereby $n$ is the number of feature vectors of the respective utterance. With this procedure we created two types of datasets $D_c$ and $D_p$. $D_c$ contains the means of each feature matrix and therefor consists of 535 feature vectors $x_i \in \mathbb{R}^{74}$. For $D_p$ we computed the means of each feature matrix partially over $l = 15$ rows as described in Sect. 4. After windowing the data with a window size of $k = 7$ and step size of $s = 1$ as described in Sect. 5 $D_p$ contains 6915 features vectors $x_i \in \mathbb{R}^{518}$.

## 6.1    Testing the Datasets with a Support Vector Machine

Because erroneous data can lead to wrong conclusion, we tested the generated data sets with a support vector machine with a radial basis kernel (SVM-RBF). Hence for a given training set $X \in \mathbb{R}^{n \times 74} \subset D$ the following classifier has to be learned:

$$f(x) = \sum_{i} \alpha_i y_i exp(-\gamma\|x - x_i\|_2) + b \tag{16}$$

The classifier is learned by solving the following optimization problem:

$$\text{minimize: } \frac{1}{2}\|\boldsymbol{w}\|^2 + C\sum_{i=1}^{n}\xi_i \tag{17}$$

$$\text{subject to: } y_i(\boldsymbol{w}^T\boldsymbol{x}_i + b) \geq 1 - \xi_i \ \ \forall i = 1,\dots,n \tag{18}$$

The tests were performed in a speaker independent scheme with leave-one-speaker-out cross-validation (LOSO) and a text independent scheme with leave-one-text-out (LOTO).

Table 1. Classification accuracies for SVM-RBF

| Dataset | LOSO accuracy (avg) | LOTO accuracy (avg) |
|---------|---------------------|---------------------|
| $D_c$ | 59.0% | 78.9% |
| $D_p$ | 65.5% | 79.0% |

To optimize the parameters $C$ and $\gamma$ of the SVM a grid-search was performed with $C = \{10^{-2}, 10^{-1}, 10^0, 10^1, \dots, 10^9\}$ and $\gamma = \{10^{-5}, 10^{-4}, \dots, 10^4\}$. Table 1 shows that with the windowed dataset $D_p$ of partially feature-means an improvement of 6.5% classification accuracy could be achieved for the LOSO. In contrast to that, for LOTO the difference between $D_c$ and $D_p$ is negligibly small. Since the SVM can be considered as model with a flat architecture, these results suggest that more time dependent information can lead to better results only for a speaker independent scheme, even without a deep architecture.

## 7 Results

In this section we describe the results of our experiments. Like in the work of Albornoz et al. [5] we evaluated the models with two kinds of cross validation. Once with leave-one-speaker-out (LOSO) and once with leave-one-text-out (LOTO). LOSO is possibly the severest challenge for a speech based emotion recognition system, because the model has to abstract from the speaker dependent properties in the speech signal. To solve this very non-linear classification problem a very complex function has to be learned. And this is exactly where deep architectures could help [20]. For LOTO a much lesser complex relation between input data and desired output has to be learned. In this case deep learning methods could provide a less benefit [5].

We investigated several neural network architectures which differ from the number of hidden layers and hidden units. In order to facilitate the examination all architectures shared the same setting of the hyper-parameters *learn rate* = 0.3, *learn rate decay* = 0.09, *momentum* = 0.9, *weight decay* = 0.007, *dropout* = 0.09, *mini batch size* = 64 and *epochs* = 55. To train the networks we used a variant of the backpropagation-algorithm based on Nesterovs Accelerated

Gradient Descent (cf. [19]). The weights for a given layer are initialized with random values $\in (-1, 1)$. We denote an architecture as $DNN_{i \times j}$ whereby i is the number of hidden layers and j the number of hidden neurons. We first used a shallow neural network $DNN_{1 \times 80}$ with one hidden layer and 80 hidden units as baseline by using the smaller Dataset $D_c \in \mathbb{R}^{535 \times 74}$. In the LOSO-scheme the $DNN_{1 \times 80}$ classified with 58.5 % accuracy (avg). With LOTO a substantially higher accuracy (avg) of 76.6 % is achieved.

## 7.1   Experiments with DNNs

In order to examine if an improvement can be obtained with more hidden layers in combination by using the larger dataset $D_p \in \mathbb{R}^{6195 \times 518}$, we investigated the following set of architectures:

$$\{DNN_{i \times j} \mid i \in \{1, 2, \ldots, 5\}, \ j \in \{100, 200, \ldots, 600\}\}$$

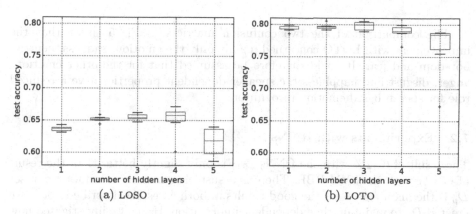

(a) LOSO

(b) LOTO

**Fig. 4.** Effect of adding more hidden layers to a DNN on classification accuracy with LOSO (a) and LOTO (b). Box plots shows the distribution of accuracies associated with 6 different numbers $j$ of hidden neurons per layer $j \in \{100, 200, \ldots, 600\}$.

Figure 4(a) shows that for LOSO additional hidden layers improved the classification accuracy (avg) until the 4th hidden layer. However with the 5th hidden layer the accuracy gets worse. Maybe the DNNs are then to big and need more data to prevent overfitting. The best result of 67.2 % with was achieved by the network $DNN_{4 \times 500}$ with 4 hidden layers and 500 neurons per layer. The confusion matrix in Fig. 5(a) shows that the best classification performance is significantly obtained whit the emotion angry. One reason for that could be the far larger number of samples of this class. Another reason could be the high energy in the speech signal of angrily spoken utterances. As assumed above and also observed in [5] with LOTO additional hidden layers did not improve the accuracy significantly as Fig. 4(b) shows. The highest result of 80.6 % is here achieved by the network $DNN_{2 \times 600}$ with 2 hidden layers and 600 hidden neurons per layer.

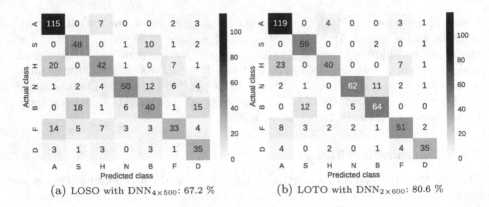

**Fig. 5.** Classification performance of the best DNNs of the individual classes illustrated as confusion matrix with A = Anger, B = Boredom, S = Sadness, H = Happiness, D = Disgust, F = Fear, N = Neutral.

The comparison of the two confusion matrices in Fig. 5 shows that the improvement with LOTO concerned mainly only the emotions sadness, neutral, boredom and fear. It could therefore be assumed that for the other emotions anger, disgust and happiness the speaker dependent properties have a reduced role for speech based emotion recognition.

## 7.2   Experiments with RNNs

The result of 67.2 % with the $DNN_{4 \times 500}$ is only slightly better than the result of 65.5 % with the SVM-RBF. There is a suspicion that the windowed dataset $D_p$ is the main reason for the good results in both cases. As described above we created $D_p$ to maintain time dependent information. Hence we investigated now Elman-Networks which are constructed to process time series. The depth of a RNN depends on the length of the time series to process. To vary the number of hidden layers we created therefor several datasets by computing the means of each features matrix over several number $l$ of rows as described in Sect. 4. Table 2 shows the smaller the averaged sections of the initial feature matrix, the longer the resulting time series and thus the more hidden layer in a respective unfolded RNN. The RNNs are trained via backpropagation through time (bptt) with the hyper-parameters *learn rate* = 0.004, *momentum* = 0.9 and *epochs* = 200. To asses the essential performance of the RNN for speech based emotion recognition no other optimization or regularization techniques are used. However to deal with the unbounded rectified linear function (cf. [21]) in the hidden layer we clipped the outputs $y = relu(x)$ of the hidden neurons as follows:

$$y = \begin{cases} y & \text{if } y \leq x_{\max} \\ x_{\max} & \text{else } y > x_{\max} \end{cases} \qquad (19)$$

In some way $x_{max}$ can also be viewed as hyper-parameters and we chose the value $x_{max} = 40$. We performed the tests again with both validation schemes LOSO and LOTO.

**Table 2.** Relationship between averaging and length of resulting time series

| Avg. over $l$ rows | $\approx$ Length of time series |
|---|---|
| 30 | 10 |
| 60 | 5 |
| 90 | 4 |
| 120 | 3 |
| 150 | 2 |
| 180 | 2 |

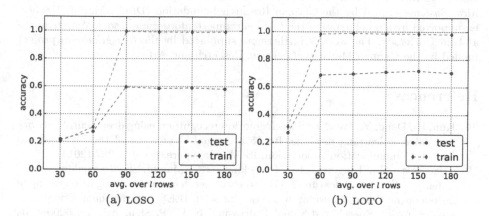

(a) LOSO                    (b) LOTO

**Fig. 6.** Effect of using longer time series as input for a RNN on classification accuracy (test, train) with LOSO (a) and LOTO (b). The X axis shows the respective number $l$ of averaged rows of the feature matrices.

Interestingly as with the DNN-experiments the highest classification accuracy (here: 59.5 %) with LOSO is again obtained by using a network with 4 unfolded hidden layers ($l = 90$) as it can be seen in Fig. 6(a). In view of the fact that only a basic version of the bptt-algorithm is used the result of 59.5 % accuracy with LOSO is highly promising. With LOTO the highest accuracy of 71.8 % is achieved by using the network with 2 hidden layers ($l = 150$). This suggests again that for LOTO deeper architectures may not improve the result significantly.

## 8    Conclusion

The performed experiments with the partially averaged and windowed time series suggest adding more hidden layers could improve the performance of feed forward neural networks for recognizing emotion in speech. The performance of the support vector machine also improved considerably when using the partially averaged feature matrices whereby the classification accuracy is only 1.7 % worse than with the DNN. Hence there is a suspicion that the method of preprocessing that we were using is the main reason for the improved results. To assess this more precisely more experiments have to be conducted.

Highly promising are the results with the RNNs, because only with a basic version of the backpropagation through time algorithm a accuracy of about 60 % is obtained. It would be interestingly to investigate if a significantly better performance can be obtained with several regularization and optimization techniques.

**Acknowledgment.** This paper is based on work done within the Transregional Collaborative Research Centre SFB/TRR 62 *Companion-Technology for Cognitive Technical Systems* funded by the German Research Foundation (DFG). Markus Kächele is supported by a scholarship of the Landesgraduiertenförderung Baden-Württemberg at Ulm University. The work is furthermore supported by the *SenseEmotion* project funded by the German Ministry of Science, Research and Arts.

## References

1. Kun, H., Dong, Y., Ivan, T.: Speech emotion recognition using deep neural network and extreme learning machine. In: 15th Annual Conference of the International Speech Communication Association, ISCA, Singapore, pp. 223–227 (2014)
2. Trigeorgis, G., Ringeval, F., Brueckner, R., Marchi, E., Nicoalou, A.M., Zafeiriou, S.: Adieu features? End-to-end speech emotion recognition using a deep convolutional recurrent network. In: 41st IEEE International Conference on Accoustics, Speech and Signal Processing, ICASSP, Shanghai, pp. 5200–5204 (2016)
3. Kim, Y., Lee, H., Provost, E.M.: Deep learning for robust feature generation in audiovisual emotion recognition. In: IEEE International Conference on Acoustics, Speech, and Signal Processing, Vancouver, pp. 3687–3691 (2013)
4. Hinton, G.E., Osinderos, S., The, Y.W.: A fast learning algorithm for deep belief nets. Neural Comput. **18**, 1527–1554 (2006). MIT Press, Cambridge
5. Albornoz, E.M., Sánchez-Gutiérrez, M., Martinez-Licona, F., Rufiner, H.L., Goddard, J.: Spoken emotion recognition using deep learning. In: Bayro-Corrochano, E., Hancock, E. (eds.) CIARP 2014. LNCS, vol. 8827, pp. 104–111. Springer, Heidelberg (2014)
6. Huang, C., Gong, W., Fu, W.: A research of speech emotion recognition based on deep belief network and SVM. Math. Probl. Eng. **2014**, 1–7 (2014). Beijing, Article ID 749604
7. Glorot, X., Bordes, A., Bengio, Y.: Deep sparse rectifier neural networks. In: Proceedings of the 14th International Conference of Artificial Intelligence and Statistics, JMLR Proceedings, Fort Lauderdale, pp. 315–323 (2011)

8. Maas, A., Hannun, A., Ng, A.: Rectifier nonlinearities improve neural network acoustic models. In: ICML Workshop on Deep Learning for Audio Speech, and Language Processing, JMLR, Atlanta (2013)
9. Goodfellow, I., Bengio, Y., Courville, A.: Deep Learning. MIT Press (2016, in preparation). http://www.deeplearningbook.org
10. Elman, J.L.: Finding structure in time. Cogn. Sci. **14**, 179–211 (1990). Wiley
11. Werbos, P.: Backpropagation through time: what does it do and how to do it. Proc. IEEE **78**, 1550–1560 (1990)
12. Koolgaudi, S.G., Rao, K.S.: Emotion recognition from speech: a review. Int. J. Speech Technol. **15**, 99–117 (2012). Springer
13. Ma, Z., Fokoué, E.: A comparison of classifiers in performing speaker accent recognition using MFCCs. Open J. Stat. **4**, 258–266 (2014). Scientific Research Publishing Inc
14. Mohino-Herranz, I., Gil-Pita, R., Alonso-Diaz, S., Rosa-Zurera, M.: MFCC based enlargement of the training set for emotion recognition in speech. Signal Image Process. Int. J. **5** (2014)
15. Cowie, R., Douglas-Cowie, E., Tsapatsoulis, N., Votsis, G., Kollias, S., Fellenz, W., Taylor, J.: Emotion recognition in human-computer interaction. IEEE Signal Process. Mag. **18**, 32–80 (2001). IEEE
16. Burkhardt, F., Paeschke, A., Rolfes, M., Sendlmeier, W., Weiss, B.: A database of German emotional speech. In: Proceedings of Interspeech, Lissabon, pp. 1517–1520 (2005)
17. Lee, H., Pham, P., Largman, Y., Ng, A.Y.: Unsupervised feature learning for audio classification using convolutional deep belief networks. In: Advances in Neural Information Processing Systems, vol. 22, pp. 1096–1104. Curran Associates Inc., Vancouver (2009)
18. Krizhevsky, A., Sutskever, I., Hinton, G.E.: ImageNet classification with deep convolutional neural networks. In: Advances in Neural Information Processing Systems, vol. 25, pp. 1097–1105. Curran Associates Inc., Nevada (2012)
19. Sutskever, I., Martens, J., Dahl, G.E., Hinton, G.E.: On the importance of initialization and momentum in deep learning. In: Proceedings of the 30th International Conference on Machine Learning, Atlanta, pp. 1139–1147 (2013)
20. Larochelle, H., Bengio, Y., Louradour, J., Lamblin, P.: Exploring strategies for training deep neural networks. J. Mach. Learn. Res. **10**, 1532–4435 (2009). JMLR.org
21. Le, Q., Jaitly, N., Hinton, G.E.: A simple way to initialize recurrent networks of rectified linear units. CoRR (2015)
22. LeCun, Y.A., Bottou, L., Orr, G.B., Müller, K.-R.: Efficient backprop. In: Orr, G.B., Müller, K.-R. (eds.) NIPS-WS 1996. LNCS, vol. 1524, pp. 9–50. Springer, Heidelberg (1998)
23. Kächele, M., Zharkov, D., Meudt, S., Schwenker, F.: Prosodic, spectral and voice quality feature selection using a long-term stopping criterion for audio-based emotion recognition. In: Proceedings of the International Conference on Pattern Recognition (ICPR), pp. 803–808 (2014)

# On Gestures and Postural Behavior as a Modality in Ensemble Methods

Heinke Hihn, Sascha Meudt$^{(\boxtimes)}$, and Friedhelm Schwenker

Institute for Neural Information Processing, Ulm University, 89069 Ulm, Germany
{heinke.hihn,sascha.meudt,friedhelm.schwenker}@uni-ulm.de

**Abstract.** Knowledge about the users emotional state is important to achieve human like, natural HCI in modern technical systems. Humans rely on body gestures and posture when communicating. We investigate the relation between gestures and human emotion, specifically when completing tasks. The main focus of this work lies on discriminating between mental overload and mental underload, which can e.g. be useful in an e-tutorial system. Mental underload is a new term used to describe the state a person is in when completing a dull or boring task. It will be shown how to select suited features, such as gestures, movement and postural behavior. Furthermore those features will be investigated regarding their discriminative power. After features are selected, a multiple classifier system will be designed, trained and evaluated.

## 1 Introduction

A fundamental part of human communication is noticing a change in the affective state of the conversational partner. Affective state refers to the experience of feelings or emotions. To elaborate on this more, consider the following scenario: A person is telling another about a rather complex topic, e.g. in an teacher-student setting. During this conversation the student starts to look a bit overwhelmed by all the new information. In this case one would expect the teacher to change his pace as the student obviously can't follow up. Let that state the student is experiencing henceforth be referred to as *mental overload*. This term is meant to describe the state one is in when being confronted with a very complex task, e.g. understanding something completely new. The opposite, i.e. completing an easy task or listening to a teacher talking about a already well known topic, shall be called *mental underload*. In terms of the student-teacher example one can consider a electronic tutorial platform which controls its pace depending on the student's behavior. A user centered system should offer possibilities for the user to express their emotions [10,16]. Based on human interaction one can imagine two ways: verbal and non-verbal. Verbal communication focuses on information retrieved from speech. These can be loudness and pitch or the words being sad. There has been a lot of research in this area [2,9,11]. Non-verbal ways of expressing feelings can be facial expressions and gestures, to name a few. While facial expressions have been researched very thoroughly in the past [6,12,13,15], the same doesn't quite hold for gestures. Even tough they play a crucial part

© Springer International Publishing AG 2016
F. Schwenker et al. (Eds.): ANNPR 2016, LNAI 9896, pp. 312–323, 2016.
DOI: 10.1007/978-3-319-46182-3_26

in human-to-human communication they have been only used little compared to other modalities in Affective Computing [17], e.g. in [7,8]. This work aims to close the gap and develop a method to employ postural behavior and gestures as a powerful additional modality. Specifically to distinguish between mental overload and underload, as described earlier.

## 2   Related Work

Kapur et al. [7] conducted a study in 2005 on gesture based affective computing. Their goal was to train a classifier to distinguish between the four basic emotions *Sad, Joy, Anger* and *Fear*. The authors equipped five actors with markers of a motion capture system and asked them to "perform" given emotions. After collecting the data 10 participants were asked to identify each emotion only by watching the moving points, i.e. the position of the markers. In a next step the authors compute mean of velocity and acceleration and the standard deviations of positions. The resulting data is then used to train and evaluate several classifiers. The participants achieved an average recognition rate of 93 %, while the classifiers achieved between 66.2 % and 91.8 %.

Kipp and Martin [8] investigate four basic gestural features and their respective relation to the emotional state. Their main goal was to create embodied conversational agents, i.e. defining a set of gestures to discriminate emotions. They introduced *lexemes* to describe gestures by a set of constraints on *handedness, hand shape, palm orientation* and *motion direction*. Data was gathered from the movie *Death of a Salesman* (1966 DS-1 and 1985 DS-2). In order to estimate the correlation between emotion category and gestures, the authors computed pairwise $\chi^2$ values. Results suggested a highly significant correlation between emotion category and handedness ($\chi^2 = 40.14$; p < .001, in film DS-1, $\chi^2 = 35.37$; p < .001, in film DS-2). They also found for the film DS-2 a correlation between emotion category and palm orientation ($\chi^2 = 42.50$; p < .05).

Bianchi-Berthouze et al. conducted a study on posture and gesture and immersion [1]. They focused on two things: is there a relationship between postural behavior and immersion and the importance of full-body control to improve user experience. High immersion occurs when the participant has the perception of being physically present in a virtual reality. Twenty participants were randomly assigned to two groups: a simple point and click game and a first person shooter. The authors hypothesized that the players in group two experienced a higher lever of immersion. After playing 10 min the players were asked to complete a *Immersion Questionnaire* [5] to quantify the level of immersion. Group 1 returned rather low immersion scores (mean 47.1, $\sigma^2 = 16.64$). They also showed many shifts in sitting position, e.g. from a very relaxed to a very attentive pose. Group 2 showed significantly higher scores (mean 68.11, $\sigma^2 = 11.95$). Movement in this group occurred fewer, with players scoring lower in immersion showing more movement. The authors argue that the results suggest that higher immersion causes fewer unnecessary movements. They also infer that the observed reduction in movement is caused by the higher engagement, i.e. players in group 2 are more focused.

## 3   Experimental Setting

The dataset is based on an experiment conducted within the Transregional Collaborative Research Centre SFB/TRR 62 "Companion-Technology for Cognitive Technical Systems". Participants were asked to play a series of games based on the interaction paradigm of Schüssel et al. [14]. The task of each game sequence was to identify the singleton element, i.e. the one item that is unique in shape and color(see Fig. 1). The participants interacted with the system by speech. The difficulty was set by adjusting the number of shapes and the time to answer. If the given answer was incorrect, the player received no reward for that particular round. After a introduction each participant completed four game sequences of decreasing difficulty. The first sequence was designed to induce overload ($6 \times 6$ board, 6 s to answer, see Fig. 1), the second was $5 \times 5$ with 10 s, the third was set to $3 \times 3$ with 100 s, sequence four was a $3 \times 3$ mode with 100 s (underload). The last sequence induces frustration, e.g. by purposely logging in a wrong answer. As the sequences 1 and 4 are explicitly designed to cause over- and underload, we focused only on those two. After each sequence the participants answered a self assessment questionnaire (SAM). The aim of those questions was to determine valence, arousal and dominance experienced in the particular sequence. A total of 52 participants were recorded. Of those were 26 male and 26 female. Their age spanned from 17 to 27 (mean 21.66, $\sigma^2 \approx 2.7$). During the experiment participants were monitored by several sensors. This work focuses on the depth data provided by a Kinect sensor to compute body movements and postural behavior. The skeleton is extracted by the Kinect itself. We do not employ any extraction algorithm.

**Fig. 1. Experimental setting. Left**: Front view. **Middle**: 3D projection of the data provided by the Kinect sensor. The depth is color coded, such that green indicates near objects and rad objects further away. **Right**: Rear view. The red rectangle indicates the Kinect. (Color figure online)

## 4    Feature Engineering

By watching the recordings a couple of mentionable static gestures were found. For each of those a set of constraints was defined:

**Arms crossed**: the right hand is near the left elbow and vice versa.
**Hands behind back**: both hands are not visible to the Kinect.
**Resting hand on hips**: the right hand is near the right hip and vice versa.
**Crossed feet**: the left foot is right of right foot or vice versa.
**Feet in front of another**: the left foot is closer to the Kinect then the right one or vice versa.

The occurrences are almost identically distributed and therefore bare only little discriminative power. To overcome this, they have to be combined with further information. We chose to enhance the features by adding the duration. One drawback of this approach is that suitable thresholds for the constraints have to be defined. Assuming a threshold for a given person is found, the threshold doesn't necessarily apply to other persons just as well. To avoid setting thresholds the mean distance of the respective joints over the set of frames is computed. Figure 4 gives an example. Again, by watching the video material and observing the participants' behavior two main linear movements have been identified: moving both hands away from the torso and scratching the head or face. The latter occurred about evenly in both sequences. "Moving both hands away from the torso" is mostly done in combination with confused facial expression as Fig. 2 shows (this might be very useful in fusion paradigms, i.e. in combination with a facial expression detector). It seems to be a rather clear indicator whether a

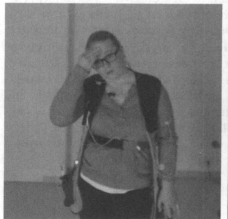

**Fig. 2. Examples of participant behavior. Left**: The participant is scratching their head. This occurred about evenly during overload and underload. **Right**: The participant is moving their hand away. This occurred far more often during overload. During underload it only occurred when the participant gave a wrong answer.

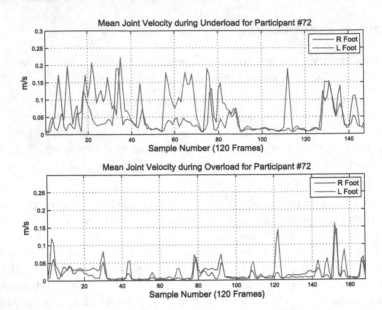

**Fig. 3. Extraced features. Top**: Mean joint velocities during underload. **Bottom**: Mean joint velocities during Overload. It can be seen that participant #72 had a rather active feet movement during underload while their feet were mostly standing still during overload.

person is experiencing over- or underload, because it occurs mostly during overload. We captured this gesture by computing mean joint distances (e.g. between both hands, hand and hip), velocity, and acceleration. For each joint the values were computed within a window of 120 frames. Figure 3 gives an example. Joint velocity is computed as the first derivative of the joint positions $(x, y, z)$ w.r.t. time: $\bar{v} = \frac{\Delta \mathbf{s}}{\Delta t}$. Joint acceleration is computed by approximating the derivative of the joint velocities, i.e. $\bar{a} = \frac{\Delta \mathbf{v}}{\Delta t}$. To account for varying movement within the frame the standard deviation of velocity and acceleration are also computed. Additionally, most participants showed a rather highly active head movement. The Kinect sensor measures the rotation angles of the head in yaw, pitch, and roll notation. The yaw angle can be used to detect whether a participant is looking at the camera or not. Pitch angle indicates movement towards the floor or the ceiling and roll angle is measuring head tilt. To capture those movements a threshold for each angle is defined.

## 5   Results

We focus on person dependent classification. This type of classification refers to training a classifier such that it fits well to a given person. A Random Forest [3,4] (RF) of 200 trees was trained and evaluated with a 10-fold cross validation for each participant. We choose RFs because they can be trained and evaluated easily and can handle large amounts of data well. Furthermore they do not

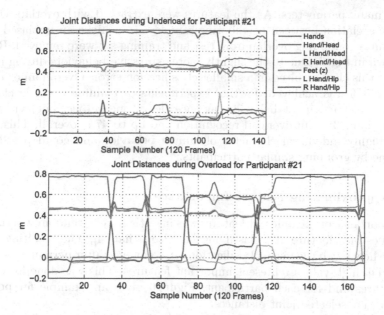

**Fig. 4. Examples for postural behavior.** Mean distances (window 120 frames, 60 frames overlap) between selected joints during underload (top) and overload (bottom). Negative values e.g. indicate that the left hand is below the left hip.

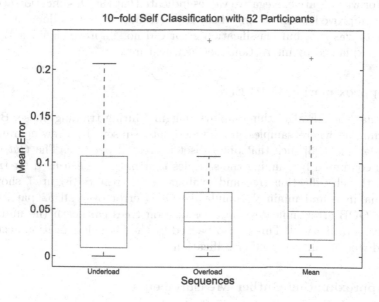

**Fig. 5. Person dependent classification results.** Each group contains 52 samples, i.e. the number of participants. Each sample represents the mean classification accuracy obtained by a 10-fold cross validation.

require many parameters. As the features were extracted with overlap, there is a high correlation between neighboring samples. To achieve an unbiased result the features containing overlap from test and training set were removed. Results are shown in Fig. 5. Due to the slightly more expressive behaviors during mental overload this class resulted in a smaller classification error. Overall an error rate of about 5 % was achieved. Further experiments of training a single classifier matching all participant behaviors at once (leave one subject out) yielded an error of about 38 % in overload recognition and up to 47 % overall. This is due to the highly individual character of human behavior and could possibly be overcome by grouping similar participants.

## 5.1   Approximating Feature Importance

The importance of the different features was investigated by randomly permuting the values of a feature and classifying based on that [4]. The resulting values indicate how much the mean classification error changed after permutation (*delta error*), i.e. high values represent important features. This was done for each of the features and for each participant. Figure 6 gives an example for postural behavior, i.e. selected joint distances.

Moving direction of the joints seems to be important for some participants and for others it's not. Tests without this feature did not yield significantly higher accuracies. Head positions yielded rather low importance measurements. In fact, the values obtained were the lowest of all features and for some participants the delta error was negative. Negative values indicate that the classification accuracy could be improved when leaving those features out. This was done and a new set of tests was run, but classification error did not improve significantly. This could be explained by interactions between features.

## 5.2   Approximating OOB Error

Recall, that the RF algorithm employs bagging during training phase. Bagging is a technique where samples are divided into subsets by drawing randomly with replacement [3] such that one subset is created per tree. The *Out Of Bag Error* is computed by running the samples that haven't been used for training through the classification tree and evaluating the results. Figure 7 shows the mean, maximal and minimal cumulative OOB error over all 52 participants. The low OOB error indicates the classification trees can learn the underlying distribution rather well. This is also backed by the high classification accuracies obtained when evaluating self classification.

## 5.3   Approximating Outlier Measurements

Outliers in RFs can be found by first computing the proximity of the data and then averaging by the number of trees. Proximity between two observations is defined as the fraction of trees in the ensemble for which these two observations

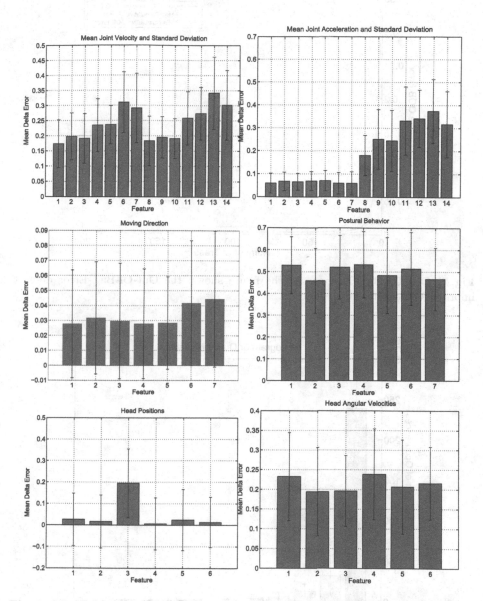

**Fig. 6. Mean feature importance and respective standard deviation.** Delta error refers to the change in classification error made when permuting values of a given feature. High values indicate important features and low values less important features.

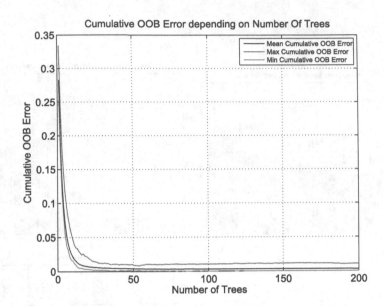

**Fig. 7. Out-Of-Bag Errors.** As the figure shows, the Out-Of-Bag Error doesn't improve further after about 50 trees.

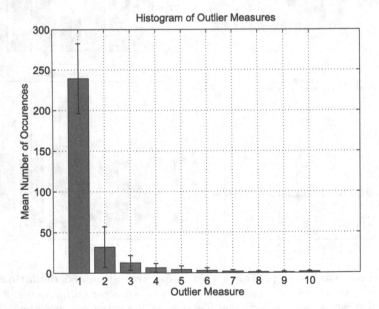

**Fig. 8. Outlier measurement and respective standard deviation.** The samples within one participant contain only little outliers. Note that the outlier value has been omitted and only the bin numbers are shown.

land on the same leaf [4]. Outliers can then be found by taking the squared inverse proximity of a given sample and compare that value with the squared inverse proximity of the remaining samples. A high value indicates this sample is an outlier. Figure 8 shows the histograms of outlier measurements over all 52 participants and the corresponding standard deviations. As most of the samples have a low measurement (first bin), the samples within each participant are very similar.

## 5.4   Comparing Different Ensemble Members

The previous section evaluated RFs. To get an idea of how well that classifier compares to others and how they influence classification accuracy, several tests were run using different classifiers. Figure 9 shows the results. A 3-NN ensemble, a Linear Discriminant Analysis (LDA) ensemble, and a mix of both were trained using the random subspace method, which is a generalization of the RF algorithm. It operates on a random subset of the feature space. In this case it was set to $m = \lfloor log_2(M) \rfloor$, as suggested by Breiman in is original paper. We chose these two classifiers because they are trained similarly to RF and don't require many parameters.

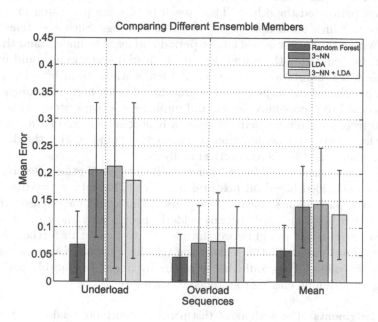

**Fig. 9. Comparing ensemble members.** As the figure shows, the RF performs well compared to the others.

# 6 Conclusion

It was shown that there is indeed a relation between postural behavior and mental over- and underload. Specifically, the findings suggested mental overload is in most cases accompanied by a rather high physical arousal, i.e. a lot of movement. This was then used as a basis to identify suitable features. These were mean joint velocity, acceleration, and distances between several selected joints. Additionally, the respective motion direction is important. Head movement has been captured by computing rotational velocities on each axis and six predefined positions. To prove the usefulness of the feature an extensive analysis was conducted. In particular, the training error (OOB error), the sample outlier measurement and the feature importance have been investigated. The analysis of the OOB error and the outlier measurement showed the features do indeed separate the samples well into mental overload and underload for each participant. For each of the features their respective discriminative power was also approximated. The results indicated mean joint velocities and accelerations bare the most information and head position and moving direction the least. In a last step the model itself has been evaluated to prove the RF algorithm is indeed the best fitting choice for this task. To achieve this the classification results were compared to ensembles of 3-NN, LDA, and a mix of both classifiers. Comparing the results revealed the RF outperformed the others. The overall results are promising in terms of HCI systems which are adaptable to the users bearings. Such a systems would interact with a single user over a longer period and learn to understand the users behavior. Remembering the initial described tutorial system one could imagine a systems which assists a student over at least a whole term by adapting the teaching pace based on the presented approaches. The theoretical findings of this work also used to successfully design and implement a live system. This system is able to record a given participant using a Kinect sensor, extract features and classify those features. Classification is achieved by training with the data from the participants used in the theoretical analysis.

For future works it could be of interest to investigate if other affective states can also be classified based on postural behavior. The data we used was from rather young participants (mean 21.66 years). This could have biased the gestures we found and analyzed, because elderly people may show different (less expressive) gestures. It would be worthwhile to investigate this further. Another drawback of our method is the rather long training phase, because we assumed a person dependent. This could be overcome by finding a suitable participant grouping and training classifier systems for each group.

**Acknowledgments.** The authors of this paper are partially funded by the Transregional Collaborative Research Centre SFB/TRR 62 "Companion-Technology for Cognitive Technical Systems" funded by the German Research Foundation (DFG).

# References

1. Bianchi-Berthouze, N., Cairns, P., Cox, A., Jennett, C., Kim, W.W.: On posture as a modality for expressing and recognizing emotions. In: Emotion and HCI Workshop at BCS HCI London (2006)
2. Breazeal, C., Aryananda, L.: Recognition of affective communicative intent in robot-directed speech. Auton. Robots **12**(1), 83–104 (2002)
3. Breiman, L.: Bagging predictors. Mach. Learn. **24**(2), 123–140 (1996)
4. Breiman, L.: Random forests. Mach. Learn. **45**(1), 5–32 (2001)
5. Cairns, P., Cox, A., Berthouze, N., Jennett, C., Dhoparee, S.: Quantifying the experience of immersion in games (2006)
6. Caridakis, G., Malatesta, L., Kessous, L., Amir, N., Raouzaiou, A., Karpouzis, K.: Modeling naturalistic affective states via facial and vocal expressions recognition. In: Proceedings of the 8th International Conference on Multimodal Interfaces, pp. 146–154. ACM (2006)
7. Kapur, A., Kapur, A., Virji-Babul, N., Tzanetakis, G., Driessen, P.F.: Gesture-based affective computing on motion capture data. In: Tao, J., Tan, T., Picard, R.W. (eds.) ACII 2005. LNCS, vol. 3784, pp. 1–7. Springer, Heidelberg (2005)
8. Kipp, M., Martin, J.-C.: Gesture and emotion: can basic gestural form features discriminate emotions? In: 3rd International Conference on Affective Computing and Intelligent Interaction Workshops, pp. 1–8. IEEE (2009)
9. Meudt, S., Zharkov, D., Kächele, M., Schwenker, F.: Multi classifier systems and forward backward feature selection algorithms to classify emotional coloured speech. In: Proceedings of the 15th ACM on International Conference on Multimodal Interaction, pp. 551–556. ACM (2013)
10. Picard, R.W.: Affective Computing, vol. 252. MIT Press, Cambridge (1997)
11. Plesa-Skwerer, D., Faja, S., Schofield, C., Verbalis, A., Tager-Flusberg, H., Dykens, E.M.: Perceiving facial and vocal expressions of emotion in individuals with williams syndrome. Am. J. Ment. Retard. **111**(1), 15–26 (2006)
12. Russell, J.A., Bachorowski, J.-A., Fernández-Dols, J.-M.: Facial and vocal expressions of emotion. Annu. Rev. Psychol. **54**(1), 329–349 (2003)
13. Schels, M., Glodek, M., Meudt, S., Scherer, S., Schmidt, M., Layher, G., Tschechne, S., Brosch, T., Hrabal, D., Walter, S., et al.: Multi-modal classifier-fusion for the recognition of emotions. In: Coverbal Synchrony in Human-Machine Interaction (2013)
14. Schüssel, F., Honold, F., Bubalo, N., Huckauf, A., Traue, H., Hazer-Rau, D.: In-depth analysis of multimodal interaction: an explorative paradigm. In: Kurosu, M. (ed.) HCI 2016. LNCS, vol. 9732, pp. 233–240. Springer, Heidelberg (2016). doi:10. 1007/978-3-319-39516-6_22
15. Shan, C., Gong, S., McOwan, P.W.: Robust facial expression recognition using local binary patterns. In: IEEE International Conference on Image Processing, ICIP, vol. 2, pp. II-370. IEEE (2005)
16. Wendemuth, A., Biundo, S.: A companion technology for cognitive technical systems. In: Esposito, A., Esposito, A.M., Vinciarelli, A., Hoffmann, R., Müller, V.C. (eds.) COST 2102. LNCS, vol. 7403, pp. 89–103. Springer, Heidelberg (2012)
17. Zeng, Z., Pantic, M., Roisman, G.I., Huang, T.S.: A survey of affect recognition methods: audio, visual, and spontaneous expressions. IEEE Trans. Pattern Anal. Mach. Intell. **31**(1), 39–58 (2009)

# Machine Learning Driven Heart Rate Detection with Camera Photoplethysmography in Time Domain

Viktor Kessler$^{(\boxtimes)}$, Markus Kächele, Sascha Meudt, Friedhelm Schwenker, and Günther Palm

Institute of Neural Information Processing, Ulm University, Ulm, Germany
{viktor.kessler,markus.kaechele,sascha.meudt,
friedhelm.schwenker,guenther.palm}@uni-ulm.de

**Abstract.** Measuring bio signals such as the heart rate in non medical applications is gaining an increasing importance. With camera based photoplethysmography (PPG) it is possible to measure the heart rate remotely with built in webcams of every tablet and laptop. Recent research with machine learning based methods showed great success compared to signal processing based methods. In this paper, we use k-nearest neighbor (kNN) and multilayer perceptron (MLP) with an alternative representation of the input vector. Estimating the quality of peaks with a Gaussian distribution could further improve the detection. Overall we could improve the root mean square error (RMSE) from 23.97 to 8.62.

**Keywords:** Photoplethysmography (PPG) · remote Photoplethysmography (rPPG) · Camera · Webcam · k-nearest neighbor · Neural network · Gaussian distribution

## 1 Introduction

Most tablets and laptops are equipped with a front camera and are often used for hours every day. For health care applications this is an interesting means to monitor the health of a person. Several works in the last ten years showed that camera based photoplethysmography (PPG) can be used to remotely measure bio signals such as the heart rate. This allows a long term monitoring system which does not interfere with the user in his/her daily work and at the same time does not need a daily scheduled, explicit measurement. Advances in signal processing based measurement methods for camera based PPG as well as camera technology enables better detection rates but until now not reliable for serious applications.

Machine learning technology methods show great success in replicating systematic occurrences. However, until now only few learning based approaches were presented. One of the early works was presented by Lamonaca et al. [7]. They used a neural network to evaluate the blood pressure from facial videos recorded with a smartphone camera and its flashlight. Hsu et al. [6] used support

© Springer International Publishing AG 2016
F. Schwenker et al. (Eds.): ANNPR 2016, LNAI 9896, pp. 324–334, 2016.
DOI: 10.1007/978-3-319-46182-3_27

vector regression (SVR) in the frequency domain to detect the heart rate. They showed three times better results than a pure signal processing based method. Maaoui et al. [8] used a support vector machine (SVM) and seven features from time and frequency domain with the aim of detecting the stress level.

The remainder of this paper is organized as follows. In Sect. 2 the generation of the signal for the detection is explained. This includes skin extraction, signal filtering and detection of the heart rate. Two machine learning algorithms, k-nearest neighbors (kNN) and multilayer perceptron (MLP), are described in Sect. 3 and analyzed in Sect. 4 on the Open_EmoRec_II dataset [11]. A conclusion follows in Sect. 5.

## 2  Signal Extraction

### 2.1  Region of Interest (ROI) Detection

The interesting content of a video for this work is the face of a participant within which we measure the heart rate. We use the Cambridge face tracker [1] implemented by Baltrušaitis et al. [2] to detect the face in each video frame. This detector returns multiple facial landmarks for the eyes, eyebrows, nose, mouth and the lower part of the contour of the face (Fig. 1(a)). Then the area corresponding to the face is estimated by mirroring the contour on an axis of reflection through the eyes. A convex hull over the lower and mirrored contour determines the outermost landmarks (Fig. 1(b)).

Parts of the face which contains skin (like cheeks and forehead) are more suitable for heart rate estimation in comparison to eyes or mouth. These parts are detected within the convex hull with the skin detection Algorithm of Mahmoud [9]. Therefore the frame is converted from RGB color space into the Y'CbCr color space and all pixels within the value space

$$Y' > 80$$
$$77 \leq C_b \leq 127$$
$$133 \leq C_r \leq 173$$

will persist in the frame (Fig. 1(c)).

### 2.2  Signal Preprocessing

Cui et al. [12] showed that the strongest heart rate signal of PPG systems is in the green light wavelength. Therefore we consider only the green color channel in this work (Fig. 1(d)). The raw pulse signal $X_{raw}^G(t)$ is extracted by computing the average green color value of the skin (the ROI, Fig. 1(d)). The discrete time $t$ is the frame rate of the video. The raw signal $\vec{X}_{raw}^G$ will be bandpass filtered with a zero-phase digital filter and third order butterworth coefficients [10]. The resulting filtered signal is denoted $\vec{X}_{filtered}^G$. The desired frequency range is set to [0.5, 3.0] Hz ([30, 180] bpm). The zero-phase filtering removes the bias and the

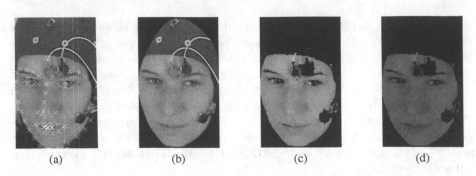

(a)              (b)              (c)              (d)

**Fig. 1.** For each frame (a) the face is detected and (b) the ROI is limited to the face. Then (c) the skin is extracted and (d) the mean green color is used for heart rate detection (Sect. 2.2). (Color figure online)

trend from the signal. Complementary information regarding the filtering can also be found in [12] (Sect. 2.3).

The ground truth signal $\vec{X}_{GT}$ is commonly recorded with a electrocardiograph or a blood volume pulse (BVP) sensor. In this work we will use a finger BVP sensor which records the photoplethysmographic signal in the fingertip. The correctness of the ground truth heart rate for our dataset is evaluated.

## 2.3  Heart Rate Detection

The time-domain based technique detects the peaks (corresponding to the heart beats) in the filtered signal $\vec{X}^G_{\text{filtered}}$. This Method will further be denoted as PEAK. For the peak detection we use the algorithm implemented by Carbajal [3]. The detected timestamps of the peaks are defined as $\vec{R}$ and the distances of adjacent timestamps as $\vec{RR}$. The average heart rate $f_{HR}$ is calculated by dividing the frame rate $fps$ through the median peak distance $\widetilde{\vec{RR}}$. In case of the average heart rate of a window segment, the peak timestamps $\vec{R}$ are restricted to the window segment:

$$f_{HR} = \frac{fps * 60}{\widetilde{\vec{RR}}_{\text{segment}}}, \quad (1\ \text{bpm} = 1/60\ \text{Hz}) \tag{1}$$

The heart rate $f_{HR}$ is measured in beats per minute (bpm) while the right side of the equation is measured in Herz (Hz). We convert Hz into bpm by multiplying with 60.

Our primary error measurement method is the root mean square error (RMSE). It compares all predicted heart rates $\vec{f}_{HR}$ against the reference heart rates $\vec{f}_{HR,GT}$ from $\vec{X}_{GT}$:

$$RMSE(\vec{f}_{HR}) = \sqrt{\text{mean}((\vec{f}_{HR} - \vec{f}_{HR,GT})^2)} \tag{2}$$

# 3  Learning

We evaluate two machine learning algorithms in our work. The first is k-nearest neighbors (kNN) and the second is a multilayer perceptron (MLP). In Sect. 3.1 we discuss the conversion of the data into a form so we can process it.

## 3.1  Data Preparation

From the filtered signal $\vec{X}^G_{\text{filtered}}$, window segments of 3 s length are generated where each segment starts from a detected peak. Each segment relates to a heart rate $f_{HR}$ calculated from the reference signal $\vec{X}_{GT}$. Therefore, the resulting segments begin at a local maximum. The signal's minimum and maximum should be located at the same positions. Our aim is to learn these similarities.

This segmentation method is extended by placing the peak at the center of the segment and using the segment length of 3 s in both directions. Further extensions consist in placing the peak at the end of the segment (online mode) or in case of the training set, creating the segments from reference BVP signal $\vec{X}^{BVP}_{GT}$ (but not from ECG signals).

Each segment is normalized independently for the learning algorithms.

In the following sections we learn the segments with kNN and MLP. As input vector we use all time steps of the 2*3 s of a segment. As target output we use the related heart rate $f_{HR}$. In Sect. 3.4 we use only the timestamps $\vec{R}$ and $\vec{R}_{GT}$ of the detected peaks from $\vec{X}^G_{\text{filtered}}$ and $\vec{X}_{GT}$.

## 3.2  k-Nearest Neighbor

One of the simplest and successful classifiers for quickly verify a concept is kNN [5]. kNN finds similar segments in the training set for each segment in the test set. The 'k' describes the counts of similar segments (nearest neighbors). For each segment we compute $k = 10$ nearest neighbors with correlation as the distance function. We assume that firstly the first nearest neighbor does not represent the target segment (outliers) so we use 10 neighbors and secondly a lot of the neighbors highly differs from the target segment and thus have a low correlation. To increase the classification quality we remove all neighbors with a correlation below 80 % from each class k. This will remove a huge part of the neighbors. The classification will be repeated for all window segments (in time). Each neighbor (respectively segment) corresponds to a heart rate; the related heart rates of the removed segments are linearly interpolated (independent of other classes). The resulting heart rate of a segment is the mean heart rate of the 10 classes.

## 3.3  Multilayer Perceptron

Some problems of kNN are that firstly a huge amount of segments is removed and interpolated and secondly we do not have an abstraction of the basic problem. Trials with different machine learning methods showed best results using an

MLP to train from the segments with one hidden layer and 20000 neurons. With increasing count of hidden neurons the performance of the system increases, as well as the execution time. In order to achieve some good balance between the execution time and the performance of the system, the number of neurons in the hidden layer is set to $\approx$ 20000. The input vector contains all time steps of the 2*3 s. A standard logistic function is used as transfer function. The bias is initialized with '1' and the weights with a Gaussian distribution. In the output layer, a linear neuron targets against the heart rate $f_{HR}$. The optimization is performed with the Adaptive Subgradient Method (Adagrad) [4]. The learning phase stops after 2000 epochs.

### 3.4    Gaussian Distribution

The filtered signal $\vec{X}^G_{\text{filtered}}$ has many detected peaks $\vec{R}$ which does not match with the reference peaks $\vec{R}_{\text{GT}}$. Reasons are overshooting of the filter during fast lighting changes (e.g. head motion) or high noise. Therefore we estimate the probability that a detected peak is a real peak. We assume that the heart rate changes slowly and the distance of a peak to its neighboring peaks is nearly constant with $\Delta t_{\text{peak}} = 60/f_{HR}$.

In an ideal case the neighboring peaks have a distance of $\Delta t_{\text{peak}}$; in case of a misdetection its neighbors should not be multiples of $\Delta t_{\text{peak}}$ (e.g. a peak between two correct peaks like in Fig. 2(a)). For a distance below $0.5\Delta t_{\text{peak}}$ or above $1.5\Delta t_{\text{peak}}$ we assume that this neighbor peak is misdetected (small distance between second and third peak in Fig. 2(a)) or corresponds to another reference peak (big distance between first and fourth peak in Fig. 2(a)). Moreover including the second and third neighbor peaks should improve the estimation (first peak considers second and fourth peak in Fig. 2(a)).

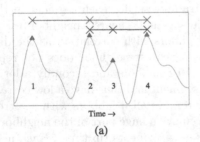
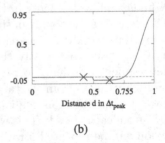

(a)                                    (b)

**Fig. 2. Left:** An example of a misdetected peak. The third peak is wrongly detected, its distances to neighboring peaks (red line) are smaller than $\Delta t_{\text{peak}}$ and this leads to low values in the distribution ($-0.05$ and $-0.049$). **Right:** the Gaussian distribution between 0 and 1 with zero for distances below $0.5\Delta t_{\text{peak}}$. The effective values for the third peak are marked with red crosses. (Color figure online)

We define a Gaussian distribution (GD) with the GD center at $d\Delta t_{\text{peak}}$ (Fig. 2(b)):

$$g_d(\vec{r}, \Delta t_{\text{peak}}) = b_d * \exp(-a_d * (\vec{r} - c_d)^2) + m_d \qquad (3)$$

$$d\Delta t_{\text{peak}} = \frac{d * 60}{f_{HR}}, \quad d = 1, 2, 3 \tag{4}$$

The probability decreases with the distance of the neighbor peak from the GD center. For every segment we calculate the distances $\vec{r}$ from the center peak to all other peaks $\vec{R}$ and exclude the center peak and all peaks with more than 3 s distance.

We experimentally discovered the following parameters for Eq. 3 (Fig. 3):

$$g_d(\vec{r}, \Delta t_{\text{peak}}) = \frac{1}{d} * \exp(-\frac{50}{\Delta t_{\text{peak}}} * (\vec{r} - d\Delta t_{\text{peak}})^2) - 0.05 \tag{5}$$

The parameter $a_d$ controls the width of the gaussian curve; a consistent value of $50/\Delta t$ for all parameters $d$ showed the best results. The maximum height $c_d$ reduces the influence with increasing $d$ by $1/d$. The center $c_d$ of the gaussian curve is fixed to $d\Delta t_{\text{peak}}$. The parameter $m_d$ controls the generic height and the minimum value; with $-0.05$ big differences have a negative influence.

In the next step neighbor peaks with a distance of more than $0.5\Delta t_{\text{peak}}$ from the GD center are excluded:

$$(d - 0.5)\Delta t_{\text{peak}} \leq g_d(\vec{r}, \Delta t_{\text{peak}}) < (d + 0.5)\Delta t_{\text{peak}} \tag{6}$$

From this it follows that each neighboring peak corresponds to only one distance $d$. We normalize each $g_d$ by dividing through the count of corresponding peaks $|\vec{p}_d|$ but minimum 2 because in ideal case each distance $d$ corresponds to two neighboring peaks and having only one is an indication for misdetection. Finally all $g_d(\vec{r}, \Delta t_{\text{peak}}) \ \forall \ \vec{r}, d$ are summarized:

$$GD = \sum_{d=1}^{3} \frac{g_d(\vec{r}, \Delta t_{\text{peak}})}{\max(2, |\vec{p}_d|)} \tag{7}$$

The probability is not a percentaged value and has values outside of $[0, 1]$.

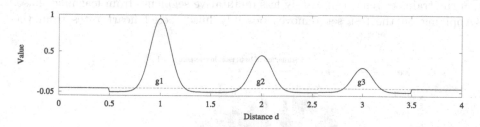

**Fig. 3.** GD for all distances $d = 1, 2, 3$. Each distribution is truncated to $d \pm 0.5$; every peak corresponds to only one $g_d$. The minimum value is $-0.05$ and the maximum height decreases with each $d$.

## 4  Experiments

### 4.1  Dataset

For our evaluation we use the Open_EmoRec_II dataset [11]. It provides 30 participants recorded in a human-computer interaction (HCI) scenario with a session length of 20 min per participant. The participants are sitting still most of the time during the session. The participants were recorded with several Pikes as well as a webcam ($720 \times 576@25$ resolution, MPEG-4 codec, 1600 kb/s bitrate, manufacturer unknown). The face region has an effective resolution of $240 \times 280$ pixels. A finger BVP sensor was recorded for the heart rate signal. The dataset combines two sessions in two different environments. The dataset consist of two parts; the second part (details can be found in [11]) is used for the evaluation. Participants 1, 4, 19 and 28 were excluded because of a bad detection rate of the peaks in the BVP signal.

### 4.2  Experimental Setting

The reference system for our comparison (denoted as PEAK) extracts the signal $\vec{X}^{G}_{\text{filtered}}$ from the video and detects the peaks as explained in Sect. 2.3. Then window segments with seven seconds length and a shift of one second are defined from which the heart rate $f_{HR}$ is determined (Eq. 1). With the shift we get a constant response time of one second. For all 26 participants 30655 segments are generated. We get a baseline RMSE value of 23.97.

The evaluation of our methods is performed with leave-one-subject-out (LOSO). The segments are generated as explained in Sect. 3.1. With heart rates higher than 60 bpm the shift of the segments is smaller than one second and a total of 44406 segments for all 26 participants are generated.

The distribution of the heart rates (Fig. 4) is between $[47, 124]$ bpm and shows a high occurrence between $[62, 92]$ bpm. Test samples outside of this area tend to produce higher error rates.

The GD can be applied on the train set as well as on the test set. Applying on the train set removes possibly less qualitative segments from learning phase. Applying on the test set removes possibly misdetected heart rates from the

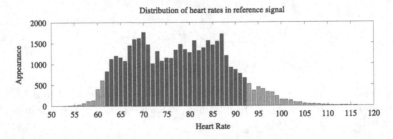

**Fig. 4.** Histogram of all heart rates $\vec{f}_{HR,GT}$. Heart rates between $[62, 92]$ bpm appear more than half as often as the most often heart rate.

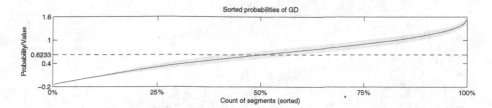

**Fig. 5.** The GD is applied on the test set. The probabilities are sorted. The red line is the mean of the probabilities computed from all participants and the light red area is the standard deviation. The probability has a value between $[-0.2, 1.6]$. (Color figure online)

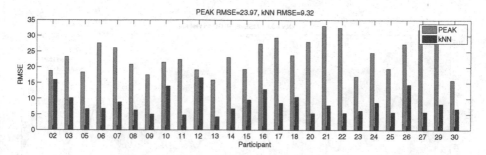

**Fig. 6.** Comparison between PEAK and kNN shows better results of kNN for all participants.

prediction. This will directly influence the result. For a more accurate comparison the removed heart rates are linearly interpolated. A threshold analysis is shown in Fig. 5. The behavior of the probabilities shows for a threshold of 0.6233 a standard deviation of approximately 20 % (blue line cutting the red area). Based on these findings we decided to use 50 % of the segments instead of a threshold.

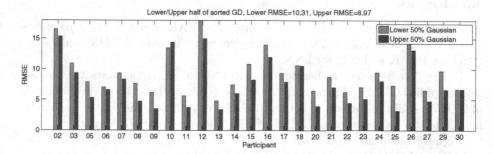

**Fig. 7.** Evaluation of GD applied on the test set. The heart rates for all participants are detected with kNN and sorted by the probability of GD. The RMSE is computed on lower (inferior) and upper (superior) half of sorted prediction. Restricting to the upper half of GD improves the detection from 10.31 to 8.97.

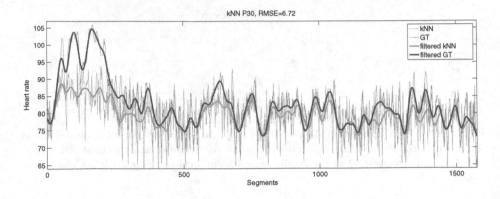

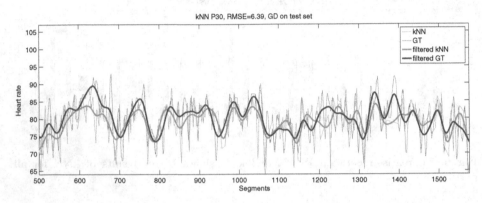

**Fig. 8.** Prediction (thin green line) with kNN for the participant 30. After removing and interpolating the lower half of GD from the test set (bottom figure, zoomed to [500, 1580]), the prediction is less noisy (thin green line). The filtered representations (thick lines) have nearly the same behavior. 30 segments are interpolated around segment 1400. This has a detection delay as consequence. (Color figure online)

## 4.3   Results

The comparison between the baseline method PEAK and kNN (Fig. 6) shows that kNN outperforms PEAK for all participants. The overall RMSE values are 23.97 for PEAK and 9.32 for kNN.

In Fig. 7 we detected the heart rates with kNN for all participants. On the test set we applied GD (Fig. 5) and sorted the detected heart rates by the probabilities of GD. We split the sorted prediction into a lower (inferior) and upper (superior) half. The two halves have an RMSE of 10.31 (lower) and 8.97 (upper).

Figure 8 compares kNN with and without GD for the participant 30. The signal duration is 18 min. For the bottom figure the heart rates of the lower half are removed from the prediction and interpolated. The error decreases from 6.7 to 6.4. Long detection gaps do not appear but shorter gaps could be problematic for live systems. For segments between [30, 250] kNN has problems following heart rates higher than 90 bpm.

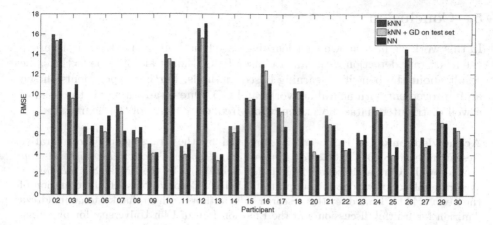

**Fig. 9.** Per participant comparison shows benefits of GD when applied on the test set. The neural network is better than kNN in most cases

Comparison of kNN, kNN with GD on the test set and MLP (Fig. 9) shows better results for GD over pure kNN for all participants. MLP is better than kNN in most cases.

Finally Table 1 summarizes the results of this work.

**Table 1.** Results of the methods.

|  | PEAK | kNN | kNN + GD | MLP |
|---|---|---|---|---|
| RMSE | 23.97 | 9.32 | 8.97 | 8.62 |

## 4.4   Discussion

Comparing the PEAK method with kNN in Fig. 6 shows the predominance of machine learning approaches for this task. The RMSE decreases from 23.97 to 9.32. Applying GD on the test set further decreases the RMSE to 8.97. Trials to apply GD on the train set degrades the RMSE both with kNN and MLP.

One big drawback of kNN and especially of GD applied on the test set is that a huge amount of segments is removed. For live systems this leads to detection delays. Therefore post processing methods needs to be designed (e.g. giving a coarse prediction and correcting it to a later time). MLP don't remove segments and with an RMSE of 8.62 it is better than kNN. Applying GD in combination with MLP should improve the detection as well but we aim for a method without interpolating segments.

## 5 Conclusion

In this work we have shown that learning based methods can extremely improve the heart rate detection with camera based PPG. Hsu et al. [6] came to the same result about the benefit of learning based methods. Further experiments shown an improvement with neural networks and GD. One disadvantage is the need of a well distributed dataset over the whole frequency range of the heart rate.

**Acknowledgements.** This work is supported by the Transregional Collaborative Research Centre SFB/TRR 62 *Companion-Technology for Cognitive Technical Systems* funded by the German Research Foundation (DFG) and by the BMBF *SenseEmotion*. We gratefully acknowledge the support of NVIDIA Corporation with the donation of the Tesla K40 GPU used for this research. The authors wish to thank Mohammadreza Amirian for fruitful discussion and the Emotion Lab of Ulm University for providing the Open_EmoRec_II dataset.

## References

1. Baltrusaitis, T., Robinson, P., Morency, L.P.: Constrained local neural fields for robust facial landmark detection in the wild. In: Proceedings of the IEEE International Conference on Computer Vision Workshops, pp. 354–361 (2013)
2. Baltrušaitis, T.: CLM-Framework. https://github.com/TadasBaltrusaitis/CLM-framework
3. Carbajal, J.P.: Findpeaks. http://sourceforge.net/p/octave/signal/ci/default/tree/inst/findpeaks.m
4. Duchi, J., Hazan, E., Singer, Y.: Adaptive subgradient methods for online learning and stochastic optimization. J. Mach. Learn. Res. **12**, 2121–2159 (2011)
5. Friedman, J.H., Bentley, J.L., Finkel, R.A.: An algorithm for finding best matches in logarithmic expected time. ACM Trans. Math. Softw. (TOMS) **3**(3), 209–226 (1977)
6. Hsu, Y., Lin, Y.L., Hsu, W.: Learning-based heart rate detection from remote photoplethysmography features. In: 2014 IEEE International Conference on Acoustics, Speech and Signal Processing (ICASSP), pp. 4433–4437, May 2014
7. Lamonaca, F., Barbe, K., Kurylyak, Y., Grimaldi, D., Moer,W.V., Furfaro, A., Spagnuolo, V.:Application of the artificial neural network for blood pressure evaluation with smartphones. In: 2013 IEEE7th International Conference on Intelligent Data Acquisition and Advanced Computing Systems (IDAACS), vol. 01, pp. 408–412, September 2013
8. Maaoui, C., Bousefsaf, F.: Automatic human stress detection based on webcam photoplethysmographic signals. J. Mech. Med. Biol. **16**(04), 1650039 (2015)
9. Mahmoud, T.M.: A new fast skin color detection technique. World Acad. Sci. Eng. Technol. **43**, 501–505 (2008)
10. Oppenheim, A.V., Schafer, R.W., Buck, J.R.: Discrete-Time Signal Processing, 2nd edn. Prentice-Hall Inc., Upper Saddle River (1999)
11. Rukavina, S., Gruss, S., Walter, S., Hoffmann, H., Traue, H.C.: OPEN_EmoRec_II- a multimodal corpus of human-computer interaction. World Acad. Sci. Eng. Technol. Int. J. Comput. Electr. Autom. Control Inf. Eng. **9**(5), 1135–1141 (2015)
12. Verkruysse, W., Svaasand, L.O., Nelson, J.S.: Remote plethysmographic imaging using ambient light. Opt. Express **16**(26), 21434–21445 (2008)

# Author Index

Abbas, Hazem M.   233
Abdoola, Rishaad   220
Abe, Shigeo   29
Amirian, Mohammadreza   269
Aswolinskiy, Witali   197

Baldassin, Alexandro   126
Bernstein, Alexander   55
Bongini, Marco   257
Breen, Vivienne   18
Bunke, Horst   185

Calder, Stefan   18
Chowdhury, Shubham   246
Costa, Kelton   117, 138

Destercke, Sébastien   92
Djouani, Karim   220
Du, Peng   18
Du, Shengzhi   220

E, Xu   80
Erdal, Mehmet   298

Fang, Yunfei   220
Fischer, Andreas   163, 185

Hennebert, Jean   163
Hihn, Heinke   312
Hou, Jian   80

Kaban, Ata   42
Kächele, Markus   269, 298, 324
Kasabov, Nikola   18
Kessler, Viktor   324
Kestler, Hans A.   105
Khalil, Mahmoud I.   233
Kuleshov, Alexander   55

Lausser, Ludwig   105
Laveglia, Vincenzo   257
Liu, Weixue   80

Lomonaco, Vincenzo   175
Lopes, Ricardo   117

Magg, Sven   150
Maltoni, Davide   175
Meudt, Sascha   285, 312, 324
Mitra, Mridul   246

Palm, Günther   285, 324
Papa, João   117, 126, 138
Pasa, Luca   3
Passos, Leandro   138
Pereira, Clayton   138
Pereira, Danillo   126

Radwan, Mohamed A.   233
Raj, Vidwath   150
Reinhart, René Felix   197
Riesen, Kaspar   185
Rosa, Gustavo   138
Roy, Kaushik   246

Sarkar, Ram   246
Schirra, Lyn-Rouven   105
Schleif, Frank-M.   42
Schmid, Florian   105
Schwenker, Friedhelm   269, 285, 298, 312, 324
Sen, Shibaprasad   246
Soullard, Yann   92
Sperduti, Alessandro   3
Steil, Jochen   197

Thiam, Patrick   285
Thouvenin, Indira   92
Tino, Peter   42
Trentin, Edmondo   68, 257

Valdenegro-Toro, Matias   209

Wermter, Stefan   150
Wicht, Baptiste   163

Yang, Xin-She   126, 138

Printed in the United States
By Bookmasters